ELECTRONIC
COMMUNICATIONS

INCLUDES _NEW_ TROUBLESHOOTING
SIMULATION DISK

Thomas A. Adamson

ELECTRONIC COMMUNICATIONS:
SYSTEMS AND CIRCUITS

INCLUDES <u>NEW</u> TROUBLESHOOTING SIMULATION DISK

Delmar Publishers Inc.®

DEDICATION: This book is dedicated to my wife—
JUDY ADAMSON
. . . whose encouragement, trust and love made it possible.

COVER PHOTOS:

National Network Operations Center of Telecom Canada—photo by Jay Freis, printed with permission of Tandem Computers Incorporated, Cupertino, CA

Photo of GRASIS® 4SHT tower and Andrew SHX® horn antenna by B. Surtz, courtesy Andrew Corporation, Orland Park, IL

Fiber Optics photo from COMSTOCK INC./Michael Stuckey

Delmar Staff
   Executive Editor: Mark W. Huth
   Managing Editor: Barbara A. Christie
   Production Editor: Eleanor Isenhart
   Design Coordinator: Susan C. Mathews
   Publications Coordinator: Karen Seebald

For information, address Delmar Publishers Inc.,
2 Computer Drive West, Box 15-015,
Albany, New York 12212

Library of Congress Cataloging-in-Publication Data

Adamson, Thomas A.
   Electronic communications.

   Includes index.
   1. Telecommunication.   I. Title.
TK5101.A34   1988   621.38'0413   87-12173
ISBN 0-8273-2640-8

CONTENTS

Chapter 4   Amplitude Modulation Systems   71

Chapter 5   Sideband Systems   109

Chapter 6   Frequency and Phase Modulation Systems   141

PREFACE

Electronic Communications: Systems and Circuits is the first truly "high-tech" approach to the presentation of electronic communications systems and circuits. All of the latest technology of integrated circuits, digital communications techniques, lasers, fiber optics, and the microcomputer as a problem-solving tool are used here. It was written to:

- Bring the latest communication technology into the classroom.
- Take advantage of classroom accessibility to microcomputers.
- Demonstrate new integration of digital technologies into the mainstream of electronic communications.
- Emphasize the integrated circuit as a major building block in communications equipment.
- Expand upon the topics of laser and fiber optics as a major method used in communication systems.
- Omit out-dated topics that tend to stay in revised editions of early communication textbooks.
- Address the importance of troubleshooting and instrumentation by including these topics as a major section in every chapter of the book.

This text is designed for a first course in electronic communications. You can employ it early in a curriculum because the only prerequisites are DC/AC theory and some elementary algebra. This is an excellent text to use in the first or second year technician programs or sophomore/junior year of engineering technology programs. It is recommended that a devices class be taken as a corequisite.

Electronic Communications: Systems and Circuits uses a systems to circuits approach. This is achieved by dividing the text into two major parts. The first part presents communication systems. Circuits are presented in the last part. This method has several advantages:

- You need not have completed a course in electronic devices.
- Attention can be focused on the theory, purpose and application, and troubleshooting of *communication systems,* without bogging you down in the details of circuit theory.
- The integrated circuit can be introduced early as a major component in communication systems. From a troubleshooting standpoint it can be treated as a circuit element.

■ Communication *circuits* are introduced when you are technically ready to see where they are used in a system. This approach allows for the similarities and differences of communications circuits to be presented to you in a clear and logical arrangement.

■ You can now focus attention on one major area at a time, first systems, then circuits.

Major Features

The major features of this text include:

Chapter Objectives. Every chapter begins with a set of objectives that you are expected to achieve after completing the chapter. These are written in a manner that makes them easy for you to understand.

Chapter Introduction. Every chapter begins with an important introduction. This is designed to motivate you and give an overview of the key chapter topics and the reason for their presentation.

Review Questions. Each chapter is divided into topical sections. The end of each section contains review questions designed to give you an opportunity to test your comprehension of the material just presented. The answers to the review questions are found at the end of the text.

Numbered Equations. Every equation presented is numbered for easy reference. This numbering system is used when the equation appears in an example so that you may make easy reference to where it is first presented.

Detailed Examples. Numerous detailed examples illustrating step-by-step solutions to problems are presented throughout the text. Each major step is presented with all variables and units clearly defined.

Troubleshooting and Instrumentation. Every chapter ends with a special section on Troubleshooting and Instrumentation. These sections contain topics that relate to the topic of the chapter. Techniques in troubleshooting communication systems and the instrumentation required are presented in the *systems* portion of the text, while circuit troubleshooting and instrumentation is presented in the *circuits* portion of the text.

Tabular Information. Numerous tables are used to present the interrelationship of important communication topics and concepts. These tables are designed to show you the commonalities and differences between communication systems, terminology, and circuits. They are an outstanding teaching aid for the classroom and valuable learning tool for the individual reader.

Illustrations. Many elaborate illustrations are used in this text with the intent of teaching concepts. These are extremely valuable in class presentations of new concepts. These illustrations are presented in a manner that removes as many learning obstacles as possible while preserving the important technical nature of the subject.

Chapter Problems. Problems are presented in enough quantity to provide ample problem solving experience in electronic communications. The intent is to build your confidence and reinforce important chapter features. The solutions to the odd-numbered problems are found at the end of the text.

Digital Appendix. An appendix that introduces fundamental digital concepts (such as binary numbers, logic gates and flip-flops) is included if you lack a digital foundation. This affords the instructor the opportunity to include this topic which is now well integrated into communications technology. The topic is a suggested prerequisite for the chapters on pulse communication techniques and digital communication techniques, as well as the chapter on phase-locked loops.

Formula Appendix. All new formulas presented in a chapter are summarized in the appendix. This feature offers a method of reviewing and reinforcing important mathematical relationships.

Ancillary Materials

The concept of the microcomputer as a patient teacher is used here. Every text comes with a floppy disk. The disk contains a series of self-instructional troubleshooting simulation programs. With the use of a special disk used only by your instructor, you may have your progress monitored. In this manner, you have the opportunity to get more individual attention than what is normally available in a standard laboratory setting. This is a powerful learning feature never before available to electronic communication students.

A coordinated laboratory manual is available with this text. The experiments use the same systems to circuits approach used by the text. Integrated circuits used in communication systems that are easy and inexpensive to purchase are emphasized in the experiments. The experiments cover the full range of communications from the VCR, optical fibers, antennas and transmission lines to phase-locked loops, active filters and integrated circuit voltage regulators. Every experiment contains an important troubleshooting section, with detailed explanations for conducting a successful experiment.

A special Instructor's Guide is available for those who teach from this text. All end of chapter problems are answered and important comments and suggestions concerning all of the lab experiments are presented. Special emphasizes is placed on areas where students may encounter problems. Here, learning activities are suggested for the instructor's use. Each program used on the accompanying floppy disk is clearly explained with its use and limitations.

Acknowledgments

The author wishes to express his deep appreciation to the reviewers and staff of Delmar Publishers for their assistance and guidance in the development of this text. The reviewers include the following: Steven Kalina, DeVry Institute of Technology; Robert Silva, Middlesex Community College; Charles Holling, El Camino College and Western Airlines; Charles Dewater, ITT Technical Institute; and especially Ronald Suptic, DeVry Institute of Technology, who provided in-depth reviews of the book and lab manual. Thanks also to Dave Dusthimer, who initiated the project, and to Dr. John Ingram, CSU Sacramento, for his valuable contributions to the computer programs used in this textbook. The author would also like to thank Jonathan Plant for his guidance, patience, and expertise of the subject area and publishing profession during the development of this text.

SECTION ONE
COMMUNICATION SYSTEMS

Courtesy of GTE Communication Systems

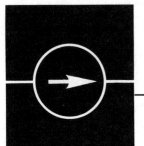

CHAPTER 1

Introduction and Review

OBJECTIVES

In this chapter, you will study:

- ☐ What this book covers.
- ☐ What you should already know.
- ☐ A review of reactance.
- ☐ A review of impedance.
- ☐ A review of resonance.
- ☐ The definitions of troubleshooting and instrumentation.
- ☐ Major categories of troubleshooting.

INTRODUCTION

The material you learned in the alternating current (AC) electrical circuits course is used in this text. Reviewing this material now will help you learn the new information presented in electronic communications. The microcomputer simulation for this chapter is an *RLC* circuit simulation. This program will determine the resonant frequency and bandwidth of an *RLC* circuit. You need only supply the values of the capacitor, inductor, and circuit resistance; the computer will do the rest.

1-1   SCOPE OF THE TEXT

This text covers all of the important concepts of electronic communications. When you complete this text, you will have studied the technical details of radio, television, satellite communications, microwaves, fiber optics, and lasers. In addition, *integrated circuits* are emphasized. Different integrated circuits (ICs) are shown in Figure 1-1.

Integrated circuits are rapidly replacing discrete components such as transistors, diodes, resistors, capacitors, and inductors. Because of the increasing use of ICs, this text is divided into two main sections. Section 1 (Chapters 1–12) introduces electronic *commu-*

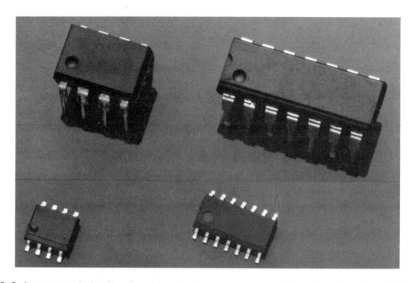

Figure 1-1 Integrated circuits used in electronic communication circuits. Copyright by Motorola, Inc. Used by permission.

nication systems. You will discover how complete systems, such as AM-FM receivers and transmitters, function. Integrated circuits are also introduced in these chapters.

Section 2 (Chapters 13–17) introduces the analysis of discrete component circuits, such as radio frequency amplifiers, oscillators, and all the other important *communication circuits* that use discrete components or act in conjunction with ICs.

Learning electronic communications in this manner also allows you the time needed to have your devices and circuits course be in-step with this text. Thus, to start this text, you need not know anything about transistor operation or ICs, a welcome relief from the more traditional methods of presenting electronic communications.

Each chapter contains a section of the highest priority: *troubleshooting* and *instrumentation*. These two areas have the greatest job potential for graduate technicians. These important areas are not treated lightly in this text, nor are they included at the expense of important electrical concepts.

This text has many other important features. You can use your microcomputer as a system and circuit simulation tool. Each text comes with a floppy-disk containing programs that relate to each chapter of the text.

A very important chapter, not usually included in traditional communications textbooks, is Digital Communication Techniques. This chapter applies these circuits to communications systems and circuits. Digital and logic circuits are widely used in home entertainment systems and industrial and military communication applications.

A laboratory manual accompanies this text. It presents practical applications of the theory given here. The manual also presents experiments in practical troubleshooting and circuit analysis techniques using ICs and discrete components. Developing these skills will help you get and keep your first job in the growing field of electronic communications.

1-2 WHAT YOU SHOULD ALREADY KNOW

Before starting this text, you should have successfully completed the following:

- A basic DC and AC electrical circuits course including laboratories
- Introductory algebra
- Some introduction to trigonometry

DC Electrical Circuits

You should be familiar with:

- Series, parallel, and series-parallel resistive circuits
- Electrical power dissipation
- Circuit loading
- Bridge circuits

AC Electrical Circuits

This chapter reviews some important AC electrical circuit topics. Areas you should know that are not covered are

- Analysis of periodic waveforms such as the sine wave and square wave
- Transformers and impedance matching
- Time constants, integrators, and differentiators
- The basic theory of operation of the analog DC and AC milliammeter
- Concepts of phasors

Laboratory Skills

You should have acquired the following laboratory skills:

- Using the VOM and electronic multimeter to measure resistance, voltage, and current
- Using the oscilloscope to measure period, frequency, peak-to-peak and instantaneous values of periodic waveforms
- Soldering, desoldering, and basic electrical assembly techniques
- Identifying and testing resistors, capacitors, inductors, transformers, and relays
- Safely operating DC laboratory power supply
- Graphing electrical circuit variables on semilog graph paper
- Constructing simple electrical circuits from schematic diagrams
- Using the laboratory function generator

Mathematics Skills

You should know how to

- Add, subtract, multiply, and divide signed algebraic expressions
- Use powers of ten, scientific notation, and metric units as they apply to electrical standards

■ Solve a linear equation in one unknown

■ Solve right triangles and use them to solve AC electrical circuits

It would also be helpful to know electrical applications of logarithms.

Computer Programming

You do not need to know anything about computer programming. To start the student disk, simply refer to the instructions on the back of the last page in the book. Once you are started, follow the instructions presented on the disk. Your progress will automatically be monitored so that you may receive valuable help from your instructor when you need it.

Devices and Circuits

It isn't necessary to have completed a course in electronic devices and circuits, where you learn about diodes, transistors, and ICs. If you are just starting a devices and circuits course, then you will find that this text will pace you so that as you cover topics in the devices course, their communication applications will be presented here. If you are not taking such a course, you will need to spend more time on the material presented in Section 2 of this text.

1-2 Review Questions

1. Explain the relationship between voltage, current, and resistance.
2. In a series DC circuit, what happens to the total circuit current if the circuit resistance increases? decreases? What happens to the total circuit resistance if one resistor in the circuit increases? decreases?
3. In a parallel DC circuit, what happens to the total circuit current if the circuit resistance increases? decreases? What happens to the total circuit resistance if one resistor in the circuit increases? decreases?
4. For a DC circuit consisting of a 5-kΩ resistor connected to a 12-V source, find (A) the circuit current, (B) the power dissipated in the resistor, and (C) the power delivered by the source.
5. A 1-kΩ resistor, a 3.3-kΩ resistor, and a 5.2-kΩ resistor are connected in series to a 12-V DC source. Calculate (A) the total circuit current, (B) the voltage drop across each resistor, and (C) the power delivered by the source.
6. A 1-kΩ resistor, a 3.3-kΩ resistor, and a 5.2-kΩ resistor are connected in parallel across a 12-V DC source. Calculate (A) the total circuit current, (B) the current in each resistor, and (C) the power delivered by the source.
7. What are the rms value and the peak-to-peak value of a sine-wave having a peak voltage of 24 V? If the voltage has a frequency of 12 kHz, what is the period? How much power would be dissipated by this source when it is connected to a 150-Ω resistor?
8. A transformer has 10 turns on the primary and 150 turns on the secondary. If a 120-V rms signal is applied to the primary, what is the secondary voltage?
9. If the transformer secondary in problem 8 is connected to a 100-Ω resistor, find the secondary current, the primary current, the power delivered by the source, and the power dissipated by the 100-Ω resistor.

1-3 | REVIEW OF REACTANCE

Overview

Recall that *reactance* is nothing more than the opposition of a capacitor or an inductor to current flow. The unit of measurement for reactance is the ohm (Ω).

Capacitive Reactance

A capacitor opposes a *change* in voltage. The more rapidly the voltage applied across a capacitor is changing, the quicker will be the current to and from the capacitor. In other words, the greater the change of voltage applied to a capacitor, the greater will be the amount of capacitive current. This can be expressed mathematically as

$$i_C = C \frac{\Delta v}{\Delta t}$$

(Equation 1-1)

where i_C = Instantaneous capacitor current in amperes
$\quad\quad\quad C$ = Capacitor value in farads
$\quad\quad\quad \Delta v$ = Rate of *change* of the applied voltage in volts
$\quad\quad\quad \Delta t$ = Rate of *change* of time in seconds

When a sine wave is applied across a capacitor, its reactance may be expressed mathematically as

$$X_C = \frac{1}{2\pi f C}$$

(Equation 1-2)

where X_C = Capacitive reactance in ohms
$\quad\quad\quad \pi$ = 3.14159 ...
$\quad\quad\quad f$ = Frequency of the applied sine wave in hertz
$\quad\quad\quad C$ = Capacitor value in farads

Example 1

Determine the peak current for the circuit in Figure 1-2.

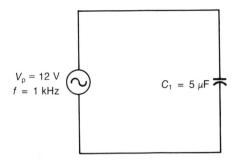

V_p = 12 V
f = 1 kHz

C_1 = 5 µF

Figure 1-2

Solution

First determine the amount of capacitive reactance:

$$X_C = \frac{1}{2\pi f C}$$
$$X_C = 31.8\ \Omega$$

(Equation 1-2)

Next, using Ohm's law for reactance, calculate the circuit current:

$$I = \frac{V_P}{X_C}$$
$$I = 377\ \text{mA}$$

Therefore, the circuit in Figure 1-2 will have a peak current of 337 mA when a sine wave with a peak voltage of 12 V and a frequency of 1 kHz is applied to the 5-μF capacitor.

Inductive Reactance

An inductor opposes a *change* in current. The more rapidly the current in an inductor tries to change, the more the inductor will create a voltage across itself to oppose the current change. In other words, the greater the change of current in an inductor, the greater will be the amount of voltage developed by the inductor to oppose the current. This can be expressed mathematically as

$$v_L = L\frac{\Delta i}{\Delta t}$$

(Equation 1-3)

where v_L = Instantaneous inductor voltage in volts
   L = Inductor value in henries
   Δi = Rate of *change* of inductor current in amperes
   Δt = Rate of *change* in time in seconds

When a sine wave is applied across an inductor, its reactance may be expressed mathematically as

$$X_L = 2\pi f L$$

(Equation 1-4)

where X_L = Inductive reactance in ohms (Ω)
   π = 3.14159 . . .
   f = Frequency of the applied sine wave in hertz
   L = Inductor value in henries

Example 2

Determine the peak current for the circuit in Figure 1-3.

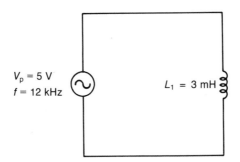

$V_p = 5$ V
$f = 12$ kHz

$L_1 = 3$ mH

Figure 1-3

Solution

First determine the amount of inductive reactance:

$$X_L = 2\pi f L \qquad \textbf{(Equation 1-4)}$$
$$X_L = 226 \ \Omega$$

Next, using Ohm's law for reactance, calculate the circuit current:

$$I_P = \frac{V_P}{X_L}$$
$$I_P = 22 \text{ mA}$$

Hence, the circuit in Figure 1-3 will have a peak current of 22 mA when a sine wave with a peak voltage of 5 V and a frequency of 12 kHz is applied to the 3-mH inductor.

1-3 Review Questions

1. What is reactance? Describe the difference between inductive reactance and capacitive reactance.
2. For a circuit consisting of a capacitor and an AC voltage source, what will happen to the circuit current if (A) the source frequency is increased? the source frequency is decreased? (B) the source amplitude is increased? the source amplitude is decreased?
3. For a circuit consisting of an inductor and an AC current source, what will happen to the inductor voltage if (A) the source frequency is increased? the source frequency is decreased? (B) the source amplitude is increased? the source amplitude is decreased?
4. A 15-pF capacitor is connected to a source with an rms voltage of 24 V. What is the capacitive reactance if the source frequency is (A) 10 kHz? (B) 10 MHz?
5. For the circuit of problem 4, what is the *peak* circuit current for the given frequencies?

6. A 120-μH inductor is connected to a source with an rms voltage of 12 V. What is the inductive reactance if the source frequency is (A) 10 kHz? (B) 10 MHz?
7. For the circuit of problem 6, what is the *peak* circuit current for the given frequencies?
8. What value of capacitor will produce an rms current flow of 10 mA when connected to a 50-V AC voltage source (peak value) at a frequency of 25 MHz?
9. What value of inductor will produce an rms current flow of 120 mA when connected to a 110-V voltage source (peak value) at a frequency of 60 Hz?

1-4   REVIEW OF IMPEDANCE

Overview

Impedance is nothing more than the total opposition to current flow in a reactive circuit. The unit of measurement for impedance is the ohm (Ω).

Series Circuit Impedance

For a series circuit consisting of a resistor, capacitor, and inductor, the total circuit impedance may be expressed mathematically as

$$Z = \sqrt{R^2 + (X_L - X_C)^2}$$ **(Equation 1-5)**

where Z = Impedance of the circuit in ohms
R = Resistance of the circuit in ohms
X_L = Inductive reactance of the circuit in ohms
X_C = Capacitive reactance of the circuit in ohms

Example 1

Find the impedance of the circuit in Figure 1-4.

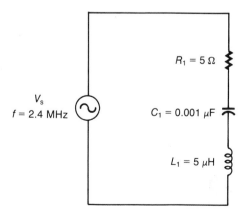

V_s
$f = 2.4$ MHz

$R_1 = 5\ \Omega$

$C_1 = 0.001\ \mu F$

$L_1 = 5\ \mu H$

Figure 1-4

Solution

First, calculate the capacitive reactance:

$$X_C = \frac{1}{2\pi f C}$$ **(Equation 1-2)**

$$X_C = 66\ \Omega$$

Next, compute the value of the inductive reactance:

$$X_L = 2\pi f L$$ **(Equation 1-4)**

$$X_L = 75\ \Omega$$

Now calculate the circuit impedance:

$$Z = \sqrt{R^2 + (X_L - X_C)^2}$$ **(Equation 1-5)**

$$Z = 10\ \Omega$$

The circuit in Figure 1-4 has an impedance of 10 Ω.

Phase Angle

The mathematical relationship of the phase angle between the circuit current and the applied voltage for a series *RLC* circuit is

$$\theta = \text{arc tan}\ \frac{X_T}{R}$$ **(Equation 1-6)**

where θ = Phase angle in degrees
X_T = Total circuit reactance ($X_L - X_C$) in ohms
R = Circuit resistance in ohms

Example 2

For the circuit in Figure 1-4, find the phase relationship between the circuit current and the applied voltage. Construct the resulting phasor diagram.

Solution

Obtain the values from Example 1-3:

$$\theta = \text{arc tan}\ \frac{X_T}{R}$$

$$\theta = \text{arc tan}\ 1.8 = 61°$$

The resulting phasor diagram is shown in Figure 1-5.

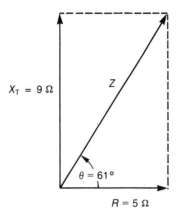

Figure 1-5

Parallel Circuit Impedance

There are several ways to find the impedance of a parallel *RLC* circuit. One method, which requires a minimum of mathematics (but not the fewest calculations) is to first find the total circuit current and then use Ohm's law to find the total impedance.

The total circuit current in a parallel *RLC* circuit can be expressed mathematically as

$$I_T = \sqrt{I_R^2 + (I_C - I_L)^2}$$ **(Equation 1-7)**

where I_T = Total circuit current in amperes

I_R = Current in resistive branch in amperes

I_L = Current in inductive branch in amperes

I_C = Current in capacitive branch in amperes

Example 3

What is the impedance of the parallel *RLC* circuit in Figure 1-6?

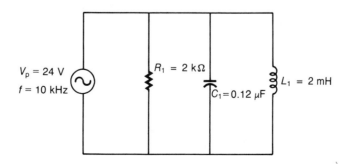

Figure 1-6

Solution

First calculate the peak current in each branch of the circuit.
For the resistive branch:

$$I_R = \frac{V_P}{R}$$
$$I_R = 12 \text{ mA}$$

For the capacitive branch, first find the capacitive reactance:

$$X_C = \frac{1}{2\pi f C}$$
$$X_C = 133 \ \Omega \qquad\qquad \textbf{(Equation 1-2)}$$

Now the current in the capacitive branch is

$$I_C = \frac{V_P}{X_C}$$
$$I_C = 180 \text{ mA}$$

For the inductive branch, first find the inductive reactance:

$$X_L = 2\pi f L \qquad\qquad \textbf{(Equation 1-4)}$$
$$X_L = 126 \ \Omega$$

Now the current in the inductive branch is

$$I_L = \frac{V_P}{X_L}$$
$$I_L = 190 \text{ mA}$$

The total current is

$$I_T = \sqrt{I_R^2 + (I_C - I_L)^2} \qquad\qquad \textbf{(Equation 1-7)}$$
$$I_T = 16 \text{ mA}$$

The circuit impedance is

$$Z = \frac{V_P}{I_T}$$
$$Z = 1.5 \text{ k}\Omega$$

Conclusion

You have completed your review of impedance. The concept of impedance is very important in electronic communication systems because inductors, capacitors, and resistors are found throughout communication circuits. If you had any difficulty with the example problems, go back and review them.

1-4 Review Questions

1. How would you define impedance?
2. What is the difference between reactance and impedance?
3. What is the phase relationship between the current and the voltage in (A) an inductor? (B) a capacitor? (C) a resistor?
4. What is the phase relationship between the current in a series *RLC* circuit and the current in (A) the inductor? (B) the capacitor? (C) the resistor?
5. What is the phase relationship between the current in a series *RLC* circuit and the voltage across (A) the inductor? (B) the capacitor? (C) the resistor?
6. Calculate the impedance and determine the phase angle between the total circuit current and the applied voltage for a series *RLC* circuit where V_S = 8 V (peak value), f = 75 MHz, C = 1.9 pF, L = 5 μH, and R = 1.2 kΩ.
7. If the components with the circuit values in problem 6 are placed in parallel, find the impedance and the phase angle between the applied voltage and the total circuit current.

| 1-5 | REVIEW OF RESONANCE |

Overview

Resonance can be defined as that condition in a series *LC* circuit when the capacitive reactance is equal to the inductive reactance. In a parallel *LC* circuit, resonance occurs when the current in the inductive branch is equal to the current in the capacitive branch.

If a series *LC* circuit is connected to an AC voltage source whose frequency is gradually increased, the value of the inductive reactance will start to increase while the value of the capacitive reactance will start to decrease. At only one frequency will the inductive reactance equal the capacitive reactance. This frequency is called the *resonant* frequency of the circuit.

The same is true for a parallel *LC* circuit. The value of the impedance of the inductive branch will equal the value of the impedance of the capacitive branch at only one frequency. This frequency is again called the resonant frequency.

Series Resonance

The resonant frequency of a series *LC* circuit can be expressed mathematically as

$$f_r = \frac{1}{2\pi\sqrt{LC}}$$

(Equation 1-8)

where f_r = Frequency of resonance in hertz
  π = 3.14159 ...
  L = Inductor value in henries
  C = Capacitor value in farads

Example 1

What is the resonant frequency for the circuit in Figure 1-7?

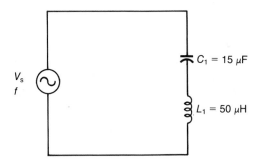

Figure 1-7

Solution

The resonant frequency is

$$f_r = \frac{1}{2\pi\sqrt{LC}}$$

$$f_r = 5.8 \text{ kHz}$$

(Equation 1-8)

The frequency where the inductive reactance equals the capacitive reactance for the circuit in Figure 1-7 is 5.8 kHz.

Frequency Response of Series LC Circuit

For the *LC* portion of a series *RLC* circuit, the voltage across the capacitor is 180° out of phase with the voltage across the inductor. Thus the total reactive voltage is equal to the difference of V_C and V_L. This relationship is shown in Figure 1-8. Note, from Figure 1-8, that as the frequency of the source is increased toward resonance, the circuit current increases.

The net result of this is that the total impedance of the circuit is decreasing as the frequency of the source is increased toward resonance. You can conclude from this that resonance in a series *RLC* circuit produces minimum circuit impedance (only the resistance *R* is left). It also produces maximum circuit current.

As the frequency of the voltage source is increased above resonance, the value of the inductive reactance becomes greater than the value of the capacitive reactance. When this

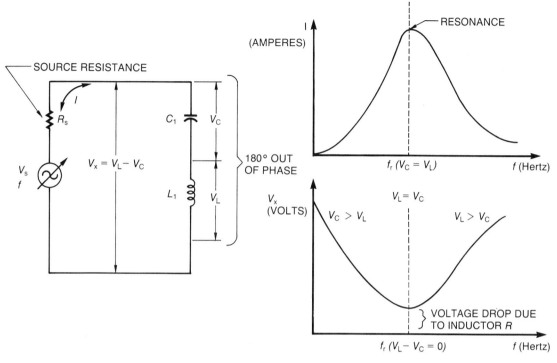

Figure 1-8 Reactive voltages in series *RLC* circuit.

happens, the total circuit impedance will begin to increase and the circuit current will start to decrease. These relatitonships are shown in Table 1-1.

Parallel Resonance

The resonant frequency of a parallel *LC* circuit may be determined by the same method as for a series resonant circuit.

Table 1-1	EFFECTS OF CHANGING FREQUENCY IN A SERIES *RLC* CIRCUIT				
f	V_L	V_C	$V_L - V_C$	*Z*	I_T
Below f_r	decr.	incr.	incr.	incr.	decr.
At f_r	$-V_C$	$-V_L$	0	min.	max.
Above f_r	incr.	decr.	incr.	incr.	decr.

Example 2

What is the resonant frequency of the circuit in Figure 1-9?

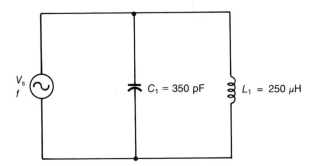

Figure 1-9

Solution

The resonant frequency is

$$f_r = \frac{1}{2\pi\sqrt{LC}}$$

$f_r = 538$ kHz

(Equation 1-8)

The resonant frequency of the circuit in Figure 1-8 is 538 kHz.

Frequency Response of Parallel RLC Circuits

For a parallel *RLC* circuit, the current in the capacitor branch is 180° out of phase with the current in the inductor branch.

It is helpful to compare the phase relationships of a series *RLC* circuit and a parallel *RLC* circuit. See Figure 1-10. Note that for the series *RLC* circuit, the current is everywhere the same.

For the parallel *RLC* circuit, voltage is everywhere the same. This means that the voltage (both phase and magnitude) across each component of the circuit is the same. Only the phase and magnitude of the current in each component of the circuit can be different. Table 1-2 summarizes this discussion.

Currents in a Parallel LC Circuit

Current behavior in a parallel *LC* circuit is summarized in Table 1-3.

Conclusion

The series *LC* circuit is sometimes referred to as the dual of the parallel *LC* circuit. You can see the reason for this. At resonance, the impedance of a series *LC* circuit is a minimum, and maximum circuit current flows. Exactly the opposite is true for the parallel *LC* circuit.

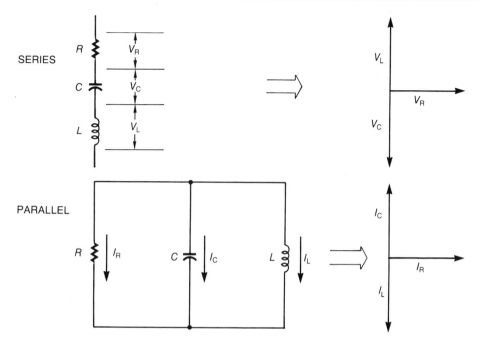

Figure 1-10 Phase comparisons of series and parallel *RLC* circuits.

Table 1-2	VOLTAGE AND CURRENT RELATIONS IN *RLC* CIRCUITS			
Series *RLC* Circuit	**Circuit Currents**			
	I_T	I_R	I_C	I_L
	All in phase and equal in value			
	Circuit Voltages			
	V_T	V_R	V_C	V_L
	Phase changes with I_T	In phase with I_T	Lags I_T by 90°	Leads I_T by 90°
Parallel *RLC* Circuit	**Circuit Voltages**			
	V_T	V_R	V_C	V_L
	All in phase and equal in value			
	Circuit Currents			
	I_T	I_R	I_C	I_L
	Phase changes with V_T	In phase with V_T	Leads V_T by 90°	Lags V_T by 90°

Table 1-3	EFFECTS OF CHANGING FREQUENCY IN A PARALLEL *LC* CIRCUIT				
f	I_L	I_C	$I_C - I_L$	Z	I_T
Below f_r	incr.	decr.	incr.	decr.	incr.
At f_r	$-I_C$	$-I_L$	0	max.	min.
Above f_r	decr.	incr.	incr.	decr.	incr.

These concepts are very important in the analysis of electronic communication systems. Learn them well.

1-5 Review Questions

1. Define resonance.
2. What are the differences between resonance in a series *LC* circuit and in a parallel *LC* circuit?
3. When is the voltage across the inductor exactly equal and opposite to the voltage across the capacitor in a series *LC* circuit?
4. Is the current in a series *LC* circuit ever equal and *opposite* in any of its reactive components? Explain.
5. When is the current in the capacitive branch equal and opposite to the current in the inductive branch of a parallel *LC* circuit?
6. Is the voltage in a parallel *LC* circuit ever equal and *opposite* in any of its reactive components? Explain.
7. What is the resonant frequency of a circuit containing only a 12-mH inductor and a 5-µF capacitor? Does it make any difference if these components are in series or in parallel?
8. What value of inductor should be used with a 1-µF capacitor in order to have a circuit with minimal current at 1 MHz? How should these components be connected?
9. What value of capacitor should be used with a 5-µH inductor in order to have a circuit with a maximum current at 500 MHz? How should these components be connected?

TROUBLESHOOTING AND INSTRUMENTATION

Discussion

This section introduces the concepts of troubleshooting. *Troubleshooting* is the act of locating and repairing in an electrical circuit a fault that causes improper operation of the circuit.

Instrumentation is the use of equipment to measure electrical quantities for the purpose of keeping these quantities within prescribed limits to ensure proper system operation. More will be said about instrumentation in the next chapter.

Troubleshooting

Several major categories of faults in electronic communications systems are as follows:

1. *Complete failures*

   This is usually the easiest kind of problem to correct. A complete failure means that the entire electrical system is inoperative. A complete failure may occur because the power cord is not connected to the wall outlet, or the on–off switch may be faulty, or the electrical power supply within the system is defective.

2. *Poor system performance*

   This area can be more difficult to analyze. Poor performance means that the system is not operating within the performance limits of its original design. Thus a communication receiver may not be receiving all of the stations it was designed to receive. A transmitter may not be conveying all of the information it was designed to handle. Poor system performance may be due to a weak component within the system. A knowledge of system operation and proper instrumentation is necessary to locate errors due to poor system performance.

3. *Tampered equipment*

   An old saying among communication technicians is, "If it works, don't fix it!" Tampering with equipment for the purpose of experiment, modification, or attempted repairs can leave the system inoperative or performing poorly.

4. *Intermittent fault*

   This problem is the most difficult. An intermittent is an inconsistant fault that causes the system to be inoperative or perform poorly. The difficulty of finding the cause of an intermittent is in keeping the problem consistent. Intermittents can be caused by mechanical defects, temperature changes, or erratic electrical behavior of components and connections.

5. *Massive traumas*

   These system failures may cause more than one part of the system to fail. A massive trauma is a system failure caused by an outside intrusion into the system that results in an inoperative system, poor system performance, or intermittents. Examples of massive traumas are fire, smoke, dropped equipment, water immersion, lightning damage, and applications of voltages with incorrect polarities or values.

You will learn how to troubleshoot from examples in the text and from the accompanying laboratory manual.

1-6 Review Questions

1. How would you define troubleshooting? instrumentation?
2. Name the five major categories of troubleshooting.
3. What is usually the easiest category of troubleshooting? What is the most difficult?
4. Define an intermittent. Give an example.

MICROCOMPUTER SIMULATION

The first troubleshooting simulation program on your disk is called **DISK INTRODUCTION**. The purpose of this program is to introduce you to the details of how your troubleshooting simulation programs work and how you can take advantage of the other disk features.

The first time you use your disk, it will ask you to enter your name as well as some other information. All of the information entered by you will become a part of a permanent record on your disk and cannot be changed. Your instructor will need this information to identify your disk from those of other students.

CHAPTER PROBLEMS

(Answers to odd-numbered problems appear at the end of the text.)

1. For the circuits in Figure 1-11, find the total current.

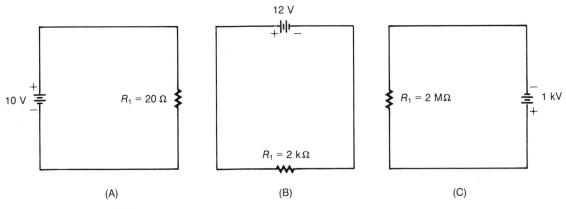

Figure 1-11

2. Find the total current for the circuits in Figure 1-12.

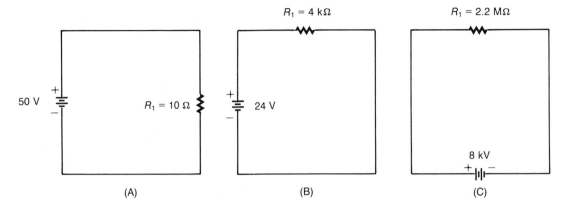

Figure 1-12

3. Using the circuits shown in Figure 1-13, find the value of the resistor.

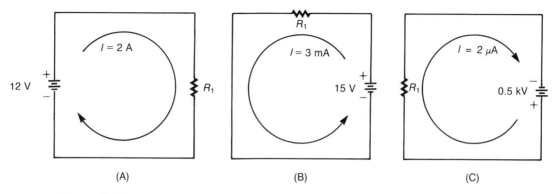

(A)   (B)   (C)

Figure 1-13

4. What is the value of the resistor for each circuit in Figure 1-14?

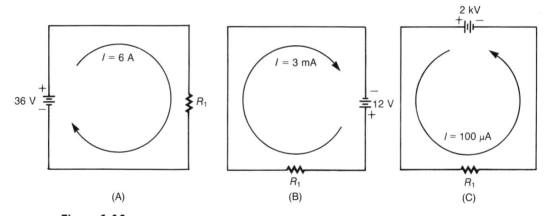

(A)   (B)   (C)

Figure 1-14

5. Calculate the applied voltage for the circuits in Figure 1-15.

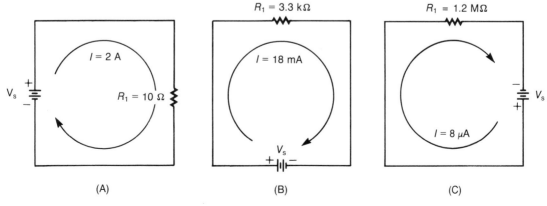

(A)   (B)   (C)

Figure 1-15

6. What is the power dissipation of each resistor in Figure 1-16?

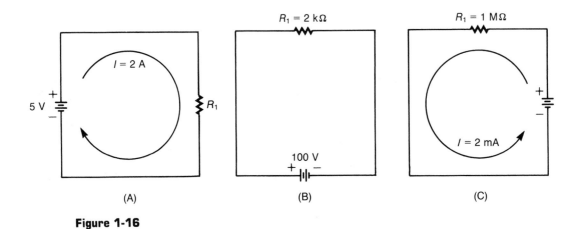

(A)   (B)   (C)

Figure 1-16

7. For the circuits in Figure 1-17, what is the power dissipation of each resistor?

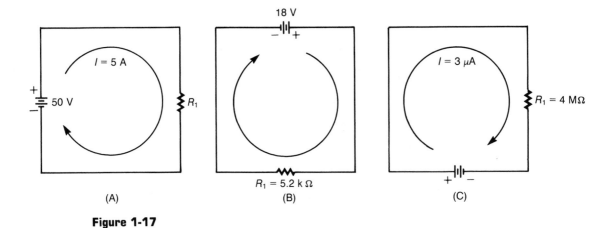

(A)   (B)   (C)

Figure 1-17

8. Calculate the total resistance for each circuit in Figure 1-18.

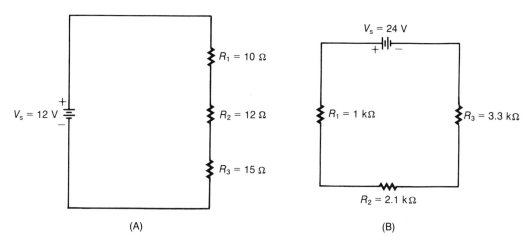

(A)   (B)

Figure 1-18

9. Solve for the total resistance for each circuit in Figure 1-19.

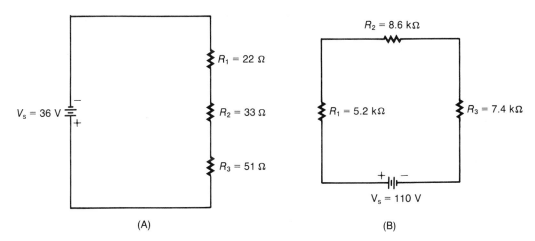

(A)   (B)

Figure 1-19

10. For each circuit in Figure 1-20, calculate the total resistance.

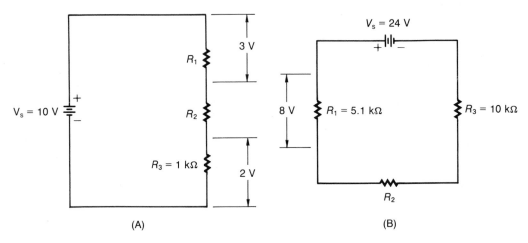

(A)   (B)

Figure 1-20

11. Find the total current for each circuit in Figure 1-19.
12. Solve for the total current in each circuit of Figure 1-20.
13. What is the voltage drop across each resistor in Figure 1-19?
14. Calculate the voltage drop across each resistor for the circuits in Figure 1-20.
15. Determine the power delivered by the source for the circuits in Figure 1-19.
16. What is the total power dissipated by the circuits in Figure 1-20?
17. For the circuits in Figure 1-21, determine the total resistance.

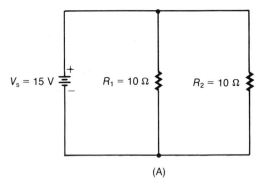

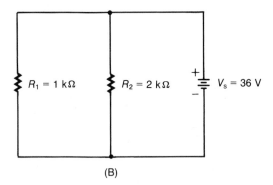

(A)   (B)

Figure 1-21

18. Find the total resistance for each circuit in Figure 1-22.

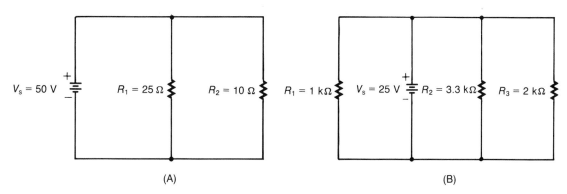

(A)   (B)

Figure 1-22

19. What is the current through each branch of the circuits in Figure 1-23?

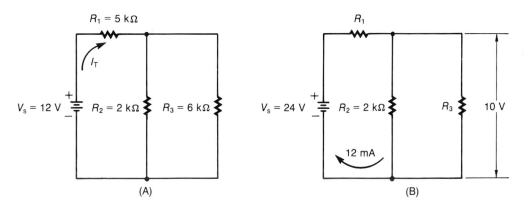

(A)   (B)

Figure 1-23

20. For the circuits in Figure 1-22, what is the current in each resistor?
21. For the circuit in Figure 1-23(B), find the value of R_3.

22. Find the indicated value for each circuit in Figure 1-24.

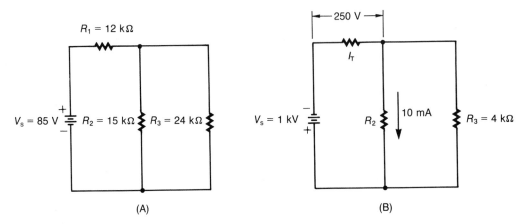

(A) (B)

Figure 1-24

23. What are the Thevenin equivalents of the circuits in Figure 1-25?

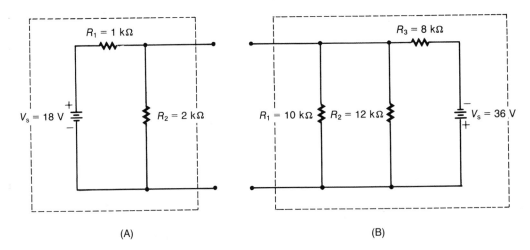

(A) (B)

Figure 1-25

24. Determine the Thevenin equivalent for each circuit in Figure 1-26.

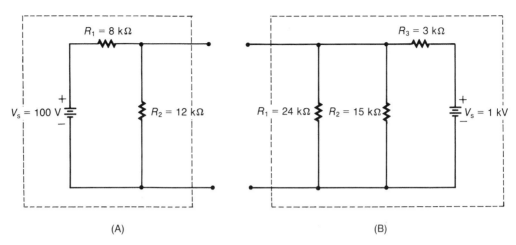

(A)   (B)

Figure 1-26

25. Find the Norton equivalent for each circuit in Figure 1-27.

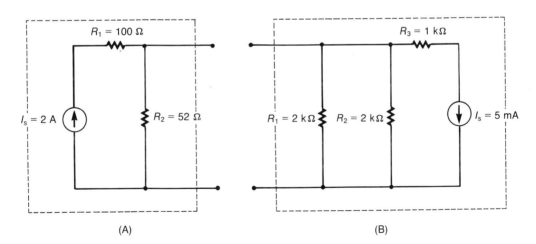

(A)   (B)

Figure 1-27

26. Solve for the Norton equivalent for each circuit in Figure 1-28.

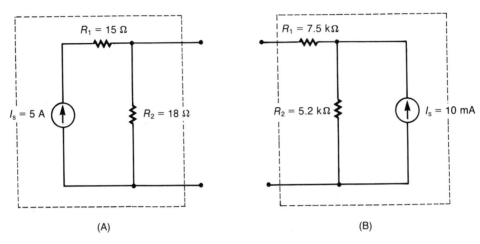

(A)   (B)

Figure 1-28

27. Determine the maximum power that can be delivered to the load for each circuit in Figure 1-29.

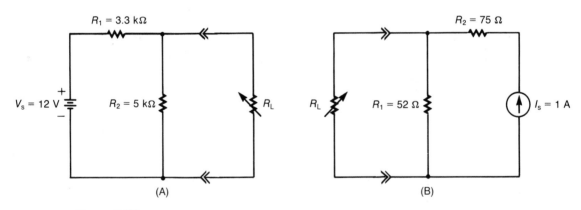

(A)   (B)

Figure 1-29

28. What is the maximum power that can be delivered to the load for the circuits in Figure 1-30?

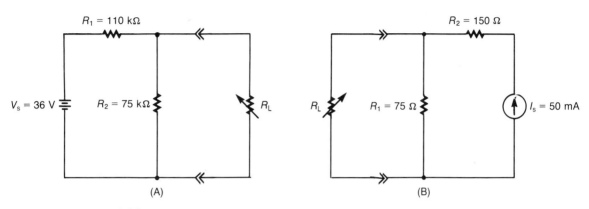

(A)   (B)

Figure 1-30

29. For the waveforms in Figure 1-31, determine (A) V_P, (B) V_{P-P}, (C) V_{rms}, (D) frequency, and (E) period.

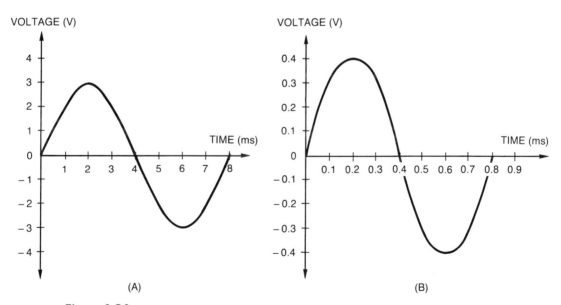

(A)   (B)

Figure 1-31

30. For each waveform in Figure 1-32, find (A) V_P, (B) V_{P-P}, (C) V_{rms}, (D) frequency, and (E) period.

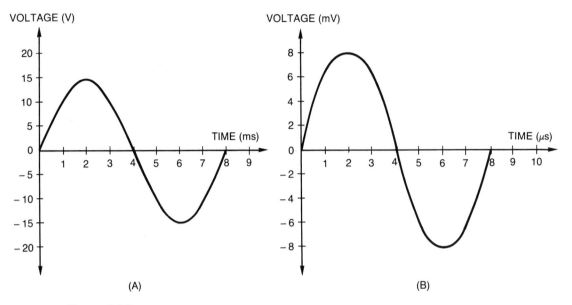

Figure 1-32

31. Find the power dissipated in each resistor in Figure 1-33.

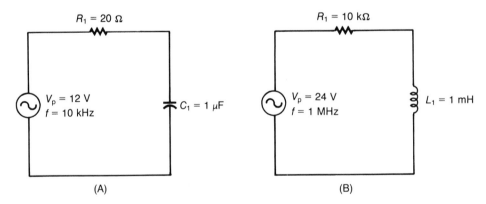

Figure 1-33

32. What is the power dissipated by each resistor in Figure 1-34?

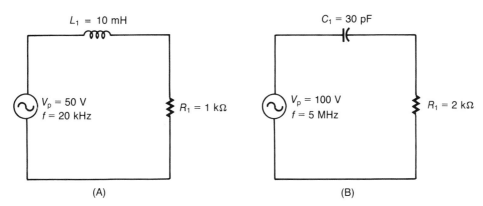

(A)   (B)

Figure 1-34

33. Determine the output voltage and output current for each transformer circuit in Figure 1-35.

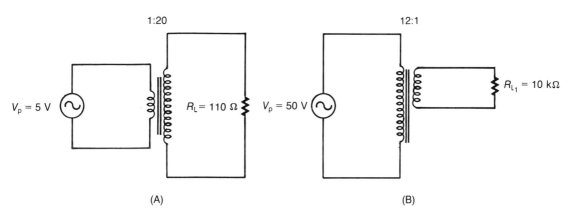

(A)   (B)

Figure 1-35

34. What are the output current and output voltage for the transformer circuits in Figure 1-36?

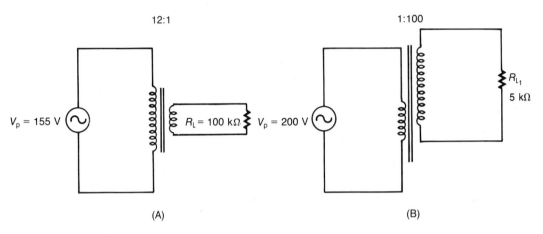

(A)   (B)

Figure 1-36

35. Determine the reflected primary impedance for the transformer circuits in Figure 1-35.
36. Calculate the reflected primary impedance for the transformer circuits in Figure 1-36.
37. Calculate the total reactance of the inductive and capacitive circuits in Figure 1-37.

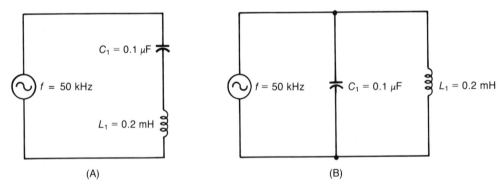

(A)   (B)

Figure 1-37

38. Determine the total reactance of the inductive and capacitive circuits in Figure 1-38.

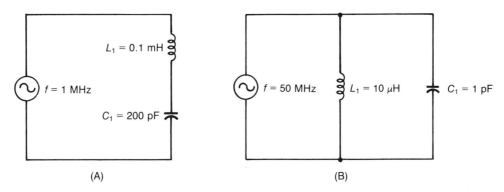

(A)   (B)

Figure 1-38

39. Find the impedance of each circuit in Figure 1-39.

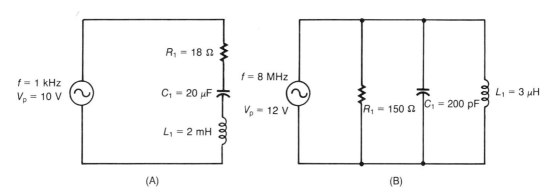

(A)   (B)

Figure 1-39

40. What is the impedance of each circuit in Figure 1-40?

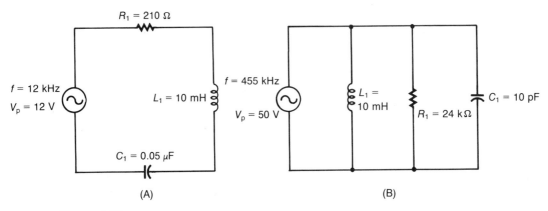

(A) (B)

Figure 1-40

41. For the circuits in Figure 1-39, determine the phase angles.
42. What is the phase angle for each circuit in Figure 1-40?
43. Calculate the actual power delivered by the source for each circuit in Figure 1-39. What is the power factor for each circuit?
44. Determine the actual power delivered by the source for each circuit in Figure 1-40.
45. Calculate the impedance of each circuit in Figure 1-41.

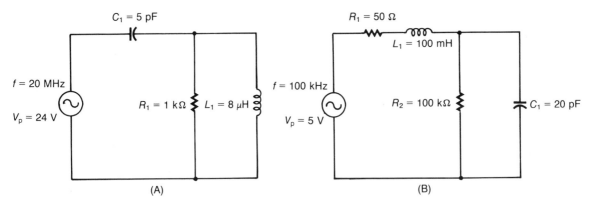

(A) (B)

Figure 1-41

46. For the circuits in Figure 1-42, what is the total impedance?

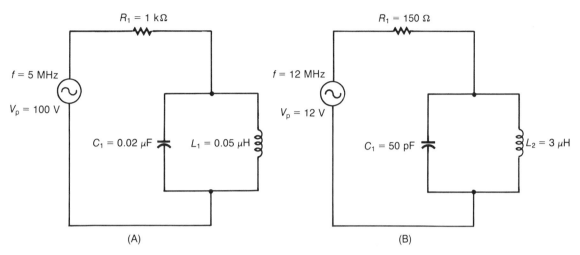

(A)   (B)

Figure 1-42

47. What is the phase angle for each circuit in Figure 1-41?
48. For each circuit in Figure 1-42, determine the phase angle.
49. What is the resonant frequency of each circuit in Figure 1-39?
50. Calculate the resonant frequency for each circuit in Figure 1-40.
51. What will be the currents at resonance for the circuits in Figure 1-39?
52. Determine the circuit current for each circuit in Figure 1-39.
53. How much current will flow at resonance for each circuit in Figure 1-40?
54. Calculate the total circuit currents at resonance for the circuits in Figure 1-41.
55. What will be the total circuit current at resonance for each circuit in Figure 1-42?

CHAPTER 2

Elements of Communication

OBJECTIVES

In this chapter, you will study:

- ☐ Some of the difficulties with early forms of communication.
- ☐ How electricity was first used to overcome some of the limitations of communications.
- ☐ What is used to make electrical currents assume the patterns of the human voice.
- ☐ How the sun's radiation differs from and is similar to the human voice.
- ☐ How radiation overcomes the disadvantages of the human voice.
- ☐ How radio stations differ.
- ☐ Some important concepts in instrumentation.

INTRODUCTION

This chapter introduces the reasons for electronic communications. The concepts here lay the foundation for all of the subsequent material. It's important to understand what a *carrier* is and why it's needed. Read this chapter carefully. It's important.

The second troubleshooting simulation program on your disk shows you how to troubleshoot a four stage amplifier using a signal generator.

2-1 | BASIC COMMUNICATION SYSTEMS

First Communication Systems

One of the most used communication systems is the human voice. It normally uses air as its *medium* for *transmission*. The mouth acts as the *transmitter*, the ear as the *receiver*.

The human voice has some major disadvantages: It cannot travel over long distances, and the louder you talk, the less private will be your message. Sound also travels relatively slowly, a little more than 1000 feet per second.

In the past, smoke signals, tribal drums, and the reflection of light off shiny objects were used to help reduce the distance problem. These methods didn't help much for privacy, but they did serve our ancestors for thousands of years.

Overcoming the Problem

If communications between people remained that way, civilization, as we know it today, could not exist. Envision living just 100 years ago. If you told someone then that someday you would be able to hear a conversation on the other side of the world while it was in progress, you would probably have been laughed at.

Suppose you told people back then that you would also be able to actually see what is happening thousands of miles away! Most certainly you would have been considered a strange person. Imagine pulling out your transistor radio or your portable TV and tuning it in (assuming someone had a transmitter)!

Everyone knows that we have overcome the limitations of the human voice. We have taken the powerful instrument of the spoken word and combined it with our intelligence to produce a combination relied upon by all but understood by few. The purpose of this text is to help you become one of the few who understand how electronic communication works and how to keep it working.

The basic communication system of the voice, ear, and eye has had its range increased to the limits of light itself.

With electronic recordings such as the audio and video cassette recorder, the sounds and visions of today can be quickly and economically preserved for the future.

2-1 Review Questions

1. What is one of the most common forms of human communication?
2. Name some disadvantages of the human voice for communications. State some advantages.
3. What were some of the methods used to overcome the limitations of the human voice?
4. How has electronics expanded your ability to acquire information?

2-2 | THE TELEGRAPH

First Telegraph System

The invention of the telegraph was a major breakthrough. It allowed people to communicate over long distances and, in theory, was private. A simple telegraph system is shown in Figure 2-1. When the electrical circuit is completed by closing the switch on the *transmitting* end, current flows, causing the relay to energize on the *receiving* end. When the relay energizes, a "clicking" sound is made, which lets the person at the receiving end know that the transmitting switch was closed. Opening and closing the switch and using a code known to sender and receiver (such as the Morse code) allowed information to rapidly exchange over long distances.

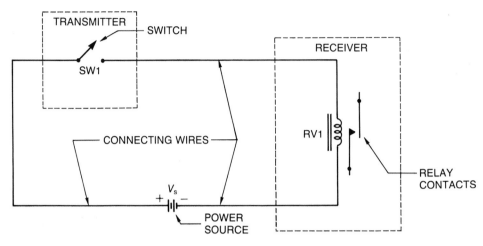

Figure 2-1 A simple telegraph system.

The advantage of that system was rapid transmission of information over very long distances. Its disadvantages were that wire was needed as a medium, and only one message at a time could be sent. In Chapter 8, you'll see how this was overcome through *multiplexing*.

2-2 Review Questions

1. What was the main advantage of the telegraph system? What were some of its disadvantages?
2. Why was a source of electrical energy needed for the telegraph system?
3. Briefly describe how a simple telegraph system operates.

2-3 | THE TELEPHONE

First Telephone

The telegraph system had another major disadvantage. Anyone wanting to use one had to know a code of dots and dashes, or pay someone who knew the code to interpret for them. The invention of the telephone opened an era of instant communications for anyone who could hear and speak the same language. The most basic features of a telephone system are shown in Figure 2-2.

The *transmitter* consists of a *microphone* that converts the sound waves of the speaker into electrical currents that essentially assume the same frequency and amplitude of the voice waves going into the microphone. These electrical currents in turn cause a changing magnetic field around the coil of the headset. This field has the same pattern of frequency and amplitude as the electrical currents in the circuit. As the magnetic field changes in frequency and amplitude, it causes the metal disk to move in the same pattern. Thus the metal disk is essentially moving in step with the voice waves of the person speaking into the microphone. The moving metal disk causes the air around it to move in much the same pattern as the original voice, and the listener hears a reproduction of what the speaker said. A device such as the microphone and the headset is called a *transducer*.

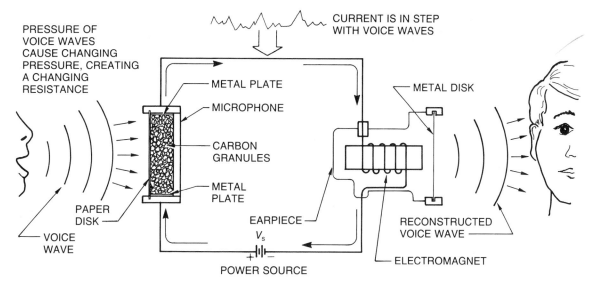

Figure 2-2 Basic telephone system.

The Importance of Electricity

The telegraph and the telephone both demonstrate the importance of electricity in overcoming the limitations of the human voice. With electricity, large distances could be covered at speeds close to 186,000 miles per second, and with some hope of a private conversation.

But the telegraph and the telephone both had one severe limitation: Wired connections needed between the transmitter and the receiver. This didn't work well for people on boats. It made communications over long distances static and sometimes unreliable. Any hope of communicating with or controlling devices in space was completely out of the picture.

Further investigation was taking place around this time concerning the exact nature of electricity. It was found that energy from the sun could influence electricity in a wire.

2-3 Review Questions

1. What was the major advantage of the telephone? State a disadvantage.
2. Why does a telephone system require electrical energy?
3. Describe how the basic telephone transmitter works. What is it called?
4. Describe how the basic telephone receiver works. What is it called?
5. What is a device such as the microphone and the headset sometimes called?

2-4   ELECTROMAGNETIC RADIATION

Definition

Electromagnetic radiation is electrical energy that can travel through space. Unlike the human voice, it doesn't require a *medium*, such as air or wires, for transmission. Light and heat can be viewed as electromagnetic radiation.

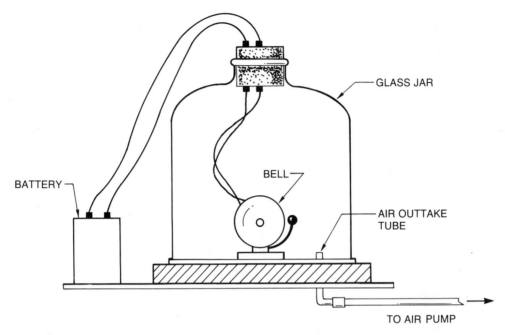

Figure 2-3 Experiment for sound and light.

Envision the energy given off by the sun. It travels at the speed of light (186,000 miles each second) in a vacuum. An experiment that illustrates this phenomenon is shown in Figure 2-3. As soon as air is removed from the glass jar, you can still see the bell ringing, but you can't hear it. This demonstrates that sound requires a medium for transmission, but light does not!

Components

Electromagnetic radiation contains electrostatic and magnetic fields. These fields can be measured in terms of frequency and intensity. What you detect as light is really a very small part of the entire frequency *spectrum*. The main parts of the frequency spectrum are shown in Figure 2-4. Note that the visible part of the spectrum is a very narrow *band* of frequencies within the electromagnetic spectrum. All of the frequencies within this spectrum travel at the speed of light and do not require a medium.

The important concept here is that since electromagnetic radiation consists of different frequencies, they can be thought of as sine waves. Thinking of electromagnetic radiation as sine waves is one of the most important concepts in electronic communications. This means that every electromagnetic wave has a frequency, period, amplitude, and phase relationship with other electromagnetic waves.

2-4 Review Questions

1. What is electromagnetic radiation?
2. How does electromagnetic radiation differ from the human voice? How is it the same?
3. Describe a simple experiment that demonstrates the difference between light and sound.

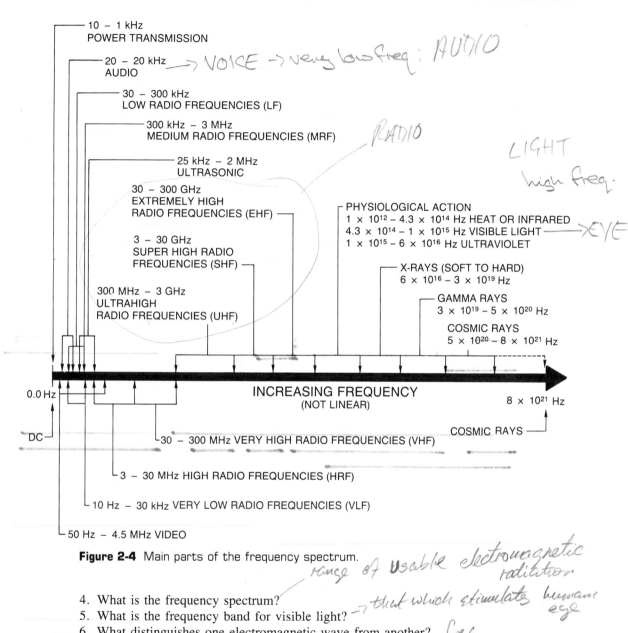

Figure 2-4 Main parts of the frequency spectrum.

4. What is the frequency spectrum?
5. What is the frequency band for visible light?
6. What distinguishes one electromagnetic wave from another?

2-5 | CONCEPT OF A CARRIER

All communication systems that can send and receive information through space, such as radio, television, and satellite communications, use electromagnetic radiation. When information, such as the human voice, uses electromagnetic radiation to transmit the infor-

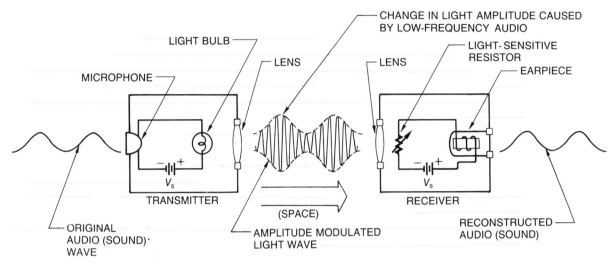

Figure 2-5 Simple "flashlight" communication system.

mation (much the same as current in a wire was used to transmit the voice patterns), then the electromagnetic radiation is called a *carrier*. It is called this because it "carries" the information.

Figure 2-5 shows a simple flashlight communications system where light is used as the carrier. If you imagine this system as waveforms, the light would be represented as a very high frequency and the voice as a very low frequency. What is happening is that the voice is causing the amplitude of the light wave to change. There is no voice in the light wave, only the change in amplitude (intensity) of the light, which represents a pattern identical to the original voice wave. Such a waveform is shown in Figure 2-6.

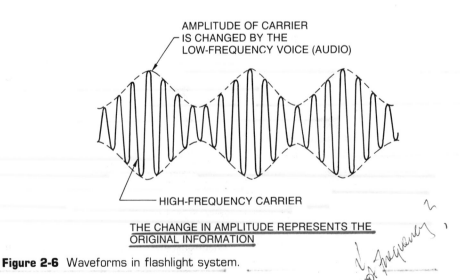

Figure 2-6 Waveforms in flashlight system.

With such a system, a voice pattern (or any other pattern of information) can now be carried by electromagnetic radiation. Here the information to be sent (in this example a voice wave) is causing a measurable characteristic of the carrier to change. This process is called *modulation.* The voice wave is causing the amplitude of the carrier wave to change. Hence, this process is called *amplitude modulation* (AM).

Information can cause a measurable characteristic of the carrier wave to change in other ways. When the information causes the frequency of the carrier wave to change, then the modulation process is called *frequency modulation* (FM). If the transmitted information causes the phase of the carrier wave to change with respect to some standard, then the modulation process is called *phase modulation* (PM).

In the simple flashlight communications system, the receiver takes the measurable changes of the carrier wave, which represents the transmitted information, and restores it to its original form. For the flashlight system, the original form was a voice wave. The process of removing the transmitted information from the carrier and restoring it to its original form is called *demodulation.*

At last, by the use of a carrier wave, information can be sent at the speed of light. No medium is needed to do this. Boats, spaceships, and star-seeking robots can respond to your commands. In return, valuable information can be returned to you just as quickly and accurately.

2-5 Review Questions

1. What kind of radiation is used by radio, television, and satellite communications?
2. What is a carrier?
3. In the simple flashlight communications system, what is the carrier?
4. What changes does the voice wave cause in the simple flashlight communications system?
5. Describe the modulation process.
6. What characteristics of a carrier can be modulated?
7. What does AM mean? Describe the process.
8. What does FM mean? Describe the process.
9. What does PM mean? Describe the process.
10. What does demodulation mean?

2-6 | FREQUENCY ALLOCATIONS

How Radio Stations Differ

The Federal Communications Commission (FCC) assigns different carrier frequencies to all transmitting radio and television stations. They do this so that you may select the one you want to receive.

This process, called *frequency-division multiplexing* (FDM), allows many different kinds of information to be transmitted at the same time without interfering with each other. Figure 2-7 illustrates this concept.

You can think of the flashlight system discussed in the last section. If frequency division multiplexing were used, it would mean that you would see the lights of many

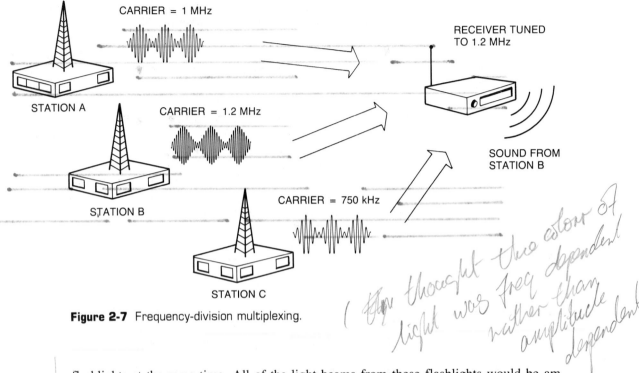

Figure 2-7 Frequency-division multiplexing.

(handwritten note: "thought this colour of light was freq dependent rather than amplitude dependent")

flashlights at the same time. All of the light beams from these flashlights would be amplitude modulated. You would be able to tell the difference between each flashlight by the color (frequency) of the transmitting light. Thus if you wanted to receive the signal from the "green" flashlight and none of the others, you would place a green filter over your receiver to block out all light except for the green light. You would then receive the information being transmitted by the green transmitter and none of the others.

Radio and Television Receivers

The preceding flashlight analogy is very similar to how radio and TV reception works. In your radio or TV, there is an electrical filter called a *tuner*. The tuner receives only one transmitting frequency and rejects all others. You will study more about tuners in the next chapter. For now, the important concept is to realize that different transmitters within your reception area all use carriers set to different frequencies.

2-6 Review Questions

1. How does the Federal Communications Commission cause one radio station to be distinguished from another?
2. What is frequency-division multiplexing?
3. Describe a simple flashlight transmitting system that uses frequency-division multiplexing.
4. What electrical device inside a communication receiver selects one station and rejects all others?

TROUBLESHOOTING AND INSTRUMENTATION

Instrumentation

When devices are used to measure electrical quantities, a general procedure is used by the experienced technician. The procedure is as follows:

1. *A decision is made to make the measurement*
   This decision may be made after you determine that a system fault exists. This determination usually results from experience with the system to be measured.
2. *A measurement procedure is selected*
   The procedure may consist of making sure that all system operating controls have been properly set, or it may be a good visual inspection. The technician may "feel" for the presence or absence of heat, based on what the technician has experienced as normal. The technician's sense of smell can detect the odor of a burning resistor or transformer.
3. *The measurement is conducted*
   Here the actual measurement is made. It is important to record what the measurement indicated. There is another old saying: "The lightest pencil mark is better than the world's best memory." There are usually good reasons for old sayings.
4. *The data is analyzed*
   The process of analyzing the data consists of comparing the recorded measurement to some standard. This step is very important. From previous measurements, experienced technicians usually know what the standard should be. If the equipment being measured comes with complete documentation, then electrical values printed on schematics or measurement tables are used as the standard. If none of these are available, then making the same measurement on an identical working system, if available, will yield a standard of reference. Without any of these standards, the technician may have to mathematically analyze the circuit.
5. *Act on the analysis*
   If a discrepancy exists between the measured data and the standard, then some action is required, such as troubleshooting to correct the fault, reporting the discrepancy to others, or simply giving the system to someone else for troubleshooting.

2-7 Review Questions

1. Before making a system measurement, what must you be sure of?
2. State a good habit when making measurements.
3. What do you do when analyzing measurement data?
4. What are some actions to be taken when a measurement discrepancy is found?

MICROCOMPUTER SIMULATION

The second troubleshooting simulation program on your disk helps you develop your troubleshooting skills for the task of troubleshooting a multi-stage amplifier using a signal generator. Be sure to activate the **instruction mode** and then the **demonstration mode** before starting the test mode. Do this for all of the simulation programs on your disk.

CHAPTER PROBLEMS

(Answers to odd-numbered problems appear at the end of the text.)

TRUE/FALSE
Answer the following questions true or false.

1. The most common form of communication is the human voice.
2. The human voice travels at the speed of light.
3. One major advantage of the telegraph system is the ability to carry information quickly over long distances.
4. An advantage of the telegraph system is that no electrical energy is required.
5. The most basic telegraph system functions by opening and closing a switch in a simple series circuit.

MULTIPLE CHOICE
Answer the following questions by selecting the most correct answer.

6. The transmitting part of a simple telephone system is the:
   (A) headset.
   (B) microphone.
   (C) battery.
   (D) none of these.
7. In a telephone system, the microphone:
   (A) converts electricity into sound.
   (B) converts sound into electrical movements.
   (C) powers the phone system.
   (D) shows the user where to speak.
8. In a telephone system the headset:
   (A) converts electrical movement into sound.
   (B) reconstructs the original voice.
   (C) converts the voice into electrical movement.
   (D) does both (A) and (B).
9. A transducer is:
   (A) a microphone.
   (B) a battery.
   (C) a headset.
   (D) both a microphone and a headset.
10. The major disadvantage of the telephone and the telegraph is:
    (A) their use of electricity.
    (B) their expense.
    (C) their need for wires.
    (D) none of these.

MATCHING

Match the correct abbreviations on the right to the statements on the left.

11. Changes the carrier frequency   (A) FM
12. Changes the carrier amplitude   (B) PM
13. Changes the carrier phase   (C) AM
14. Modulates the signal strength   (D) None of these
15. Modulates the color of the light

FILL IN

Fill in the blanks with the most correct answer(s).

16. Energy from the sun is called _____ radiation.
17. Unlike the human voice, electromagnetic radiation does not require a _____ .
18. Electromagnetic radiation contains an electrostatic field and a/an _____ field.
19. The entire range of _____ of electromagnetic radiation is called the frequency spectrum.
20. A small range of frequencies is called a _____ of frequencies.

OPEN ENDED

Answer the following questions as indicated.

21. Describe what properties of an electromagnetic wave can be measured.
22. How can one electromagnetic wave be distinguished from another?
23. Describe the concept of a carrier. Why is a carrier necessary?
24. What is the difference between the carrier and the modulating signal? How are they the same?
25. What is frequency-division multiplexing? How does this distinguish one transmitter from another? What unit in the receiver selects the signal?

CHAPTER 3

Elements of Radio Communication

OBJECTIVES

In this chapter, you will study:

- [ ] The purpose of an antenna ground system.

- [ ] The function of a tuner and how it receives different frequencies.

- [ ] Why the reproducer cannot function directly from the carrier.

- [ ] What detection is and why it is necessary.

- [ ] What all communication receivers have in common.

- [ ] What a spectrum analyzer does and how it operates.

- [ ] The fundamentals of waveform analysis and the meaning of the Fourier series.

- [ ] What block diagrams mean and how to troubleshoot them.

INTRODUCTION

This chapter is a departure from the systems to circuits approach of this text. There is a good reason for this. Every circuit component used in this chapter, with the exception of the diode, should already be familiar to you. What a diode does will be explained here. Understanding what all communication receivers must have to function will give you a good overview for the remainder of the systems section of this text.

The third troubleshooting simulation program on your disk will help you improve your troubleshooting efficiency. Here the program will keep track of the steps you use in trouble-shooting and advise you accordingly.

3-1 | ANTENNA GROUND SYSTEM

Discussion

This section starts with the *front end* of a communication receiver. You are introduced to the reception part of a communication system first because most readers are more familiar with receivers than transmitters. The next chapter introduces what happens to the transmitted signal once it is received by the antenna ground system.

Antenna Ground System

All electronic communication systems require an *antenna ground* system. An antenna is any device that converts electromagnetic radiation into electrical signals. An antenna could be a piece of wire or your body. You'll learn more about the important details of antennas in Chapter 10. For the example here, the antenna will be a 6-foot length of wire.

Giving the antenna a return path to the earth (earth ground) gives it a source and a storage place for electrons. This makes it easy to convert radio waves into electrical currents. The schematic of an antenna ground system is shown in Figure 3-1(A). What is happening to it is shown in Figure 3-1(B).

An antenna ground system will convert *all* radio signals (and anything else that comes along with them) into electrical signals. So if you were trying to receive the AM carrier wave of one transmitting station (say it had a carrier frequency of 1 MHz), your antenna ground system would pick it up and also all of the other radio stations in your area.

Recall from the last chapter that frequency-division multiplexing (FDM) is used to distinguish one station from another. Assume that there are three commercial AM stations in your area. Your antenna ground system would be converting each of their carrier waves into electrical currents, as shown in Figure 3-2. As you discovered in the last chapter, a tuner was used to select one station and reject all the others. Tuners are the subject of the next section.

Conclusion

In this section you were introduced to that part of all communication receivers that converts the radio waves into electrical energy. The chapter on antennas will present a detailed discussion about antenna theory. For now, know what an antenna does and why it's needed.

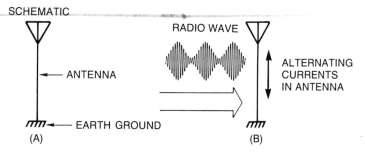

Figure 3-1 Antenna ground system.

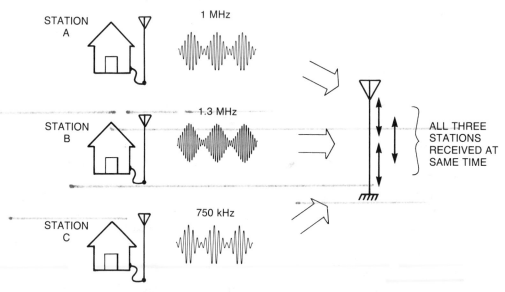

Figure 3-2 Problems with antenna ground system.

3-1 Review Questions

1. What is an antenna?
2. What does an antenna do?
3. What is the difficulty in working with an antenna?

3-2 | THE TUNER

Discussion

This section will demonstrate a most practical application of the resonant circuits you studied in AC. These circuits are used as tuners; without them, modern communications would not be possible. This is what makes frequency division multiplexing work.

The Tuner

A *tuner* is an electrical circuit capable of selecting one frequency and rejecting all others. The most basic tuner is a resonant circuit. Figure 3-3 shows the two basic types of resonant circuits.

Recall from your basic electrical circuits course that a parallel resonant circuit has maximum impedance at resonance, and a series resonant circuit has minimum impedance at resonance. If you connected a parallel resonant circuit into the antenna ground system and selected the values of L and C so that the resonant frequency of the circuit were 1 MHz, then you would have maximum voltage across the circuit at 1 MHz and no other frequency. Figure 3-4 shows this arrangement.

With the addition of a parallel resonant circuit, your AM receiver will be able to *tune* in one radio station (the one at 1 MHZ) and reject all others. Using a variable inductor

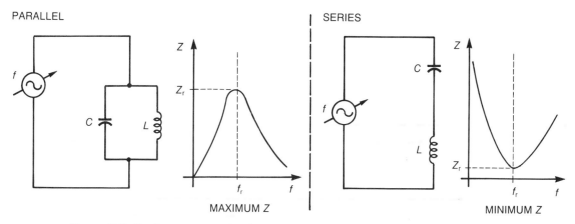

Figure 3-3 Basic resonant circuits.

or capacitor allows you to change the resonant frequency of the tuner, thus selecting different stations. For now, a variable capacitor will be shown to indicate that different radio stations may be selected. The signal you've captured contains the exact image of the transmitted AM radio wave. A way must be devised to convert this into something you can hear.

Conclusion

All communication receivers use tuners to select one station. It is important that you understand what the tuner does and its place in the total communication system.

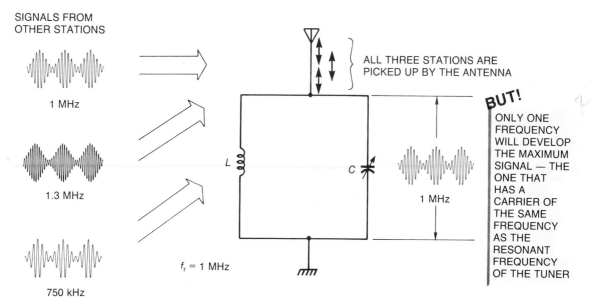

Figure 3-4 Effects of adding a parallel resonant circuit.

3-2 Review Questions

1. What is a tuner?
2. What is the purpose of a tuner?
3. What determines the station that will be selected by a tuner?
4. How can tuners be made to select different stations?

3-3 THE REPRODUCER

Discussion

In this section, you will discover why the headset will not work when connected directly to the tuner. Understanding why something doesn't work is helpful in understanding what does work.

Reproducer

In Chapter 1, you were introduced to the basic construction of a headset used in telephone communications. This is shown in more detail in Figure 3-5. For the telephone system, all currents flowed in one direction because a battery was used. These fluctuating direct currents produced a changing magnetic field that made the metal disk move in step with the changing current and thus reproduce sound.

The simplest radio receiver does not have a battery. All of the energy to operate this kind of receiver comes from the radio wave itself. As an example, the radio wave for the given tuner could be changing at a rate of 1 000 000 times each second!

Putting the headset across the tuner, as shown in Figure 3-6, doesn't cause the metal disk to produce any sound. This arrangement doesn't work for many reasons, one of which

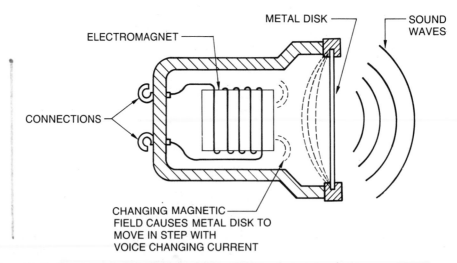

Figure 3-5 Construction of earpiece for headset.

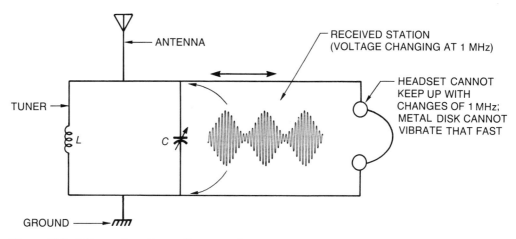

Figure 3-6 AM receiver that will not work.

is that the metal disk in the headset cannot vibrate at 1 MHz. Even if it could, you couldn't hear it!

What is needed is some way of causing only the *changes* in the *amplitude* of the carrier wave not the carrier wave itself, to move the metal disk. The changes in the amplitude of the carrier wave represent the sound; that's what you want to get. You don't need the carrier frequency any more; it's already done the job of getting from the station to you. The device that will restore the audio from the changes in the amplitude of the carrier wave is called a *detector*.

Conclusion

You are now ready to tackle the reason for a detector. As you will see in Chapter 16, there are many kinds of detectors. For now, it's important that you know why a detector is needed and that all communication receivers require some kind of detection.

3-3 Review Questions

1. What is the purpose of a reproducer?
2. Describe the basic construction of a headset.
3. What causes the headset to create sound?
4. Why isn't sound created when the headset is connected directly to the tuner?

3-4 | NEED FOR DETECTION

Discussion

Now that you know why a reproducer doesn't work when connected directly to a tuner, you will see what it takes to make it work. What you are trying to do is to reconstruct the original audio from the *amplitude changes* of the radio wave. Keep this in mind as you study this section.

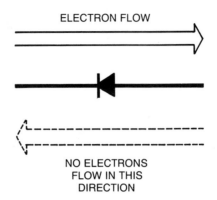

ELECTRON FLOW

NO ELECTRONS
FLOW IN THIS
DIRECTION

Figure 3-7 Schematic of a diode and direction of electron flow.

The Detector

A detector is an electrical device or circuit that detects the information imposed on a carrier wave. For the simplistic AM receiver, the detector is a *diode*. A diode is an electrical device that allows current to flow in only one direction. The schematic symbol of a diode is shown in Figure 3-7.

When a diode is added to the simple AM receiver, the current waveform of the received signal is converted from AC to pulsating DC. Figure 3-8 shows what happens. The diode has caused the currents in the headset to flow in the same direction, but they are still pulsating at the alarming rate of 1 000 000 times each second. This is still too fast for the metal disk or your ear.

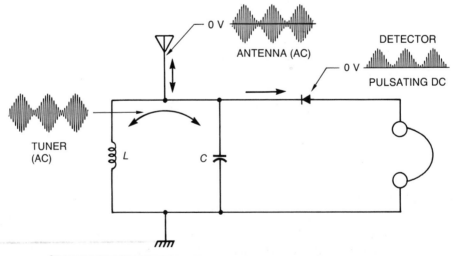

Figure 3-8 Electrical currents in a simple AM receiver with diode added.

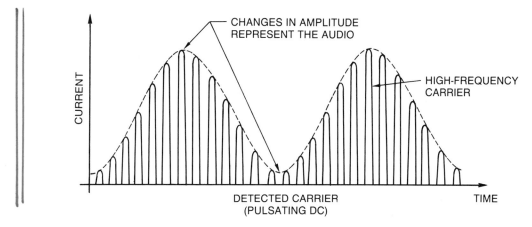

Figure 3-9 Audio on the detector carrier.

Restoring the Audio Information

If you analyzed the resultant waveform caused by the diode, the audio information has been preserved. Look at Figure 3-9. The audio can be reconstructed if a capacitor is added across the headset. See Figure 3-10 for what happens when the capacitor is added.

Conclusion

You now have a complete receiver system. This is something that you can build in the lab and it will actually work. These types of receivers were very popular at the beginning of the radio age. They were called *crystal* receivers. The crystal was the detector. These receivers do not need a battery; all of their energy came from the received signal. The next section is a summary of what is needed by all communication receivers.

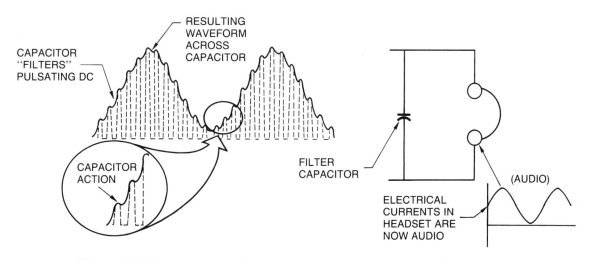

Figure 3-10 Reconstructing the audio.

3-4 Review Questions

1. Define a detector.
2. What is a diode?
3. When a diode is used as a detector in the simple AM radio, what happens to the waveform of the carrier?
4. What device other then a diode is required to reconstruct the modulating signal?

3-5 | THE BASIC RECEIVER

Discussion

This section brings together all of the new information covered in these last two chapters. It is the foundation section for the rest of the information that will follow in receiver systems.

The Basic Receiver

Figure 3-11 shows the most basic AM receiver and the waveforms of each section. *All* radio communication systems require the four sections in the most basic radio receiver:

1. An antenna ground system
2. A tuner
3. A detector
4. A reproducer

This is true for your stereo FM radio, your TV set and the satellite around Pluto taking orders from earth. All of them must have at least these four basic sections.

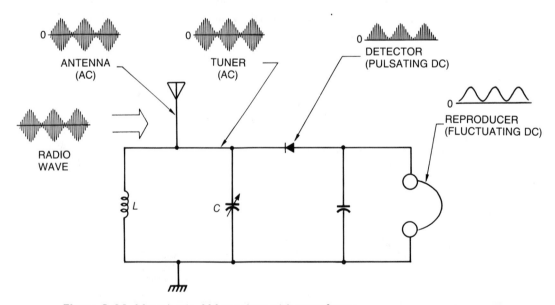

Figure 3-11 Most basic AM receiver with waveforms.

TROUBLESHOOTING AND INSTRUMENTATION

Discussion

Schematics of complex communication circuits that show all the resistors, capacitors, inductors, diodes and other circuit components are sometimes too detailed. Occasionally, a more simplified representation is desired, especially when trying to determine the signal flow through a communications system. This can be done with *block diagrams*.

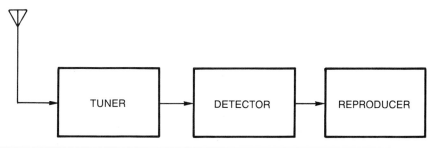

Figure 3-23 Block diagram of simple AM receiver.

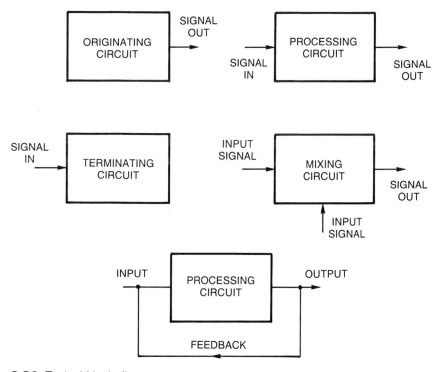

Figure 3-24 Typical block diagrams.

Table 3-1	BLOCK DIAGRAMS AND TROUBLESHOOTING TECHNIQUES		
Type of Circuit	**Function**	**Troubleshooting Method**	**Example(s)**
Originating	Creates own signal	Check output with scope or meter to see if signal is present	Oscillator
Processing	Changes the signal or impedance matching	Check input and output and compare the two signals for desired results	Amplifier Tuner
Terminating	Performs a final function	If input signal is OK and terminating circuit does not produce desired output, then problem is in terminating circuit	Headset CRT Relay
Mixing	Combines two signals to produce a single out	Check both input signals; then check for desired output	Mixer
Feedback	Output effects the input	Break open feedback loop and test as a processing circuit	Automatic gain control

Block Diagrams

The simple AM receiver introduced in this chapter could be represented by the block diagram in Figure 3-23. The signal flow is from the antenna, through the tuner, through the detector, and to the reproducer. If the detector were not functioning (suppose it were open), then the received signal would still be present in the antenna and tuner, but there would be nothing at the input of the reproducer. Figure 3-24 shows the common types of block diagram representations for different circuits in communication systems. Table 3-1 shows what each block in Figure 3-24 could represent and presents a troubleshooting strategy.

Each block in a block diagram can be considered as a *stage*. The first step in troubleshooting a defective system is to use a block diagram and to first find the defective stage. Once the defective stage is located, the system schematic can then be referenced to find the exact cause of the problem.

In the next chapter, the troubleshooting and instrumentation section will introduce you to equipment used for signal injection and signal tracing. Both methods are widely used in troubleshooting communication systems to isolate a defective stage.

3-7 Review Questions

1. What is a block diagram?
2. How does a schematic differ from a block diagram?
3. When are block diagrams used? When are schematics used?
4. Describe the function of the (A) originating circuit, (B) processing circuit, (C) terminating circuit, (D) mixing circuit, (E) feedback circuit.
5. Give examples of each circuit in question 4.
6. State the troubleshooting procedure for each circuit in question 4.

MICROCOMPUTER SIMULATION

The third troubleshooting simulation on your disk is called **SIGNAL GENERATOR II**. This is similar to the previous troubleshooting simulation program with one major difference: the computer will keep track of the steps you actually use in troubleshooting! You will again be presented with a signal generator for troubleshooting a four stage amplifier, but this time you will be encouraged to maximize your troubleshooting efficiency. As you will see, when troubleshooting a four stage amplifier, only three tests are needed to find a single defective stage using signal generator techniques.

The skills you acquire in this troubleshooting simulation can be used in future troubleshooting simulations. More importantly, these are the foundation skills necessary to build your future as an electronics technician.

CHAPTER PROBLEMS

(Answers to odd-numbered problems appear at the end of the text.)

TRUE/FALSE
Answer the following questions true or false.
1. An antenna receives one station and rejects all others.
2. An antenna ground system converts radio waves into electrical currents.
3. The most basic tuner is a resonant circuit.
4. A tuner selects one station and rejects all others.
5. You can select different radio signals by changing the value of the capacitor in the tuner.

MULTIPLE CHOICE
Answer the following questions by selecting the most correct answer.
6. Connecting a headset directly across the tuner in a simple AM radio receiver:
   (A) produces sound in the headset.
   (B) will not produce sound in the headset.
   (C) will burn out the tuner.
   (D) will not work until a battery is added.
7. To make the headset reproduce the original modulating signal, the metal disk must:
   (A) move in step with the frequency of the received carrier.
   (B) move in step with the changes in amplitude of the received carrier.
   (C) vibrate fast enough to keep up with the high-frequency radio wave.
   (D) have a source of energy from a battery.
8. A diode detector:
   (A) converts AC into pulsating DC.
   (B) allows current to flow in one direction only.
   (C) detects the modulating signal.
   (D) does all of the above.
9. The audio is reconstructed from the radio wave by:
   (A) the action of the headset.
   (B) having the audio removed from the carrier.

(C) the action of a filter capacitor connected across the headset.

(D) none of the above.

10. Which of the following do not require a tuner:

(A) TV receiver?

(B) CB receiver?

(C) communications satellite?

(D) All communication receivers require a tuner.

MATCHING

Match the block diagram in Figure 3-25 to the correct function.

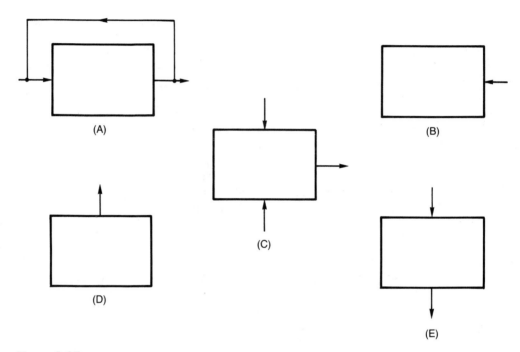

Figure 3-25

11. Processing circuit with feedback
12. Mixing circuit
13. Reproducer
14. Oscillator
15. Amplifier

FILL IN

16. Any periodic waveform consists of combinations of pure _____ waves.

17. A _____ analyzer is an instrument that indicates what frequencies are present and their amplitudes.

18. A square wave is made up of an infinite number of _____ harmonics.

19. A periodic waveform made up of an infinite number of all harmonics is the _____ .

20. The _____ _____ is a mathematical formula showing that all periodic functions are made up of pure sine waves.

OPEN ENDED

Answer the following questions as indicated.

21. For the frequency spectrum in Figure 3-26(A), what frequencies are present and what are their amplitudes?

22. For the frequency spectrum in Figure 3-26(B), what are the frequencies and their amplitudes?

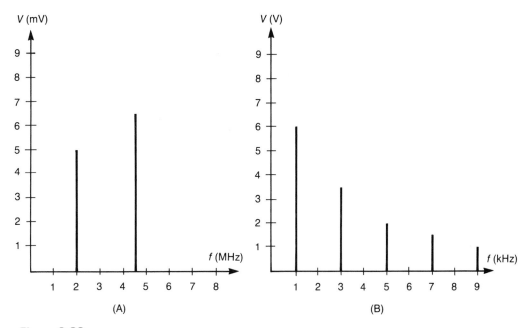

Figure 3-26

23. Determine the amplitude of the first three harmonics for the waveform in Figure 3-27(A).
24. For the waveform in Figure 3-27(B), determine the amplitude of the first three harmonics.
25. Construct a graph of the spectrum for each waveform in Figure 3-27.

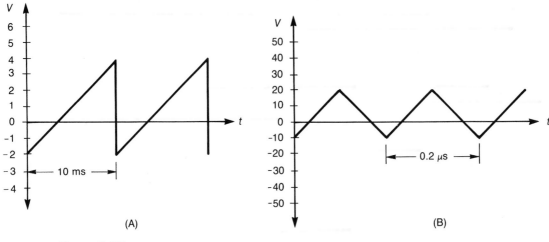

(A)

(B)

Figure 3-27

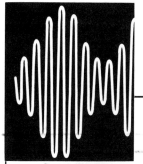

CHAPTER 4

Amplitude Modulation Systems

OBJECTIVES

In this chapter, you will study:

- [ ] How to analyze an AM waveform.
- [ ] The definition of noise and how noise is measured.
- [ ] How an AM waveform is measured using the modulation index.
- [ ] What spectrum analysis of an AM wave reveals.
- [ ] How the AM waveform is transmitted and what is required to make modulation occur.
- [ ] What a superheterodyne receiver is and why it is superior to other receivers.
- [ ] The definition of gain and sensitivity and how they are measured.
- [ ] How to use dBm to calculate power in a communication system.
- [ ] Troubleshooting techniques in signal tracing and signal injection.

INTRODUCTION

There is a lot of new information in this chapter. When you complete it, you will have a detailed understanding of an AM communication system. You will encounter your first block diagrams of a communication system. Recall that block diagrams were introduced in the troubleshooting and instrumentation section of Chapter 3. Make sure you understand that section before starting on this chapter.

The fourth troubleshooting simulation on your disk demonstrates the use of a signal tracer. Here you have the opportunity to develop troubleshooting skills using signal tracing techniques. Having this skill is a valuable addition to the troubleshooting skills you have acquired from the previous troubleshooting simulations on your student disk.

4-1 | AMPLITUDE MODULATION ANALYSIS

Discussion

In this section you will see that the AM wave consists of frequency components other than the carrier wave, because an AM wave is not a pure sine wave. From the discussion in the last chapter, you may suspect that it is made up of more than one sine wave. Your suspicion is correct.

Producing the AM Wave

If you simply combine two frequencies, a low-frequency audio with a high radio frequency (RF), the result will be as in Figure 4-1. You can see that the output waveform results in two sine waves: one riding on the other. The familiar AM waveform is not present.

What is needed is some way of making the sine waves unpure, such as a square wave or pulsating DC. Doing this will create harmonics. Harmonics are needed in order to produce a waveform that is not a pure sine wave, such as an AM carrier. One simple method of doing this is by introducing a diode as shown in Figure 4-2. The diode causes *distortion* of the input waveform. Distortion makes the resultant signal differ from the input signal.

The circuit in Figure 4-2 is called a *modulator*. A modulator is an electronic device or circuit that affects the process of modulation. The word "modulate" means to change. The diode is a *nonlinear* device. A nonlinear device is one that produces an output that does not rise and fall with the input. Clearly, the output of the modulator is AM pulsating DC. This waveform is the same as that in the diode detector of the simple AM receiver.

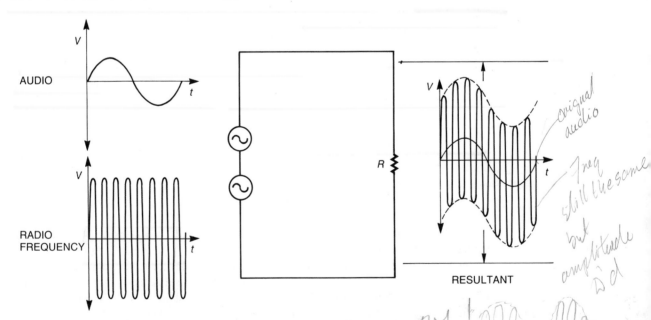

Figure 4-1 Result of the simple combining of two frequencies.

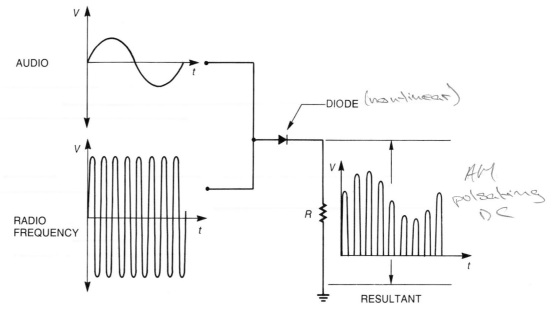

Figure 4-2 Introducing a diode to cause distortion.

Restoring the AM Wave

What is now needed is a way of converting the AM pulsating DC into the familiar AM AC waveform. You can accomplish this by using a resonant circuit that has the same resonant frequency as the RF carrier. Figure 4-3 shows the result.

Conclusion

In this section, you discovered that it takes a nonlinear device to produce an AM wave. This concept is important for the information to follow. You also saw that the circuit used to generate an AM wave from a low-frequency signal and an RF carrier is called a modulator.

4-1 Review Questions

1. What happens when you combine two sine waves of different frequencies?
2. What is distortion? Why is distortion necessary?
3. Describe the operation of a simple diode modulator.
4. What does "modulate" mean?

4-2 MODULATION FACTOR

Overview

In the last section, you saw how to create an AM waveform. In this section, you will see how much the carrier wave should be modulated by the modulating signal.

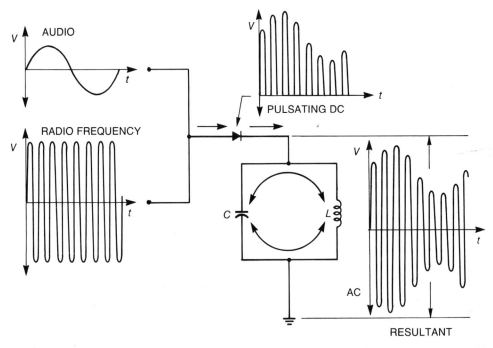

Figure 4-3 Restoring the AM waveform.

Modulation Factor

Look at the three AM waveforms in Figure 4-4. Waveform A has a small amount of modulation, waveform B a larger amount of modulation, and waveform C has so much modulation that the carrier disappears. As indicated in Figure 4-4, there are three classes of modulation:

1. Less than 100% modulation
2. 100% modulation
3. Greater than 100% modulation

Modulation of the carrier at greater than 100% is undesirable because it causes distortion of the recovered audio signal at the receiver. This distortion is shown in Figure 4-5.

The amount of modulation can be measured, and this measurement is called the modulation factor. This is expressed as

$$m = \frac{B}{A}$$

(Equation 4-1)

where m = Modulation factor (no units)
   B = Peak value of modulating signal
   A = Peak value of unmodulated carrier

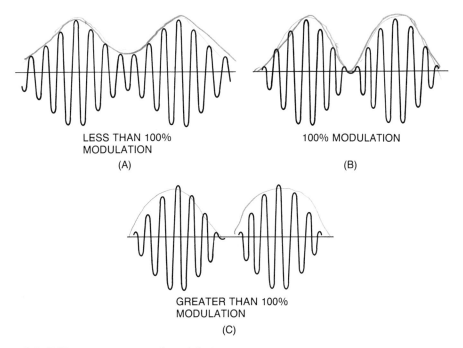

LESS THAN 100%
MODULATION
(A)

100% MODULATION
(B)

GREATER THAN 100%
MODULATION
(C)

Figure 4-4 Different amounts of modulation.

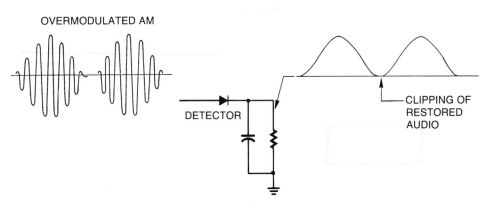

OVERMODULATED AM

DETECTOR

CLIPPING OF
RESTORED
AUDIO

Figure 4-5 Resulting distortion from overmodulation.

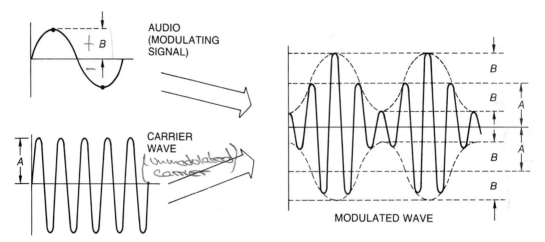

Figure 4-6 Graphical representation of modulation factor.

Note: B and A must be in the same units. This is shown graphically in Figure 4-6. When the modulation factor is expressed as a percentage, it is called the *percent modulation:*

$$m_p = \frac{B}{A} \times 100\%$$

(Equation 4-2)

Example 1

What is the percent modulation of the AM transmitter in Figure 4-7?

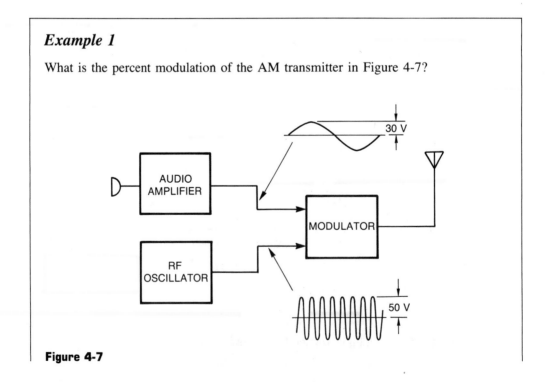

Figure 4-7

Solution

Using the formula for percent modulation:

$$m_p = \frac{B}{A} \times 100\%$$

(Equation 4-2)

$$m_p = \frac{30\ V}{50\ V} \times 100\%$$

$$m_p = 60\%$$

A sketch of the resulting waveform is shown in Figure 4-8.

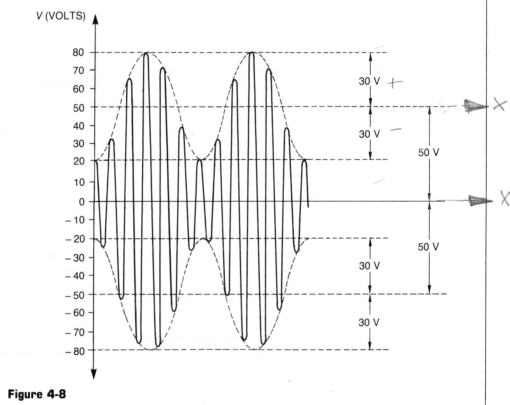

Figure 4-8

Modulation Measurement

When you want to measure the percent modulation from an existing AM waveform, use the following direct measurement procedure: Observe the AM waveform on a laboratory oscilloscope. A typical waveform is shown in Figure 4-9. The percent modulation may be found from

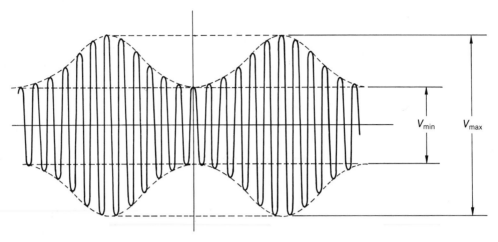

Figure 4-9 Typical AM waveform.

$$m_p = \frac{V_{max} - V_{min}}{V_{max} + V_{min}} \times 100\%$$ **(Equation 4-3)**

where m_p = Percent modulation
V_{max} = Maximum value of AM waveform
V_{min} = Minimum value of AM waveform

Example 2

What is the percent modulation of the AM waveform in Figure 4-10.

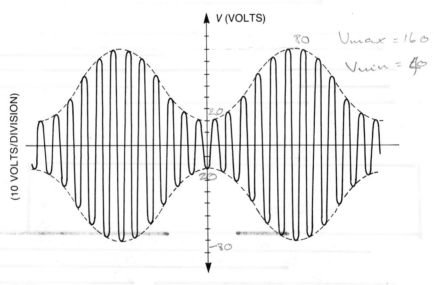

Figure 4-10

Solution

Using the relationship:

$$m_p = \frac{V_{max} - V_{min}}{V_{max} + V_{min}} \times 100\% \qquad \textbf{(Equation 4-3)}$$

$$m_p = \frac{160\text{ V} - 40\text{ V}}{160\text{ V} + 40\text{ V}} \times 100\% = \frac{120}{200} \times 100\%$$

$$m_p = 60\%$$

$\frac{80 - 20}{80 + 20} = \frac{60}{100} = 60\%$

Conclusion

In this section, you discovered that the amount of modulation of an AM waveform is measurable. You also saw that overmodulation is undesirable and that the amount of modulation can be calculated from laboratory measurements. The next section will present important information concerning what happens when you mix two frequencies together to get an AM waveform. As you will see, harmonics are produced because the resulting waveform is not, in itself, a pure sine wave.

4-2 Review Questions

1. Describe the following waveforms: less than 100% modulation, 100% modulation, more than 100% modulation.
2. What is the modulation factor? How is it different from percent modulation?
3. What is the problem with modulation of more than 100%?
4. Describe a method of measuring the amount of modulation on your oscilloscope.

4-3   AM SPECTRUM AND BANDWIDTH

Discussion

As you discovered in the last section, in order to produce an AM waveform you needed to introduce distortion. The resulting waveform, since it was no longer a pure sine wave, contained harmonics. This section introduces what these harmonics mean, why they are important, and how to measure and predict them.

Sidebands

Whenever a modulating frequency (f_M) amplitude modulates a carrier frequency (f_C), the result is four different sine waves:

1. The original modulating frequency f_M
2. The original carrier frequency f_C
3. The sum of the two frequencies: $f_C + f_M$
4. The difference between the two frequencies: $f_C - f_M$

   If the modulating frequency is a much lower frequency than the carrier (which is usually the case), then the low-frequency modulating signal is never transmitted. The transmission finally received by the receiver is therefore three sine waves: f_C, $f_C + f_M$, and $f_C - f_M$.

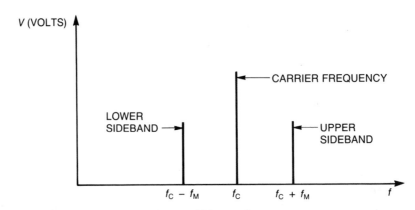

Figure 4-11 Frequencies contained in an AM waveform.

When a spectrum analyzer is used to observe a transmitted AM wave that has been modulated with a low-frequency audio sine wave, the result looks like Figure 4-11. Observe that the center frequency is the carrier, and on either side of it are the other two frequencies. Because of this relationship, the frequencies on either side of the carrier are called the sidebands. The sideband at a higher frequency than the carrier is called the *upper sideband,* and the sideband that is at a frequency lower than the carrier is called the *lower sideband.*

Hence the upper sideband frequency is

$$f_{USB} = f_C + f_M$$

(Equation 4-4)

and the lower sideband frequency is

$$f_{LSB} = f_C - f_M$$

(Equation 4-5)

When a carrier wave is modulated by a nonsinusodial waveform, such as a square wave, the resulting sidebands contain many different frequencies, as shown in Figure 4-12. This frequency spectrum occurs because the modulating square wave actually consists of an infinite number of odd harmonics. Each of these harmonics adds and subtracts from the carrier wave and causes the multiple frequencies in each sideband.

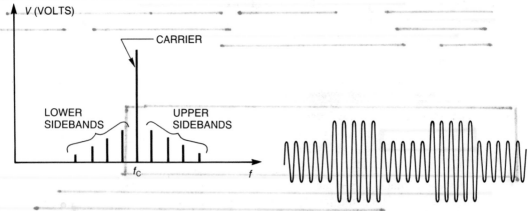

Figure 4-12 Frequency spectrum when carrier is modulated by a square wave.

Example 1

A carrier frequency of 750 kHz is modulated by a 3-kHz sine wave. State the values of the transmitted frequencies. Which frequency is the upper sideband? Which frequency is the lower sideband?

Solution

The resultant transmitted frequencies are

$$f_C = 750 \text{ kHz}$$
$$f_C + f_M = 750 \text{ kHz} + 3 \text{ kHz} = 753 \text{ kHz} \qquad \textbf{(Equation 4-4)}$$
$$f_C - f_M = 750 \text{ kHz} - 3 \text{ kHz} = 747 \text{ kHz} \qquad \textbf{(Equation 4-5)}$$

The upper sideband is 753 kHz.   Bandwidth 753−747 = 6 kHz
The lower sideband is 747 kHz.

Bandwidth

Bandwidth is defined as the range of frequencies from the lower sideband to the upper sideband. This is expressed as

$$BW = f_{USB} - f_{LSB} \qquad \textbf{(Equation 4-6)}$$

where   BW = Bandwidth in hertz
f_{USB} = Upper sideband in hertz
f_{LSB} = Lower sideband in hertz

Here is another method of calculating bandwidth for a transmitted AM signal. Start with the formula

$$BW = f_{USB} - f_{LSB}$$

Since $f_{USB} = f_C + f_M$ and $f_{LSB} = f_C - f_M$, substitute in the original equation to get

$$BW = (f_C + f_M) - (f_C - f_M)$$
$$BW = f_C + f_M - f_C + f_M$$
$$BW = f_C - f_C + f_M + f_M$$
$$BW = 2f_M \qquad\qquad 2 \times 3 \text{ kHz} = 6 \text{ kHz} \qquad \textbf{(Equation 4-7)}$$

Equation 4-7 states that the bandwidth of a transmitted AM wave is twice the value of the modulating frequency. In the United States, the allowed bandwidth of commercial AM radio stations is 10 kHz. This means that the highest modulating frequency that can be used by the transmitting station is 5 kHz (2 × 5 kHz = 10 kHz).

carrier freq fc = 10   modulating freq fm = 5

Example 2

A 10-MHz carrier is modulated with a 5-kHz sine wave. What is the bandwidth of the transmitted AM signal?

Solution

$$BW = 2f_M$$

(Equation 4-7)

$$= 2 \times 5 \text{ kHz} = 10 \text{ kHz}$$

Conclusion

The bandwidth of a transmitted AM wave plays an important role in the design and alignment of AM receivers. As you will see in the next section, the bandwidth of the resonant circuits used to select the received station must be close to that of the transmitted signal. For those of you who remember your AC well, this means that the Q of the tuner will play an important role in the quality of your radio.

4-3 Review Questions

1. What sine waves are produced as a result of mixing two sine waves of different frequencies?
2. What are sidebands? Define an upper sideband and a lower sideband.
3. Describe the resulting spectrum of modulating a carrier with a square wave.
4. What is bandwidth? What electrical circuit determines bandwidth?
5. What determines the bandwidth of a transmitted AM waveform?
6. What is the allowable bandwidth of commercial AM radio in the United States?

4-4 │ AMPLIFIERS AND NOISE

Definition

An amplifier is a circuit that increases the *amplitude* of a signal. The idea is to put a small AC signal on the input and get out a larger AC signal of the same frequency from the output. Devices such as transistors are used to construct amplifier circuits. Amplifier circuits used in communications will be covered in detail in the circuits section of this text. For now, an amplifier will be treated from a block-diagram point of view, as shown in Figure 4-13.

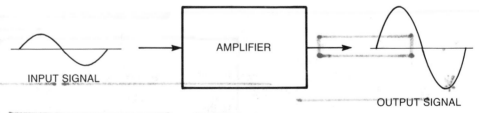

INPUT SIGNAL   AMPLIFIER   OUTPUT SIGNAL

Figure 4-13 Block diagram of an amplifier.

Amplifier Gain

Amplifier gain is defined as the ratio of the output signal strength to the input signal strength. Amplifiers can have their gains measured in terms of signal voltage, current, or power. This is expressed mathematically as

Voltage Gain

$$A_V = \frac{V_{out}}{V_{in}}$$

(Equation 4-8)

where A_V = Voltage gain (no units)
V_{out} = Output signal voltage
V_{in} = Input signal voltage

Current Gain

$$A_I = \frac{i_{out}}{i_{in}}$$

(Equation 4-9)

where A_I = Current gain (no units)
i_{out} = Output signal current
i_{in} = Input signal current

Power Gain

$$A_P = \frac{P_{out}}{P_{in}}$$

(Equation 4-10)

where A_P = Power gain (no units)
P_{out} = Input signal power in watts
P_{in} = Output signal power in watts

Example 1

What is the voltage gain of the amplifier in Figure 4-14?

Figure 4-14

Solution

From the formula for voltage gain,

$$A_V = \frac{V_{out}}{V_{in}}$$ **(Equation 4-8)**

$$A_V = \frac{1\ V}{5\ mV} = \frac{1}{(5 \times 10^{-3})}$$

$$A_V = 0.2 \times 10^3 = 200$$

Amplifiers are used in communication systems to increase the strength of signals. In this chapter, you will see how they are used in communication receivers.

Noise

Noise is any undesired signal. When a signal is transmitted from one point to another, noise is introduced. The noise distracts from the modulating information and may be so bad that the received signal cannot be understood.

Some sources of noise are:

I. External Noise
  A. Natural
    1. Weather conditions
    2. Energy from the sun (sun spots)
    3. Radiation from space
  B. Manufactured
    1. Industry
    2. Vehicles
    3. Other communication equipment

II. Internal Noise
  A. Passive components: resistors, wires, capacitors, etc.
  B. Active components: transistors, FETs, etc.

Signal-to-Noise Ratio

When a weak radio signal is amplified, the noise accompanying the signal is amplified along with it. The ratio of the signal strength to the noise strength is an important measurement in communication work. If you have a communication receiver with amplifiers that do not introduce any noise, then the ratio of the amount of signal to the amount of noise is

$$SNR = \frac{S}{N}$$ **(Equation 4-11)**

where SNR = Signal-to-noise ratio
      S = Signal strength
      N = Noise strength

Noise Figure

Practical amplifiers introduce some noise into the signal as the signal is being amplified. The *noise figure* (NF) is a measure of the degradation of the signal caused by the receiving system and is given by

$$NF = \frac{S_{in}/N_{in}}{S_{out}/N_{out}} \qquad \text{(Equation 4-12)}$$

A perfect receiver or amplifier would have a noise figure of 1.

Frequently, noise figure is expressed in dB (decibels) as

$$NF(dB) = 10 \log(NF) \qquad \text{(Equation 4-13)}$$

An ideal receiver or amplifier has a noise figure of 0 dB.

Example 2

The input signal to a receiver is 100 μV, and the internal noise at the input is 10 μV. The signal of the output, after being amplified, is 2 V, and the noise at the output is 0.5 V. What is the noise figure?

Solution

Using the relationship for NF,

$$NF = \frac{S_{in}/N_{in}}{S_{out}/N_{out}} \qquad \text{(Equation 4-12)}$$

$$NF = \frac{100 \ \mu V/10 \ \mu V}{2 \ V/0.5 \ V} = \frac{10}{4}$$

$$NF = 2.5$$

Conclusion

In this section, you saw what an amplifier is, how amplifier gain is measured, and the role of noise in communication systems. In the next section, you will learn how amplifiers are used to make the basic AM receiver even more powerful.

4-4 Review Questions

1. Define an amplifier. Give an example.
2. Define gain. Can gain be less than 1?
3. What is the similarity between the formulas for power gain, voltage gain, and current gain? What are the differences?
4. Define noise. Name some causes of noise.
5. Explain signal-to-noise ratio. What is the definition of noise figure?
6. What is the noise figure of an ideal amplifier? Explain.

4-5 | RECEIVER SENSITIVITY AND dB

Sensitivity

Receiver sensitivity is a measure of the weakest signal that can be put into a useful form by the receiver. One problem with the most basic AM receiver introduced in the last chapter was poor sensitivity. This meant that you could hardly hear the received signal from the strongest radio station. Weaker stations could not be heard at all. This radio suffered from poor sensitivity.

Radio frequency amplifiers are used in communication receivers to increase their sensitivity. A typical RF amplifier will have tuned resonant circuits on the input and on the output to select the desired signal. See Figure 4-15.

Cascaded Amplifiers

Cascaded amplifiers have the output of an amplifier connected to the input of another amplifier. This arrangement greatly increases a receiver's sensitivity. See Figure 4-16.

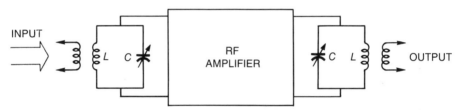

Figure 4-15 Block diagram of a typical RF amplifier with tuned circuits.

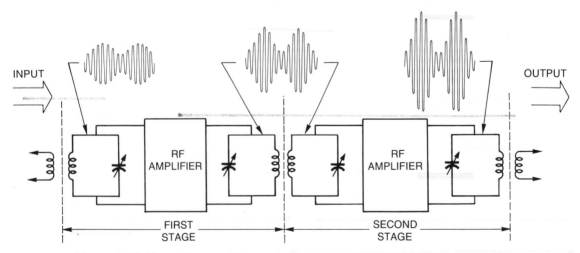

Figure 4-16 Cascading amplifiers to increase receiver sensitivity.

Sometimes, cascaded amplifiers are referred to as a *multiple-stage* amplifier, since more than one amplifier is used in the overall system. This system of RF amplifiers can take very weak radio signals (from long distances) and amplify them into larger signals. They also amplify the input noise and add their own internal noise to the received signal.

The total gain of such a system is the product of the gain of each stage:

$$A_T = A_1 A_2 A_3 \ldots A_N \qquad \text{(Equation 4-14)}$$

where A_T = Total gain of the system

   $A_1 \ldots A_N$ = Gain of each amplifier

Decibel Gain

When working with multiple-stage amplifiers it's convenient to use *decibel gains*. The power gain in dB is

$$A_P(\text{dB}) = 10 \log(A_P) \qquad \text{(Equation 4-15)}$$

The advantage of using power gains in dB is that the overall system gain of a multiple-stage amplifier is the *sum* of each of the individual amplifier gains:

$$A_T(\text{dB}) = A_1(\text{dB}) + A_2(\text{dB}) + A_3(\text{dB}) + \ldots + A_N(\text{dB}) \qquad \text{(Equation 4-16)}$$

Example 1

A two-stage RF amplifier has a power gain of 100 for its first stage and 10 for its second stage. What is the overall system power gain? Express this in dB.

Solution
There are two methods of working this problem.
   Method 1
Find the total power gain by multiplication:

$$A_T = A_{P1} A_{P2} \qquad \text{(Equation 4-14)}$$
$$A_T = 100 \times 10 = 1\,000$$

Convert to dB:

$$A_P(\text{dB}) = 10 \log(A_P) \qquad \text{(Equation 4-15)}$$
$$A_P(\text{dB}) = 10 \log(1\,000) = 10(3) = 30 \text{ dB}$$

   Method 2
Convert each power gain to dB:

$$A_P(\text{dB}) = 10 \log(A_P) \qquad \text{(Equation 4-15)}$$

For first stage:

$$A_{P1}(\text{dB}) = 10 \log(100) = 10(2) = 20 \text{ dB}$$

For second stage:

$$A_{P2}(dB) = 10 \log(10) = 10(1) = 10 \text{ dB}$$

Now add the dB power gains:

$$A_T(dB) = A_1(dB) + A_2(dB) \qquad \text{(Equation 4-16)}$$
$$A_T(dB) = 20 \text{ dB} + 10 \text{ dB} = 30 \text{ dB}$$

Either method yields the same answer.

The preceding problem is illustrated graphically in Figure 4-17.

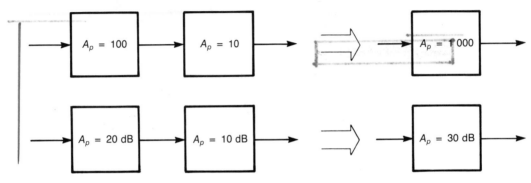

Figure 4-17 Graphic illustration of system gain.

Helpful Aid

Many communication technicians use the following rule when using power gains in dB. Every doubling of the power gain represents an increase of 3 dB. This is shown in Table 4-1.

Table 4-1	RELATIONS OF POWER GAIN TO dB
Power Gain	**dB**
1	0
2	3
4	6
8	9
16	12

Negative Decibels

Sometimes the power gain is less than 1. When a signal is transmitted through space, it loses power. This power loss (attenuation) is usually expressed in dB by the same formula for power gain in dB. As an example, if the transmitted signal is 5 W and the received signal is 10 mW, then the power gain is

$$A_P = \frac{p_{out}}{p_{in}}$$ **(Equation 4-10)**

$$A_P = \frac{10 \text{ mW}}{5 \text{ W}} = \frac{10 \times 10^{-3}}{5}$$

$$A_P = 0.002$$

Converting to dB gives

$$A_P(\text{dB}) = 10 \log(A_P)$$ **(Equation 4-15)**

$$A_P(\text{dB}) = 10 \log(0.002) = 10(-2.69) = -26.9 \text{ dB}$$

Table 4-2 gives an approximation of power gains less than 1 and their dB relations.

Table 4-2	POWER GAINS LESS THAN UNITY	
	Power Gain	**dB**
	1	0
	0.5	−3
	0.25	−6
	0.125	−9
	0.062 5	−12

dBm

The measurement of dBm is used to indicate the power level with respect to 1 mW. The "m" is a reminder that the milliwatt reference is being used. dBm is expressed mathematically as

$$P(\text{dBm}) = 10 \log\left(\frac{P}{1 \text{ mW}}\right)$$ **(Equation 4-17)**

For example, a power of 2 W in dBm is

$$P(\text{dBm}) = 10 \log\left(\frac{2 \text{ W}}{1 \text{ mW}}\right) = 10 \log(2\,000) = 33 \text{ dBm}$$

The advantage of dBm is that it simplifies power measurements. Many meters have a dBm scale. Using this scale to measure the input and output power in dBm makes amplifier gain calculations much easier. This is shown in the following example.

Example 2

An amplifier measures -9 dBm of input signal and 0 dBm of output signal, as shown in Figure 4-18. Determine the gain of the amplifier in dB.

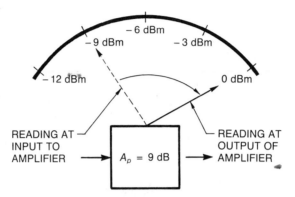

Figure 4-18

Solution

Since the needle moved from -9 dBm to 0 dBm, the amplifier has a decibel power gain of 9 dB.

Decibel Voltage Gain

You will find that in many cases voltage gain measurements are more common than power gain measurements. Decibels are also used to express voltage gain. Mathematically, the voltage gain in dB is

$$A_{V(\text{dB})} = 20 \log(A_V) \qquad \text{(Equation 4-18)}$$

If $A_V = 10$, then

$$A_{V(\text{dB})} = 20 \log(10) = 20(1) = 20 \text{ dB}$$

As the voltage gain doubles, the dB gain increases by 6; the converse is true with a decrease. Tables 4-3 and 4-4 illustrate.

Table 4-3	dB VOLTAGE GAIN	
	Voltage Gain	**dB**
	1	0
	2	6
	4	12
	8	18

Table 4-4	dB FOR VOLTAGE GAINS LESS THAN 1	
	Voltage Gain	dB
	1	0
	0.5	−6
	0.25	−12
	0.125	−18

Conclusion

The sensitivity of a receiver is greatly increased by amplifiers. The changes in the voltage and power levels of communication signals are usually measured in dB or dBm. Knowing how to use these measurements is an important step in becoming a qualified communications technician.

4-5 Review Questions

1. Explain receiver sensitivity.
2. How is a radio's sensitivity increased?
3. What are cascaded amplifiers? How does this arrangement improve receiver sensitivity?
4. How is the total gain of several amplifiers computed?
5. What is the advantage of computing gains in dB?
6. Define a dBm. What is the reason for using dBm?

4-6 | RECEIVER SELECTIVITY

Introduction

The ability of a receiver to select one station and reject all others is of prime importance. Many frequency allocations to different transmitters require that the receiver be *selective* enough to receive only one station at a time.

Factors

One factor that affects the *selectivity* of a receiver is the *bandwidth* of the tuned circuits used in the RF amplifiers. Recall from the last section that the RF amplifier used to increase the receiver *sensitivity* used tuned circuits to select the desired signal. Also recall from Section 4-3 that the transmitted AM wave contained a carrier and two sidebands. The receiver must be selective enough to select only one radio station, but not so selective that it rejects the sidebands of the received signal.

As discussed in basic AC courses, the bandwidth of a resonant circuit is

$$BW = f_U - f_L$$

(Equation 4-19)

where   BW = Circuit bandwidth in hertz
      f_U = Upper cutoff frequency in hertz
      f_L = Lower cutoff frequency in hertz

The Q (quality) of a resonant circuit is related to the bandwidth by

$$Q = \frac{f_r}{BW}$$

(Equation 4-20)

where Q = Quality of the circuit (no units)
f_r = Resonant frequency of the circuit in hertz
BW = Circuit bandwidth in hertz

Frequency Response

In the design of an RF amplifier, the frequency response is very important. Figure 4-19 illustrates the relation of *circuit bandwidth* to receiver *selectivity*.

The *frequency response* curves represent the response of three different tuned circuits. Each circuit can be analyzed as follows:

Circuit A

This circuit's bandwidth is too wide. Stations A and B will be amplified by the same amount. Hence, for circuit A, the selectivity is poor.

Circuit B

Circuit B has just the right amount of bandwidth. Only station B is being amplified. This circuit has just the right amount of selectivity.

Circuit C

This circuit has too much *selectivity*. Note that the bandwidth is so narrow that the sidebands of the received signal are not being amplified. Since the information is contained in the sidebands, no useful output will occur.

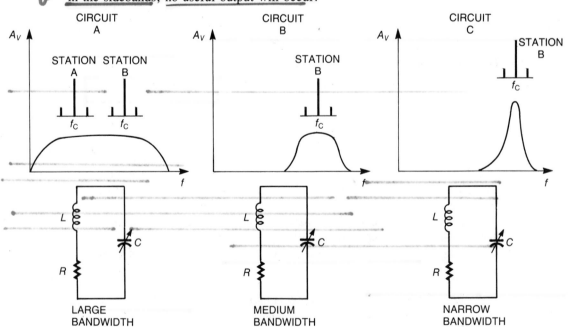

Figure 4-19 Relation of circuit bandwidth and receiver selectivity.

Coil Resistance

As explained in basic AC texts, the Q of a resonant circuit is related to the resistance of the inductor (coil) by

$$Q = \frac{X_L}{R_L}$$

(Equation 4-21)

where $Q = Q$ of the inductor (no units)
   $X_L = $ Inductive reactance in ohms
   $R_L = $ Resistance of inductor in ohms

From equations (4-20) and (4-21) the relationship between selectivity and the resonant circuit used in the receiver can be developed: Since

$$Q = \frac{f_r}{BW}$$

(Equation 4-20)

and

$$Q = \frac{X_L}{R_L}$$

(Equation 4-21)

then

$$\frac{f_r}{BW} = \frac{X_L}{R_L}$$

Solving for BW,

$$BW = \frac{R_L f_r}{X_L}$$

(Equation 4-22)

The following example illustrates an application of this relationship.

Example 1

A tuned radio frequency (TRF) receiver is shown in Figure 4-20. The transmitting AM stations all have a bandwidth of 10 kHz. The TRF is shown tuned to 550 kHz. Determine the bandwidth of any single stage.

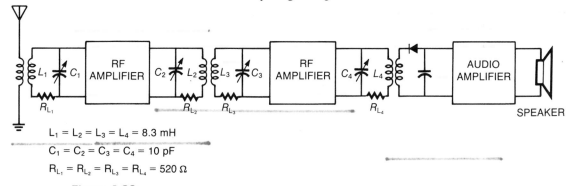

$L_1 = L_2 = L_3 = L_4 = 8.3$ mH
$C_1 = C_2 = C_3 = C_4 = 10$ pF
$R_{L_1} = R_{L_2} = R_{L_3} = R_{L_4} = 520\ \Omega$

Figure 4-20

Solution

First compute X_L:

$$X_L = 2\pi fL \qquad \text{(Equation 1-4)}$$
$$X_L = 6.28 \times 550 \text{ kHz} \times 8.3 \text{ mH}$$
$$X_L = 28.7 \text{ k}\Omega$$

Now compute the bandwidth:

$$BW = \frac{R_L f_r}{X_L} \qquad \text{(Equation 4-22)}$$

$$BW = \frac{520 \ \Omega \times 550 \text{ kHz}}{28.7 \text{ k}\Omega}$$

$$BW = 9.97 \text{ kHz}$$

Many problems occurred with the TRF receiver, which led to the development of a new type of receiver, presented in the next section.

Conclusion

You should now be familiar with the terms "selectivity" and "sensitivity." This section demonstrated what factors influenced selectivity.

4-6 Review Questions

1. Explain the meaning of receiver selectivity.
2. What determines the selectivity of a receiver?
3. What is the relationship between the bandwidth of the transmitted signal and the receiver selectivity?
4. Explain what is meant by "too much selectivity."
5. What problems would a receiver with poor selectivity experience?
6. Explain the difference between selectivity and sensitivity.

4-7 SUPERHETERODYNE RECEIVER

Discussion

In 1927, the TRF receiver was replaced by the *superheterodyne* receiver. The word comes from "heterodyne," which means to mix or combine frequencies. The word "super" was used for commercial effect to help sell the radio when it first appeared on the market. The main advantage of the superheterodyne receiver is its constant bandwidth while tuning to different radio stations.

The sensitivity and selectivity of a superheterodyne receiver are so good that this form of receiver is the most common in use today.

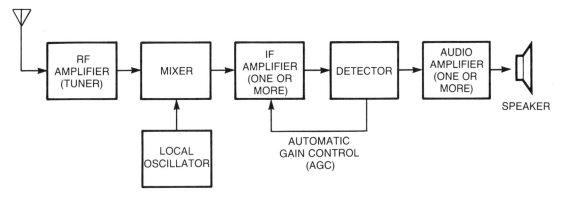

Figure 4-21 Block diagram of AM superheterodyne receiver.

Block Diagram

The block diagram of a superheterodyne receiver is shown in Figure 4-21. The *automatic gain control* (AGC) is a method used to help keep the gain of the receiver constant as stations of different signal strengths are received. The circuits section of this text will present the important details of AGC.

The purpose of a superheterodyne receiver is to keep the receiver bandwidth constant. This is done by keeping the tuned circuits used by amplifiers tuned to the same frequency. This in turn is accomplished by *intermediate frequency* (IF) amplifiers. These amplifiers are *always* tuned to the same frequency regardless of the frequency of the incoming signal. They are called IF amplifiers because the frequency they are amplifying is between the frequency of the received signal and the frequency of the audio that will eventually operate the loudspeaker.

Heterodyning

Recall that when two frequencies are mixed together in a nonlinear device the result is the two original frequencies, the sum of the two frequencies, and the difference between the two frequencies. This principle makes the superheterodyne receiver practical.

For most commercial AM receivers, the IF amplifiers are tuned to 455 kHz. If the received radio signal is 1 000 kHz, then the local oscillator will have a frequency of 1 455 kHz.

An *oscillator* is a circuit that creates its own frequency. In the receiver, an oscillator is called *local* because it is contained locally inside the receiver. It is usually referred to as the LO.

The received signal and the LO signal will come together in the *mixer*. The output of the mixer will be the original two frequencies (1 000 kHz and 1 455 kHz) along with the sum and difference frequencies (2 455 kHz and 455 kHz). Since the IF amplifier has tuned circuits that are *fixed* at 455 kHz, it is the 455-kHz signal that will pass through the receiver. This is shown in Figure 4-22.

The point is that the LO frequency must always be 455 kHz more than the frequency of the received station. This is done by physically connecting (ganging) the variable capacitor in the tuner to the variable capacitor in the tuned circuit of the oscillator. As you

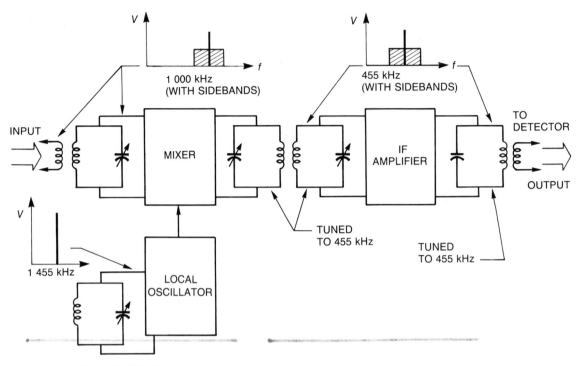

Figure 4-22 Heterodyning principle.

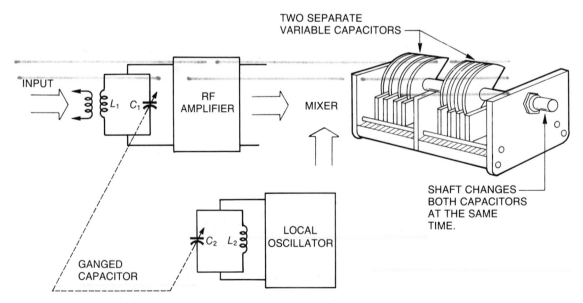

Figure 4-23 Ganged capacitors in superheterodyne receiver.

will see in the circuits section of this text, one method of controlling the frequency developed by an oscillator is a resonant circuit. This arrangement is shown in Figure 4-23.

If you now wished to receive a radio station at a different frequency, say 750 kHz, the LO would then have to be at a frequency of 1 205 kHz. This would give the required difference frequency of 455 kHz for the IF amplifier.

The LO frequency can be determined from the relationship

$$f_{LO} = f_R + f_{IF}$$
(Equation 4-23)

where f_{LO} = Frequency of the LO in hertz

f_R = Frequency of the received signal in hertz

f_{IF} = Frequency of the IF amplifier in hertz

Example 1

What is the LO frequency if the IF frequency of an AM superheterodyne receiver is 455 kHz and the received signal is 830 kHz?

Solution

Using the relationship:

$$f_{LO} = f_R + f_{IF}$$
(Equation 4-23)

$$f_{LO} = 830 \text{ kHz} + 455 \text{ kHz} = 1 \text{ } 285 \text{ kHz}$$

Image Frequency

There is a potential problem with superheterodyne receivers: They can receive more than one station at a time. One station is the one you wanted; the other, called the *image,* is not. Here is how this happens.

Suppose you want to get a radio station at 600 kHz. This means that the LO will be at 1 055 kHz. This is fine as long as there is not another radio station at 455 kHz *above* the LO frequency (1 055 kHz + 455 kHz = 1 510 kHz). If there is, then the LO will mix with it as well as the desired station. The result is as follows:

Mixing with 600 kHz station

Difference frequency: $f_{LO} - f_R = 1 \text{ } 055 \text{ kHz} - 600 = 455 \text{ kHz}$.

Mixing with 1 510-kHz station

Difference frequency: $f_R - f_{LO} = 1 \text{ } 510 \text{ kHz} - 1 \text{ } 055 \text{ kHz} = 455 \text{ kHz}$

The *image frequency* can be determined from the formula

$$f_I = 2f_{IF} + f_R$$
(Equation 4-24)

where f_I = Image frequency in hertz

f_{IF} = IF frequency in hertz

f_R = Frequency of received signal in hertz

The *image frequency rejection* is a measure of how well a receiver can reject an image frequency. The use of an RF amplifier greatly reduces image frequency interference. Many low-cost superheterodyne receivers do not use an RF amplifier.

Example 2

What is the image frequency for an AM superheterodyne receiver with an IF frequency of 455 kHz when tuned to a carrier with a frequency of 550 kHz?

Solution
Using the relationship:

$$f_I = 2f_{IF} + f_R \qquad \text{(Equation 4-24)}$$
$$f_I = 2 \times 455 \text{ kHz} + 550 \text{ kHz}$$
$$f_I = 910 \text{ kHz} + 550 \text{ kHz} = 1\ 460 \text{ kHz}$$

Conclusion

In this section, you saw how the most common form of AM receiver is constructed. The next section shows two methods for troubleshooting the superhet receiver.

4-7 Review Questions

1. Explain the meaning of superheterodyne.
2. Describe the block diagram of a superheterodyne receiver.
3. What is the purpose of AGC?
4. What is an advantage of a superheterodyne receiver?
5. Describe the heterodyning principle. State the purpose of the LO and mixer.
6. Explain what an IF amplifier does.
7. What is an image frequency? How can image frequencies be reduced?

TROUBLESHOOTING AND INSTRUMENTATION

Troubleshooting Block Diagrams

Troubleshooting with a block diagram is a convenient way of isolating a faulty stage so that a detailed analysis of that stage can be made. In this section, you will analyze the superheterodyne (superhet) receiver from a block-diagram standpoint. You will then see how to troubleshoot it using the two most common methods of isolating a faulty stage: *signal injection* and *signal tracing*.

Superhet Analysis

Consider the block diagram of a superheterodyne receiver in Figure 4-24. Table 4-5 summarizes the purpose of each stage and the input and output signals you should expect.

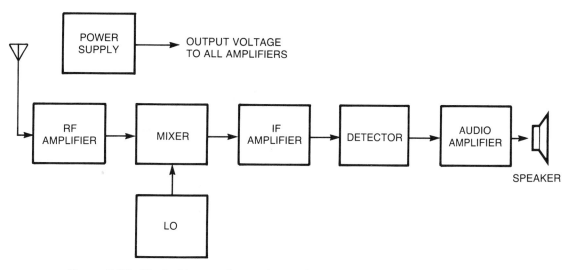

Figure 4-24 Block diagram of superheterodyne receiver with power supply.

Table 4-5	SUPERHETERODYNE FUNCTIONS		
Stage	**Function**	**Input Signal**	**Output Signal**
Power supply	Provides power to all stages of receiver	AC power 120 VAC, 60 Hz	6 to 12 VDC
RF amplifier	Selects desired frequency and amplifies it	Radio signals	Amplified radio signal
LO	Creates a sine wave to be mixed with the incoming received signal	None	AC that is 455 kHz more than f_R
Mixer	Mixes the LO signal with RF to produce a difference frequency	LO and RF	$f_{LO} + f_R$ $f_{LO} - f_R$ f_{LO} and f_R
IF amplifier	Amplifies the 455-kHz signal from mixer	Same as output of mixer	Amplified 455 kHz AM
Detector	Restores the audio and bypasses the RF	455-kHz AM	Restored audio
Audio amplifier	Amplifies the audio signal from the detector	Audio from detector	Amplified audio
Speaker	Converts the audio waveform into sound	Audio signal	Sound waves

Signal Tracing

In the troubleshooting examples that follow, assume that the problem with the receiver is no sound from the speaker. When troubleshooting using the *signal tracing* method, use an instrument like the oscilloscope. No matter which method is used, always check the power supply voltage first, using a voltmeter. The reason is that if the power supply is not functioning properly, then nothing in the receiver will work correctly, since the power supply supplies power to the amplifiers.

Once you have checked the power supply, go to the "middle" of the receiver and see if a signal is present on the output of the IF amplifier. If there is, this tells you that all the stages to the left of the IF amplifier output are working. This means that the problem must be in one of the stages to the right of the IF amplifier output. If you do not see a signal here, then the problem must be to the left of the IF amplifier output. See Figure 4-25.

Continue this procedure until you find a stage with an input signal but no output signal. That stage is the defective one. To check the LO, simply look for an output signal.

Signal Injection

The signal injection method of troubleshooting involves using an RF generator that can be modulated with an audio tone. The output frequency of the RF generator is variable and can be tuned to represent that of a radio station, LO, or IF amplifier. Most RF generators contain an audio signal output for injecting an audio signal into the audio amplifier. A laboratory RF signal generator is shown in Figure 4-26.

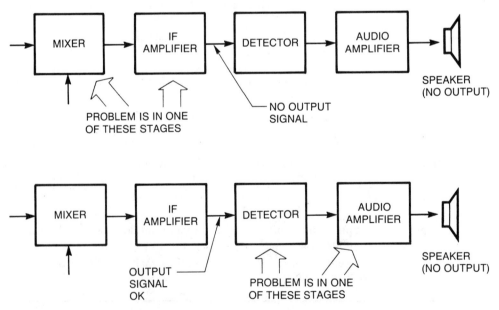

Figure 4-25 Using signal tracing in superhet.

Figure 4-26 Laboratory RF generator. *Courtesy* of Hewlett-Packard Company.

To troubleshoot a superhet receiver using the *signal injection* method, inject an audio tone at the input of the audio amplifier. If you hear a tone at the speaker, then the problem must be to the left of the audio amp. If a tone is not heard, then the problem must be from the audio amp to the speaker. As before, always check the power supply first. Figure 4-27 illustrates the signal injection troubleshooting method.

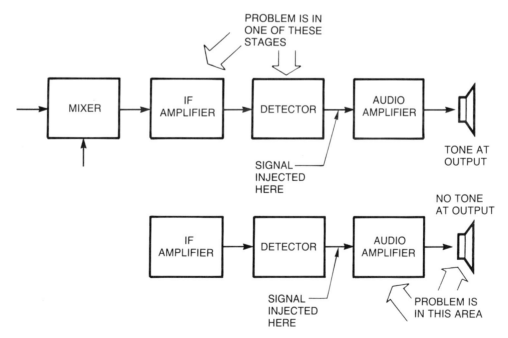

Figure 4-27 Troubleshooting by signal injection.

Conclusion

This fourth troubleshooting simulation introduces you to a signal tracer. Unlike the signal generator used in the previous two troubleshooting simulations, the signal tracer typifies a general class of instruments that depend upon the presence of a signal for troubleshooting purposes. The instructions for this simulation will show you how this instrument is used and the demonstration mode will allow you to practice with this new instrument until you are confident that you can easily use it. The testing mode will then give you an opportunity to experience some 'real' troubleshooting using a signal detector.

4-8 Review Questions

1. State the function of each section of a superheterodyne receiver.
2. Describe the input and output signals for each stage of a superheterodyne receiver.
3. Describe the signal tracing troubleshooting method. Give an example. What instrument can be used for signal tracing?
4. What is a signal generator?
5. Describe the signal injection troubleshooting method. Give an example.

MICROCOMPUTER SIMULATION

The microcomputer simulation on the diskette for this chapter introduces you to a signal tracer. This program gives you a challenging opportunity to use the signal tracer as often as you wish. The random-number generating feature is again used to simulate the practical situation where you don't know what problem you may encounter. The computer's memory capabilities are used to keep score of your troubleshooting progress.

CHAPTER PROBLEMS

(Answers to odd-numbered problems appear at the end of the text.)

1. What happens when two different frequencies are combined across a resistor?
2. Is an AM carrier a pure sine wave? Explain.
3. Explain the meaning of distortion. Give an application where distortion is useful.
4. What is a nonlinear device? Is a resistor a nonlinear device?
5. Describe the process of modulation.
6. Sketch the diagram of a simple AM modulator. Explain how it works.
7. Which of the waveforms in Figure 4-28 have less than 100% modulation?
8. In Figure 4-28, which waveforms have more than 100% modulation? Which have 100% modulation?
9. What is the modulation factor for waveforms A and B in Figure 4-28?
10. Determine the modulation factor for waveforms B and C in Figure 4-28.
11. Express the percent modulation for waveforms A and B in Figure 4-28.
12. Comment on the detected waveform of Figure 4-28(C).

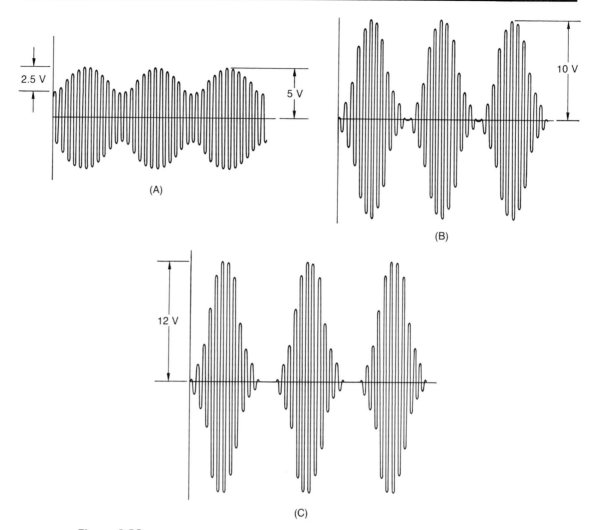

(A)

(B)

(C)

Figure 4-28

13. Sketch the resultant AM waveform of the two waves in Figure 4-29(A).
14. Sketch the resultant AM waveform of the two waves in Figure 4-29(B).

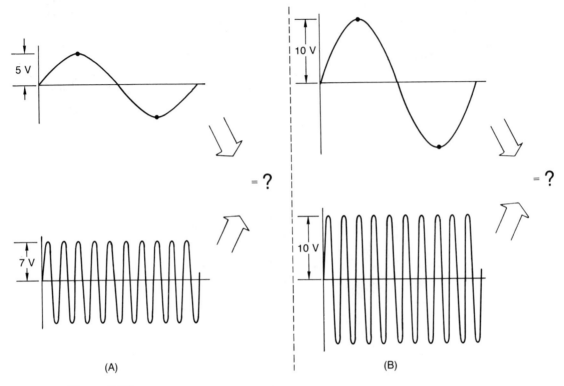

(A)   (B)

Figure 4-29

15. What is the percent modulation of the oscilloscope waveform in Figure 4-30(A)?
16. Determine the percent modulation of the oscilloscope waveform in Figure 4-30(B).

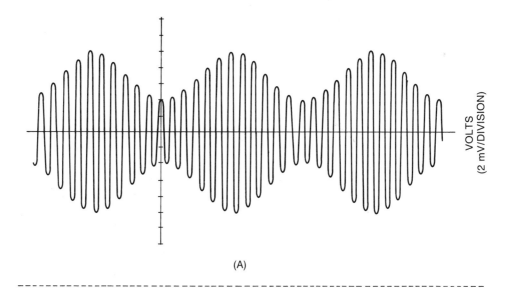

(A)

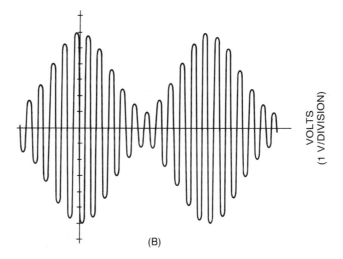

(B)

Figure 4-30

17. What are the resulting frequencies when a 12-kHz sine wave modulates a 550-kHz carrier?
18. Calculate the resulting frequencies when a 150-MHz carrier is modulated by a 25-kHz signal.
19. What is the bandwidth of the resultant AM waveform in problem 18?
20. Determine the bandwidth of the AM waveform in problem 17.

21. Sketch the resultant spectrum from modulating a 2-MHz carrier with a 10-kHz sine wave. Identify the upper and lower sidebands. What is the resultant bandwidth?
22. For each of the following sets of frequencies, the higher frequency is the carrier, the lower is the modulating frequency. Determine the upper and lower sidebands and the bandwidth. Sketch the frequency spectrum. (A) 15 MHz and 5 kHz (B) 120 MHz and 15 kHz
23. Determine the gain of each amplifier in Figure 4-31.

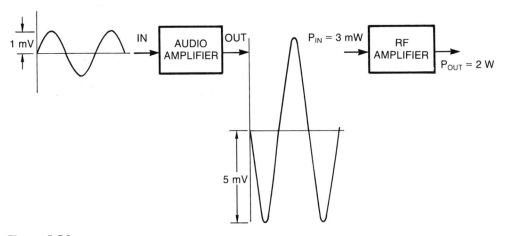

Figure 4-31

24. Determine the signal-to-noise ratio of the following: (A) $S_{IN} = 3$ mV, $N_{IN} = 20$ μV; (B) $S_{OUT} = 12$ V, $N_{OUT} = 2$ mV.
25. Determine the signal-to-noise ratio of the following: (A) $S_{IN} = 24$ mV, $N_{IN} = 0.09$ mV; (B) $S_{OUT} = 4.6$ V, $N_{OUT} = 0.5$ mV.
26. What is the noise figure of the amplifier in Figure 4-32(A)?
27. What is the noise figure of the amplifier in Figure 4-32(B)?

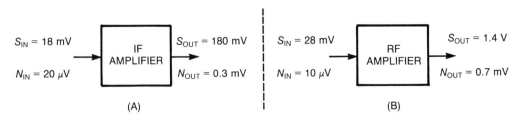

Figure 4-32

28. What is the total gain of the TRF receiver in Figure 4-33(A)?
29. Compute the total gain of the TRF receiver in Figure 4-33(B).
30. Convert the gain of each stage in the TRF receiver in Figure 4-33(B) to dB. Compute the total gain in dB.
31. Express the gain of each stage of the TRF receiver in Figure 4-33(A) in dB. What is the total gain in dB?

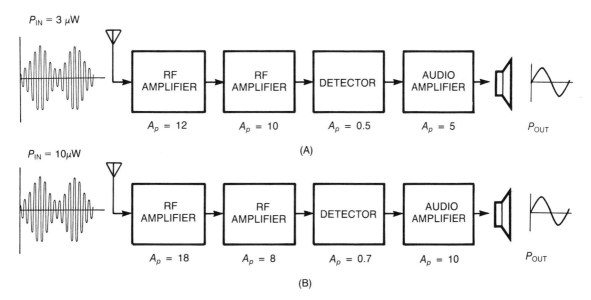

(A)

(B)

Figure 4-33

32. Express the following gains in dB: (A) 2, (B) 4, (C) 16.
33. If the input signal to an amplifier is 3 mW, compute the output signal for the following power gains: (A) 3 dB, (B) 9 dB, (C) −3 dB, (D) −12 dB.
34. For an amplifier with an input signal of 12 mW, what would be the output signal if the amplifier gain were (A) 4 dB? (B) 6 dB? (C) −2 dB? (D) −6 dB?
35. Express the signal input power of problem 34 in dBm.
36. Convert the input signal power of problem 33 to dBm.

37. Which curves in Figure 4-34 have too much selectivity if the received signal has a bandwidth of 20 kHz?
38. Determine the response curves in Figure 4-34 that have the correct selectivity for an AM signal with a bandwidth of 5 kHz.

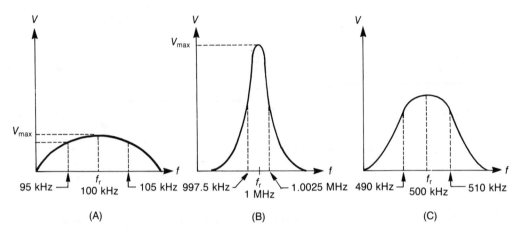

Figure 4-34

39. For an IF frequency of 455 kHz, what must be the LO frequency when receiving the following stations: (A) 580 kHz? (B) 800 kHz? (C) 995 kHz?
40. If the IF frequency is 455 kHz, what must be the LO frequency if the following stations are being received: (A) 660 kHz? (B) 1.05 MHz? (C) 775 kHz?
41. What is the image frequency for each station in problem 40?
42. Determine the image frequency for each station in problem 39.

CHAPTER 5

Sideband Systems

OBJECTIVES

In this chapter, you will study:

- [ ] The building blocks of AM transmitters.
- [ ] What factors determine transmission power.
- [ ] Different types of sideband systems.
- [ ] How sidebands are produced and how to measure them.
- [ ] How to distinguish between the following sideband systems: (A) vestigial, (B) suppressed carrier, (C) single sideband.
- [ ] The building blocks of SSB receivers.
- [ ] How to measure carrier rejection in SSB transmitters.

INTRODUCTION

As you will see in this chapter, transmitting the carrier and both sidebands is not the most efficient way to transmit information. Since there is no modulating information in the carrier, the energy used to transmit it is wasted. Intelligence is contained only in the sidebands. Putting the transmitting energy in the sidebands and not in the carrier can greatly increase the overall system efficiency.

The fifth troubleshooting simulation on your student disk will again present the use of the signal tracer as a troubleshooting tool, but this time, your troubleshooting efficiency will be encouraged. Here the computer will again keep a record of the measurements you have made and offer suggestions as to how you may improve.

5-1 | AM TRANSMISSION

Discussion

The creation and transmission of an AM radio wave is quite straightforward. An AM transmitter consists of two basic sections: the RF section and the audio section. These two sections handle the carrier and the modulating information. They are then combined in a circuit called a *modulator*. The result is then transmitted as a radio wave.

Block Diagram

Two major arrangements are used in the construction of an AM transmitter. One is for *high-level* modulation, and the other is for *low-level* modulation. The block diagrams are shown in Figure 5-1. The difference between high-level and low-level modulation will be explained shortly. For now, Table 5-1 summarizes the purpose of each stage of an AM transmitter, including the kind of input and output waveforms to expect.

High-Level Transmission

High-level AM transmission means having the carrier wave modulated at the power amplifier stage. The advantage of high-level transmission is greater system efficiency. The main disadvantage is increased system cost. The reason for the increased cost is that the audio signal must have circuits that bring this signal to a high power level. Most high-power AM transmitters use high-level transmission.

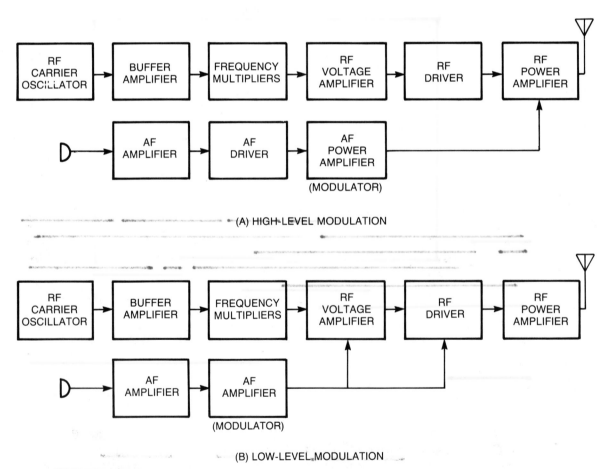

Figure 5-1 High-level and low-level AM transmitters.

Table 5-1	PURPOSE OF EACH SECTION IN AM TRANSMITTER (FOR HIGH-LEVEL MODULATION SYSTEMS)		
Stage	**Purpose**	**Input Signal**	**Output Signal**
RF carrier oscillator	Creates RF signal	None	Pure unmod. sine wave
Buffer amplifier	Isolates oscillator from next stage, helps keep oscillator stable	Pure unmod. sine wave	Pure unmod. sine wave
Frequency multiplier	Increase oscillator frequency	Pure unmod. sine wave at osc. frequency	Pure unmod. sine wave at higher frequency
AF amplifier	Amplifies audio signal from microphone	Audio signal from microphone	Audio signal with voltage gain
AF driver	Increases audio signal power	Audio signal from AF driver	Audio signal with power gain
AF modulator	Changes amplitude of RF carrier	Audio signal	Audio signal
RF voltage amplifier	Increases volt. gain of RF carrier	Pure RF sine wave*	Pure RF sine wave*
RF driver amplifier	Prepares RF to operate PA	Pure RF sine wave*	Pure RF sine wave*
RF power amplifier	Drives antenna circuit to produce transmitted AM radio wave	Pure RF sine wave and audio*	AM waveform

*Note that this is different for a low-level AM system.

Low-Level Transmission

Low-level AM transmission means having the carrier wave modulated at stages preceding the RF power amplifier stage. This method is more economical than high-level transmission. This saving comes because the audio signal can be at a low power level. The disadvantage of this method is lower system efficiency. Low-level transmitters are usually found in low-power, low-cost systems.

Conclusion

In the circuits section of this text, you will see how each of the transmitter circuits discussed here is constructed. For now, you saw what it takes to produce the fundamental AM waveform.

5-1 Review Questions

1. Name the two major sections of an AM transmitter.
2. What is the name of the stage where the audio signal is combined with the RF carrier for (A) high-level modulation? (B) low-level modulation?
3. What is the difference between high-level modulation and low-level modulation? What are the advantages and disadvantages?

4. What stage in an AM transmitter creates its own sine wave?
5. What is the main difference between an AF amplifier and an RF amplifier?
6. Describe the purpose of a frequency multiplier.

5-2 | TRANSMISSION POWER

Overview

When the AM wave is transmitted, it can be shown that all of the information is contained in the sidebands. Recall that an AM waveform consists of a carrier with upper and lower sidebands. Under optimum conditions, two thirds of the transmitter power is put into the carrier. The remaining third goes to the sidebands.

Power Content

Figure 5-2 shows the voltage and the power content of a transmitted AM wave. Since power is proportional to the square of the voltage, the power content of an AM waveform is expressed as

$$P_{SB} = \left(\frac{mV_C}{2}\right)^2 = \frac{m^2V_C^2}{4} = \frac{m^2P_C}{4}$$

$$P_{SB} = \frac{m^2P_C}{4}$$

(Equation 5-1)

where P_{SB} = Power content in one sideband in watts
         m = Modulation factor (no units)
         P_C = Power content of the carrier in watts

$$P_T = \frac{m^2P_C}{4} + \frac{m^2P_C}{4} + P_C$$

(Equation 5-2)

where P_T = Total transmission power in watts
         m = Modulation factor (no units)
         P_C = Carrier power in watts

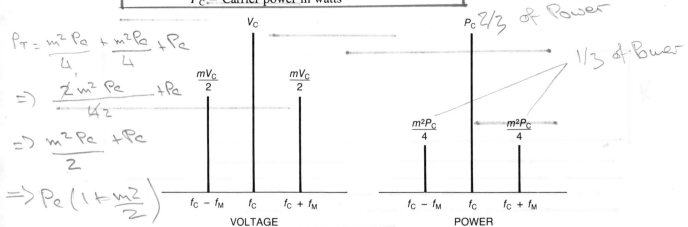

Figure 5-2 Voltage and power content of AM wave.

Equation 5-2 can be simplified to

$$P_T = \frac{m^2 P_C}{2} + P_C$$

Factoring P_C gives

$$P_T = P_C\left(1 + \frac{m^2}{2}\right)$$

(Equation 5-3)

Example 1

Determine the power being transmitted for the carrier and the sidebands when the percent modulation is 100% and the total power content of the AM signal is 1 000 W.

Solution

From the equation for total power:

$$P_T = \frac{m^2 P_C}{4} + \frac{m^2 P_C}{4} + P_C$$

(Equation 5-2)

$$P_T = P_C + \frac{m^2 P_C}{2}$$

Hence

$$1\,000 \text{ W} = P_C + \frac{(1.0)^2 P_C}{2}$$
$$1\,000 = P_C + 0.5 P_C$$
$$1\,000 = 1.5 P_C$$

Solving for P_C gives

$$P_C = \frac{1\,000}{1.5}$$
$$= 666.67 \text{ W}$$

The power in the sidebands (shared equally between them) is

$$P_{USB} + P_{LSB} = 1\,000 - 666.67 = 333.33 \text{ W}$$

Since $P_{USB} = P_{LSB}$, then

$$2P_{LSB} = 333.33$$
$$P_{USB} = P_{LSB} = \frac{333.33}{2} = 166.66 \text{ W}$$

This example shows that even under ideal conditions (100% modulation), the power in the sidebands is only one third of the total power in the transmitted radio wave.

Carrier Power

The power of the carrier remains the same no matter what the percentage of modulation is. The following example uses this concept.

Example 2

The percent modulation of an AM wave changes from 30% to 60%. At the 30% modulation level, the power content of the carrier was 500 W. Calculate the sideband and the carrier power when the percent modulation is 60%.

Solution

Since the power content of the carrier in an AM wave remains the same regardless of the percent modulation,

$$P_{C(60\%)} = P_{C(30\%)} = 500 \text{ W}$$

Using the relationship for power content in each sideband:

$$P_{SB} = \frac{m^2 P_C}{4} \qquad \text{(Equation 5-1)}$$

$$P_{SB} = \frac{(0.60)^2(500 \text{ W})}{4}$$

$$P_{SB} = \frac{0.36 \times 500}{4} = 45 \text{ W}$$

Hence

$$P_{USB} = P_{LSB} = 45 \text{ W}$$

Another Technique

Here are some important facts about AM transmission:

- Only the sidebands contain the information being transmitted.
- The upper sideband and the lower sideband are identical. You only need one of them to extract the modulating information.
- The RF carrier, which does not contain any information, requires two thirds of the total transmission power.

These facts led to the practice of transmitting only one sideband called *single-sideband* (SSB) transmissions. By not transmitting the carrier and one of the sidebands, the transmission efficiency would be greatly increased and the bandwidth would be decreased. These improvements allowed for more transmitters of different frequencies on the same band and lower transmission power requirements.

Consider the number of stations that can be accommodated in a specified band of frequencies for standard AM transmission

$$N_S = \frac{BW_S}{2f_M}$$

where did this come from

(Equation 5-4)

For single-sideband (SSB) transmission,

$$N_S = \frac{BW_S}{f_M}$$

(Equation 5-5)

where N_S = Number of stations

   BW_S = Spectrum bandwidth in hertz

   f_M = Frequency of the modulating signal in hertz

Example 3

Determine how many AM stations can be accommodated in a 100-kHz spectrum if the highest modulating frequency is 5 kHz for (A) standard AM transmission, (B) SSB transmission.

Solution

For standard AM transmission,

$$N_S = \frac{BW_S}{2f_M}$$

(Equation 5-4)

$$N_S = \frac{100 \text{ kHz}}{2 \times 5 \text{ kHz}} = \frac{100 \times 10^3}{10 \times 10^3}$$

$$N_S = 10 \text{ stations}$$

For SSB transmission,

$$N_S = \frac{BW_S}{f_M}$$

(Equation 5-5)

$$N_S = \frac{100 \text{ kHz}}{5 \text{ kHz}} = \frac{100 \times 10^3}{5 \times 10^3}$$

$$N_S = 20 \text{ stations}$$

Note: This example assumes that there are no *guard bands* between each of the transmitting stations. A guard band is a range of frequencies between each station where no transmission is allowed. Guard bands keep stations from interfering with each other.

Conclusion

In this section, you saw what led up to the development of *single-sideband* systems. These systems will be discussed in the next section.

5-2 Review Questions

1. In a transmitted AM wave, state where the information is contained.
2. For a station transmitting music, state how much music information is contained in the carrier.
3. Determine when there is maximum power in the sidebands.
4. What is the maximum power that can be contained in the sidebands compared to the carrier power?
5. What is the difference between the upper sideband and the lower sideband in terms of information content?
6. What is the relationship between the carrier power and the percent modulation?
7. Explain how SSB transmission allows more stations to be located within the same frequency spectrum when compared to standard AM transmission.
8. What is a guard band? What effects do guard bands have on the number of stations within a given frequency spectrum?

5-3 TYPES OF SIDEBAND SYSTEMS

Overview

This section presents an overview of the four standard types of AM transmission. After you understand this, then study the detailed analysis of each transmission system that follows.

Standard AM Transmission

Up to now, you have been dealing with standard AM transmission. It consists of a carrier with an upper sideband and a lower sideband. A more detailed analysis of this kind of transmission will be helpful. Suppose you want to transmit music from a standard piano. The frequency range of its keyboard is from below 30 Hz to above 4 kHz.

To take a closer look at what happens to the frequency spectrum, three cases will be studied:

1. The spectrum when there is no sound at all
2. The spectrum when a piano note of around 30 Hz is to be transmitted
3. The spectrum when a piano note of around 4 kHz is to be transmitted

In all three cases, assume that the carrier frequency is 1 MHz. Figure 5-3 shows the resulting frequency spectrum for the three cases.

Note that in all three cases, the carrier is always present, *even when no music is being transmitted.* When there are no sidebands present, the resultant waveform is a pure RF sine wave. No matter how many sidebands are present, the carrier is *always* a pure sine wave. Figure 5-4 shows how the familiar AM waveform is actually constructed from three sine waves of three different frequencies: The lower sideband ($f_C - f_M$), the carrier (f_C), and the upper sideband ($f_C + f_M$).

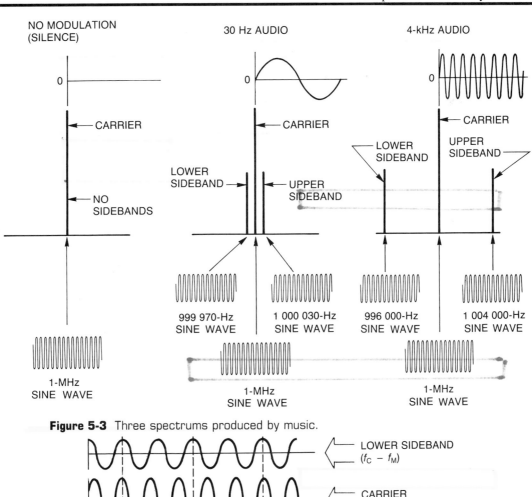

Figure 5-3 Three spectrums produced by music.

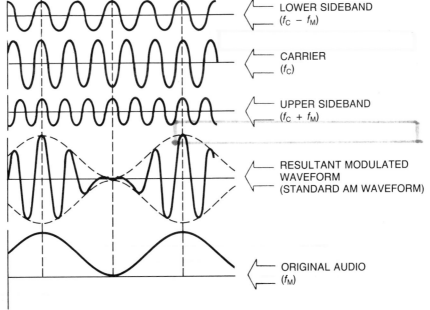

Figure 5-4 Actual construction of AM waveform.

The figure shows how critical the *phase* relationship of the sidebands is in the outcome of the resultant AM waveform. Because of atmospheric conditions, the phase of one sideband may often shift during transmission of the radio wave. This condition is sometimes called "fading"; even though the signal is still strong, the reconstructed audio seems to disappear. Fading can result from changes in the phase relationships of the sidebands.

When the waveform is put into a detector (like the simple diode detector), the result is the sum and difference of the sidebands and carrier. Recall that when two or more sine waves are combined in a nonlinear device, the result is new frequencies that are the sum and difference of the original frequencies. Hence, the difference between each sideband and the carrier is the transmitted audio frequency.

Eliminating the Carrier

If the radio transmitter could be constructed so that the carrier were not transmitted, only the sidebands, then considerable transmitter power would be saved. The basic idea of eliminating the carrier and transmitting only the sidebands is shown in Figure 5-5 for the same three cases of modulation.

As the figure shows, when there is no sound, there is no carrier. The point is that the only time power is being transmitted is when there is some information. This type of modulation is called *suppressed-carrier* modulation. This type of modulation will be studied in more detail later in the chapter. For now, just get the basic idea of what suppressed carrier modulation means—no carrier, just sidebands.

Eliminating One Sideband and the Carrier

Since both sidebands contain the same information, one sideband could be eliminated, along with the carrier, and the required transmission bandwidth would be only half as wide. The advantage here is the ability to get more different stations within a frequency range.

The resulting frequency spectrum for transmitting only one sideband (in this case the upper sideband) is shown in Figure 5-6 for the three cases of modulation. This kind of transmission is called *single-sideband* (SSB) modulation. The transmitted waveform looks just like a pure sine wave, which is what you would expect, because what is being transmitted is only one of the sine waves that was a part of the original complex AM waveform. Therefore during transmission, there is no fading. Since there is only one sideband, there is no other sideband or carrier to be interfered with.

Vestigial Sideband Transmission

Suppressed-carrier and SSB transmission have one big disadvantage. The receiver needs to *reinsert* the missing carrier. This can be expensive, lack some fidelity, and be prone to other problems.

Instead of completely eliminating the carrier vestigial sideband transmission will partially reduce only one sideband and keep the carrier. This means a carrier need not be reinserted by the receiver. This transmission is called *vestigial sideband* transmission; its frequency spectrum is shown in Figure 5-7.

The main advantage of vestigial sideband transmission is reduced bandwidth of television stations. This reduction is necessary because much information must be contained in a TV signal. Many TV stations occupy a given frequency allocation along with required guard bands.

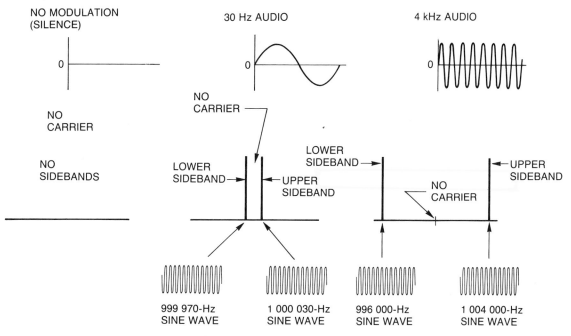

Figure 5-5 AM transmission without carrier.

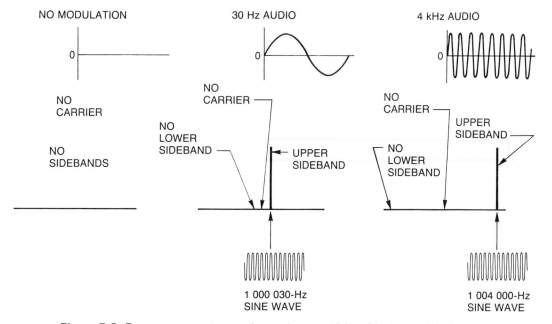

Figure 5-6 Frequency spectrum when only one sideband is transmitted.

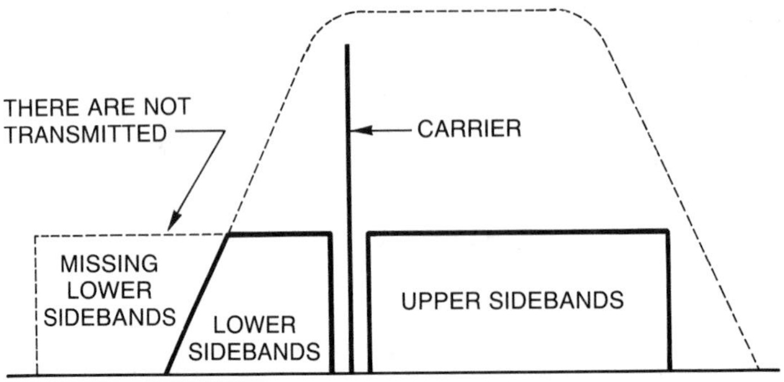

Figure 5-7 Frequency spectrum of vestigial sideband transmission.

Conclusion

This section presented the four major types of sideband systems. The next three sections will present *suppressed-carrier, single-sideband,* and *vestigial sideband* systems in greater detail.

5-3 Review Questions

1. Describe what happens in standard AM transmission when no modulating signal is being transmitted.
2. What is fading? How is it caused?
3. Explain the difference between (A) standard AM transmission and suppressed carrier; (B) suppressed-carrier transmission and single sideband; (C) single sideband and vestigial sideband.
4. Compare the transmission of a 30-Hz tone to that of a 4-kHz tone using (A) standard AM transmission, (B) suppressed carrier, (C) single sideband, (D) vestigial sideband.
5. Describe the form of AM transmission that always produces a single pure sine wave. Why does this happen? What would happen if you tried to detect this waveform with a diode detector?
6. State the main problem in receiving a suppressed-carrier transmission or a single-sideband transmission.
7. What is an application for vestigial sideband transmission?

5-4 | SUPPRESSED CARRIER

Overview

Suppressed-carrier modulation is the elimination of the carrier and the transmission of only the upper and lower sidebands. This kind of transmission is sometimes referred to as *double sideband* (DSB). In the last section, the advantages of suppressing the carrier were introduced. In this section, some of the details of how this is accomplished are presented.

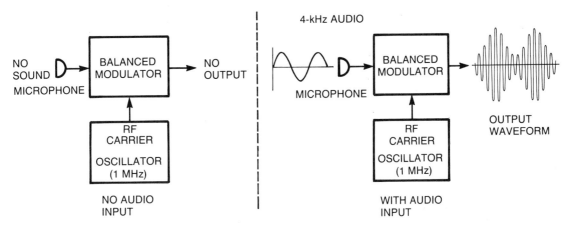

Figure 5-8 Effects of a balanced modulator.

Balanced Modulator

Recall that a *modulator* was that section of a transmitter that combined the audio information with the RF carrier. A specially constructed modulator, called a *balanced modulator,* is used in suppressed-carrier transmission. What makes a balanced modulator unique is that the effects of the carrier are "balanced out" by the internal circuit, and the resulting output contains only the sidebands. The effect of a balanced modulator is shown in Figure 5-8. Note that when there is no modulating signal, there is no output. This conforms with the spectrum of a suppressed-carrier wave shown in the last section—no audio, no output.

Example 1

A DSB signal contains 1 kW. How much power is contained in the sidebands and the carrier frequency?

Solution

In a DSB signal there is no carrier. Therefore the carrier power is zero. Since $P_{USB} = P_{LSB}$, then

$$2P_{SB} = 1 \text{ kW}$$

$$P_{SB} = \frac{1 \text{ kW}}{2} = 0.5 \text{ kW}$$

Hence, the power in each sideband is 0.5 kW.

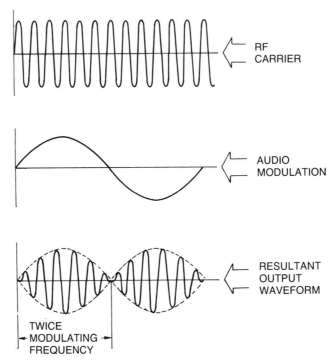

Figure 5-9 Resultant waveform from balanced modulator.

The resultant waveform of the suppressed-carrier modulator is shown in Figure 5-9. Note that the resultant "envelope" is twice the modulating frequency. This is what you would expect to get if you used a simple diode detector at the receiver end. When you pass upper and lower sidebands through a nonlinear device, the result is the sum and difference of the two signals. The difference is an audio signal that is twice the original audio.

This can be expressed mathematically as

$$f_R = f_{USB} - f_{LSB}$$

(Equation 5-6)

Example 2

A diode detector is used to reconstruct the audio from a DSB transmitter. If the transmitting frequency is 2 MHz and the modulating frequency is 5 kHz, determine the frequency of both sidebands and the reconstructed audio.

Solution

The frequency of both sidebands is the sum and the difference of the carrier frequency and the modulating frequency:

$$f_{\text{USB}} = f_C + f_M \qquad\qquad\qquad \textbf{(Equation 4-4)}$$
$$f_{\text{USB}} = 2 \text{ MHz} + 5 \text{ kHz} = 2\ 005\ 000 \text{ Hz}$$

$$f_{\text{LSB}} = f_C - f_M \qquad\qquad\qquad \textbf{(Equation 4-5)}$$
$$f_{\text{LSB}} = 1 \text{ MHz} - 5 \text{ kHz} = 1\ 995\ 000 \text{ Hz}$$

Since the envelope of a DSB wave is twice the frequency of the modulating wave, a diode detector would produce twice the modulating frequency. Using the relationship:

$$f_R = f_{\text{USB}} - f_{\text{LSB}} \quad = \left(f_c + f_m \right) - \left(f_c - f_m \right) = 2 f_m = 2 \times 5 kHz \quad \textbf{(Equation 5-6)}$$
$$f_R = 2\ 005\ 000 - 1\ 995\ 000 = 10\ 000 \text{ Hz} \qquad = 10 \, kHz$$
$$f_R = 10 \text{ kHz}$$

This is double the original modulating frequency.

A detailed discussion of balanced modulator circuits is given in the circuits section of this text. For now, know what role a balanced modulator plays in electronic communications.

IC Balanced Modulator

There are many *integrated circuits* available that perform the function of a balanced modulator. One is the LM1596/LM1496 IC balanced modulator. Note that the internal workings of the IC are not shown in a conventional schematic; only the external connections are shown. Figure 5-10(A) is the internal construction of a typical IC, and Figure 5-10(B) shows the wiring diagram of the balanced modulator.

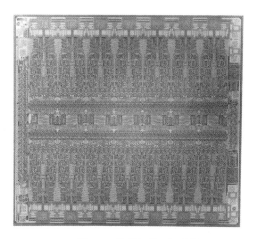
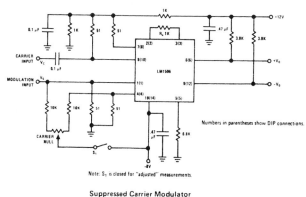

Typical Application and Test Circuit

Note: S₁ is closed for "adjusted" measurements.

Suppressed Carrier Modulator

Figure 5-10 Typical IC and IC balanced modulator. *Courtesy* of National Semiconductor Corporation

National Semiconductor

Audio/Radio Circuits

LM1596/LM1496 Balanced Modulator-Demodulator

General Description

The LM1596/LM1496 are double balanced modulator-demodulators which produce an output voltage proportional to the product of an input (signal) voltage and a switching (carrier) signal. Typical applications include suppressed carrier modulation, amplitude modulation, synchronous detection, FM or ᴾM detection, broadband frequency doubling and chopping.

The LM1596 is specified for operation over the $-55^{\circ}C$ to $+125^{\circ}C$ military temperature range. The LM1496 is specified for operation over the $0^{\circ}C$ to $+70^{\circ}C$ temperature range.

Features

- Excellent carrier suppression
  65 dB typical at 0.5 MHz
  50 dB typical at 10 MHz

- Adjustable gain and signal handling

- Fully balanced inputs and outputs

- Low offset and drift

- Wide frequency response up to 100 MHz

Schematic and Connection Diagrams

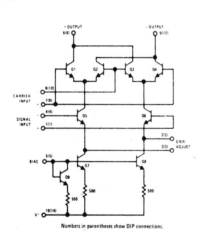

Numbers in parentheses show DIP connections.

Metal Can Package

TOP VIEW
Note: Pin 10 is connected electrically to the case through the device substrate.

Order Number LM1496H or LM1596H
See NS Package H08C

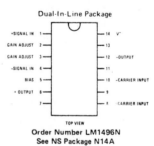

Dual-In-Line Package

TOP VIEW
Order Number LM1496N
See NS Package N14A

Figure 5-11 Specification sheet for LM1596/LM1496. *Courtesy* of National Semiconductor Corporation

A specification sheet for the LM1596/LM1496 balanced modulator is shown in Figure 5-11. The IC comes in different packages: metal can and dual-in-line.

Disadvantages of DSB

Since the carrier has been removed in DSB, the receiver must *reinsert* the carrier. This is very crucial, since the carrier must be not only at the correct frequency but also have the correct phase. These relationships are required if the AM waveform is to be restored for proper detection. This requires expensive and very accurate circuitry. Later in the chapter, you will be introduced to a receiver that reinserts the carrier.

Conclusion

The balanced modulator is the key to suppressed-carrier transmission; the bandwidth has not been reduced. The next section will introduce SSB transmission.

5-4 Review Questions

1. Describe the main feature of DSB transmission.
2. What is the difference between a standard AM modulator and a balanced modulator?
3. Explain the difference between the envelope of a standard AM transmission and that of a DSB transmission.
4. Why is the envelope of a DSB transmission twice the frequency of the modulating signal?
5. Describe the major disadvantage of DSB modulation.
6. Discuss the bandwidth of DSB transmission compared with standard AM transmission.

5-5 | SINGLE SIDEBAND

Discussion

This section presents the details of eliminating one of the sidebands from the output of a balanced modulator and reproducing the original audio at the receiver. This technique presents some major problems. Single sideband (SSB) transmission has a wide range of applications.

Eliminating a Sideband

For the purpose of discussion, it makes no difference which sideband is eliminated. In the following example, the lower sideband will be eliminated. Recall the example of transmitting piano music. Consider what must be done when 30 Hz is the modulating frequency. Assume that the carrier frequency is 10 MHz. The desired result is shown in Figure 5-12.

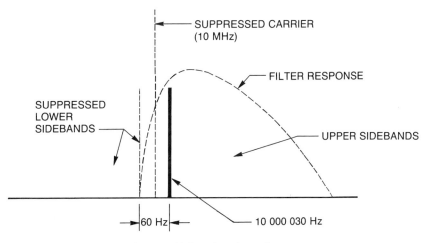

Figure 5-12 Suppressing the lower sideband and carrier.

Only 60 Hz separates the two sidebands (the result of the balanced modulator receiving the 30-Hz audio tone). Keep in mind that you are trying to filter out a frequency of 9 999 970 Hz (LSB) from a frequency of 10 000 030 Hz (USB).

The required circuit Q to do this can be computed as follows:

$$Q = \frac{f_R}{BW}$$

$$Q = \frac{10 \text{ MHz}}{60 \text{ Hz}} = 166\ 667$$

A circuit Q of this magnitude is not practical. As you will see in the chapter on filters, practical crystal filters have Q's of about 50 000. There is another method to suppress one sideband, without having to resort to impractical high-Q circuits.

Conversion Frequency

If the audio signal is first mixed with a low radio frequency, say 100 kHz, the result will be a lower Q for filtering. See Figure 5-13. Note that the separation of the sidebands is still 60 Hz, but now the required Q of the highpass filter circuit is greatly reduced:

$$Q = \frac{100 \text{ kHz}}{60 \text{ Hz}} = 1\ 666$$

For voice, where the lower limit of 100 Hz is acceptable for intelligence, the bandwidth separation would be 200 Hz, and the required Q using the same conversion frequency is

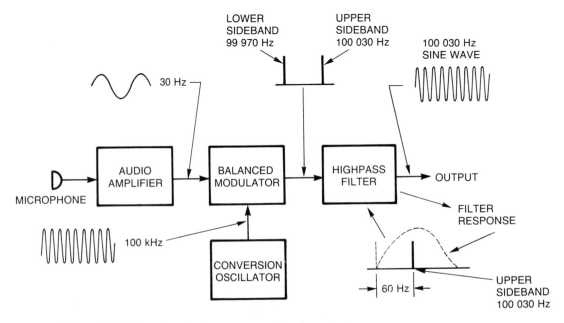

Figure 5-13 Results of using a conversion frequency.

$$Q = \frac{100 \text{ kHz}}{200 \text{ Hz}} = 500$$

This value is practical for economical highpass filters.

Power Savings

Single-sideband transmission saves considerable transmission power, as the following example illustrates.

Example 1

If a 10-kW SSB transmitter were to be replaced by a standard AM signal with the same power, what would be the power content of the carrier and each sideband when the percent modulation is 100%? Compare the sideband power of the SSB to the standard AM transmitter.

Solution

The total power content of a standard AM wave is

$$P_T = P_C + \frac{m^2 P_C}{4} + \frac{m^2 P_C}{4} \qquad \textbf{(Equation 5-2)}$$

$$10\,000 \text{ W} = P_C + \frac{(1.0)^2 P_C}{4} + \frac{(1.0)^2 P_C}{4}$$

$$10\,000 \text{ W} = P_C + \frac{P_C}{2} = 1.5 P_C$$

Solving for P_C gives

$$P_C = \frac{10\,000}{1.5} = 6\,666.67 \text{ W}$$

The power content of both sidebands is

$$P_{SBT} = P_T - P_C$$
$$P_{SBT} = 10\,000 - 6\,666.67 = 3\,333.33 \text{ W}$$

The power in one sideband is

$$P_{SB} = \frac{3\,333.33}{2} = 1\,666.67 \text{ W}$$

Thus for the same transmission power, one sideband in standard AM would contain only 1 666.67 W. This compares with 10 000 W of power contained in the same sideband of an SSB transmitter.

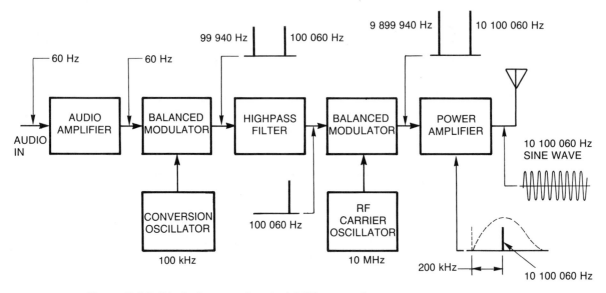

Figure 5-14 Block diagram of typical SSB transmitter.

SSB Transmitter

The block diagram of a typical SSB transmitter is shown in Figure 5-14. Note what now happens. The first filter is a highpass filter with a practical Q of about 1 000. This easily rejects the lower sideband for audio frequencies down to 50 Hz. The resulting upper sideband is then fed into another balanced modulator. This time, the two new sidebands are at least 200 kHz apart, each containing the original audio frequency.

Now, a second filter, inherent in the power amplifier, is used to remove the new lower sideband. This again can be done with a practical Q value. For the given example where $f_C = 10$ MHz:

$$Q = \frac{10 \text{ MHz}}{200 \text{ kHz}} = 50$$

The power amplifier acts as a highpass filter. For this kind of amplifier, a Q of 50 is easy to obtain.

Conclusion

This section introduced you to the concept of using a lower frequency carrier to "bring up" the audio frequency. This was done so it became practical to filter out one of the sidebands. You will see this principle used in many other applications. Understanding its use in SSB systems will help you recognize it in other systems.

5-5 Review Questions

1. In SSB transmission, which sideband can be eliminated?
2. Describe why a very high Q filter is needed to filter out a sideband from its carrier.
3. What is a conversion frequency?

4. State the purpose of a conversion frequency oscillator.
5. Describe why it is easier to filter out the desired sideband after it is mixed with a conversion frequency.
6. How many filters are used in the SSB transmitter presented in this chapter?

5-6 | VESTIGIAL SIDEBAND

Discussion

For television, vestigial sideband transmission is used. The TV frequency range is quite extensive, since visual (picture) information along with synchronizing and voice must be included in the sidebands of the transmitted wave. For color TV systems, color information adds even more requirements to the sidebands.

Vestigial Sideband Generation

Figure 5-15 is a block diagram of a portion of a vestigial sideband transmitter. The main feature of vestigial sideband generation is the presence of a highpass filter. Note that the filter allows all of the USB, carrier, and part of the LSB to pass. The advantages of this compared to standard AM are:

■   The transmission bandwidth is reduced.

■   Some power requirements are reduced.

The advantage of this compared to single sideband is:

■   Since the carrier is still present, the signal is easy to detect at the receiver end.

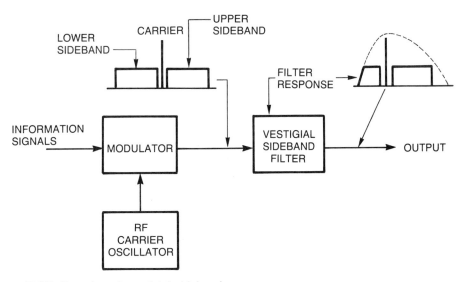

Figure 5-15 Creation of vestigial sideband.

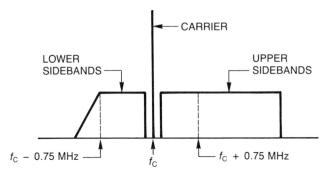

Figure 5-16 Transmitted TV spectrum.

The main disadvantage of vestigial sideband modulation is that the receiver must make up for the part of the sideband that was suppressed. The resulting spectrum for the transmitted TV signal is shown in Figure 5-16.

As required by the Federal Communications Commission (FCC), both sidebands of the TV signals below 750 kHz are transmitted. Only the upper sideband contains required picture frequencies above 750 kHz. It will work out at the receiver end that since both sidebands for frequencies of 750 kHz and below are present, these frequencies will be stronger than the higher frequencies present in only the upper sideband. Thus the IF amplifiers in TV receivers have their tuned circuits tuned to emphasize the visual frequencies above 750 kHz. The response curve for the TV IF is shown in Figure 5-17. As you will see in the chapter on television, a TV uses a superheterodyne receiver.

Conclusion

You have explored four methods of sideband transmission. The next section presents a receiver used to make sense from SSB transmission. Recall that, unlike vestigial sideband where the carrier is still preserved, SSB must have the carrier reinserted before the modulating information can be restored.

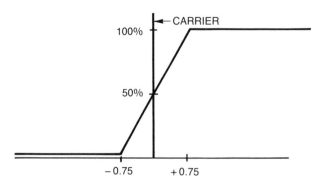

Figure 5-17 TV IF response curve.

5-6 Review Questions

1. State what information other than sound must be transmitted in TV transmission.
2. Describe what circuit is used to create vestigial transmission.
3. State some of the advantages of vestigial sideband transmission.
4. Describe a disadvantage of vestigial sideband reception.
5. Explain how the TV IF is tuned to compensate for vestigial sideband reception.

5-7   SSB RECEIVERS

Overview

Single-sideband receivers must reinsert the carrier before detection can take place. The design of SSB receivers requires very stable oscillators. This stability is necessary because the reconstructed modulating signal is the result of the relationship of the phase and frequency of the reinserted carrier to the received SSB signal.

Carrier Reinsertion

Figure 5-18 is a block diagram of a method to reinsert the carrier of an SSB signal. If the original modulating signal were 4 kHz and the received sideband were 10 004 000 Hz, then a carrier of 10 MHz would have to be generated by the oscillator in Figure 5-18. The output of the mixer (a nonlinear device) would be the original two frequencies and their sum and difference. If the result were put through a lowpass filter, the output would be the difference frequency of 4 kHz.

The oscillator used to reinsert the carrier *must* be very stable. If its frequency changes by even a small amount, the result would not be an accurate reconstruction of the original

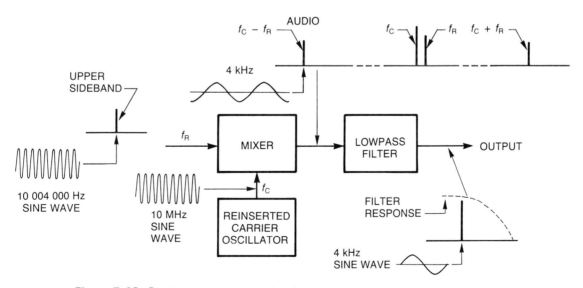

Figure 5-18 Carrier reinsertion technique.

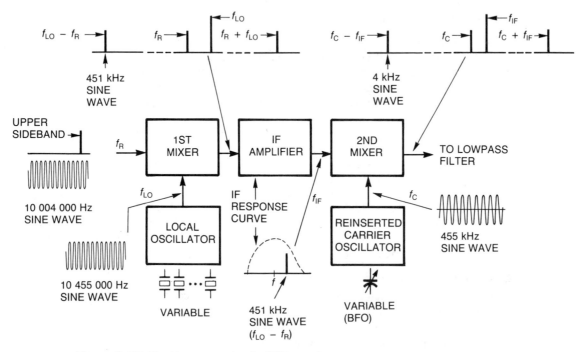

Figure 5-19 Double conversion in SSB receiver.

modulating signal. For this reason, the incoming signal is first reduced to an IF frequency using a superheterodyne receiver. See Figure 5-19.

Usually the desired accuracy is achieved by an *automatic frequency control* (AFC) circuit. Sometimes, a *pilot carrier* signal is sent along with the sideband. This pilot carrier signal is generated at the transmitter and is used as a reference frequency for the carrier reinsertion oscillator.

If the receiver is to receive SSB signals from transmitters with different frequencies, then some method of changing the frequency of the reinsertion oscillator is necessary. This can be done with different crystals for the local oscillator (LO) and a variable frequency oscillator for the carrier reinsertion.

If the pilot carrier method is used, the variable frequency oscillator can have its frequency adjusted by the radio operator. In this case, it is called a *beat frequency oscillator* (BFO), because the operator adjusts the BFO frequency until a low-frequency "beat" is heard. This beat indicates that the frequency difference between the BFO and the pilot carrier is almost zero.

Receiver Block Diagram

The block diagram of an SSB receiver is shown in Figure 5-20. Observe that the RF amplifier, LO, mixer, and IF amplifiers are the same as in a conventional AM radio. The difference is in the second mixer and BFO. Through the BFO and second mixer, the carrier is reinserted and detection can take place.

Because of the conversions required by the received signal, tuning an SSB receiver requires some patience from the operator. Many parts of the SSB receiver are made up

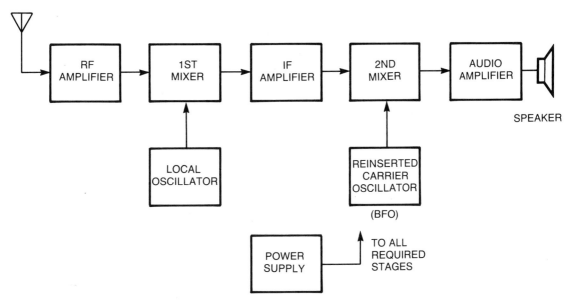

Figure 5-20 SSB receiver block diagram.

of ICs rather than discrete components. As an example, the second mixer can be constructed using an inexpensive IC. Mixers will be presented in the circuits section of the text.

Conclusion

SSB receivers require more stages than the standard AM receiver. You saw that very stable oscillators were required in order to accurately reinsert the carrier. When you start the circuits section of this text, you will see how discrete components such as transistors, resistors, and diodes are used to accomplish the task required by each of the receiver blocks.

5-7 Review Questions

1. State what an SSB must do before detection can take place.
2. State the major requirement of the oscillators used in an SSB receiver.
3. Describe what is meant by AFC.
4. What is the purpose of a pilot carrier?
5. What is a BFO? Describe how it is used.
6. How many mixers are in an SSB receiver? Describe the purpose of each one.

TROUBLESHOOTING AND INSTRUMENTATION

SSB Instrumentation

The alignment and maintenance of SSB systems is different from other types of communication systems. This section introduces the techniques used in SSB transmission testing.

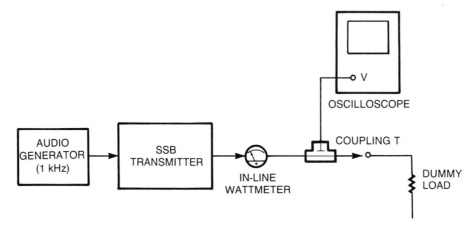

Figure 5-21 Basic carrier suppression test.

Carrier Suppression Testing

Recall that for a true SSB transmitter, the carrier should not appear at the output of the transmitter. If the SSB transmitter is not properly maintained, a carrier can appear at the output, with the result that transmitter power is wasted. This condition defeats one of the main advantages of SSB transmission.

A simple test setup to measure the relative degree of carrier suppression is shown in Figure 5-21. The *dummy load* is a device that simulates the transmitting antenna but does not allow any signal to be transmitted. Dummy loads are necessary because you don't

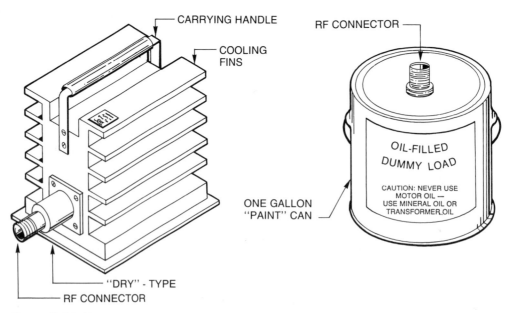

Figure 5-22 Typical dummy load.

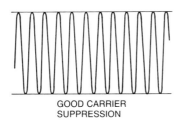 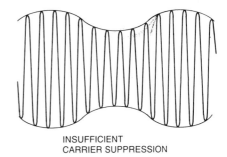

Figure 5-23 Carrier suppression scope patterns.

want to transmit a radio signal while adjusting the transmitter. That could disrupt other communication signals because your transmitting frequency is changing from the adjustments. A typical dummy load is shown in Figure 5-22.

The *in-line* wattmeter ensures that the output power of the transmitter is adjusted properly. The *coupling T* is an electrical device that extracts a small amount of RF energy for use by the oscilloscope.

The adjustment procedure in the maintenance manual for the particular SSB transmitter should be followed for proper adjustment of carrier suppression. In general, no matter what the adjustment procedure, here is what you look for.

Figure 5-23 shows the scope signal for good and poor carrier suppression. Note that the presence of a carrier begins to produce an output waveform that is similar to the standard AM transmitted wave.

Carrier-Suppression Measurement

For some systems, it is impossible to completely suppress the carrier. In these cases, the technician must adjust the transmitter to bring the carrier down to certain specifications. Carrier suppression is usually measured in dB and is defined as

$$S_C dB = 20 \log \left(\frac{V_{p\text{-}p}}{V_R} \right)$$ **(Equation 5-7)**

where $S_C dB$ = Carrier suppression in decibels

$V_{p\text{-}p}$ = Peak-to-peak envelope voltage in volts

V_R = Voltage of envelope ripple in volts

Figure 5-24 illustrates where the measurement is made on the output signal.

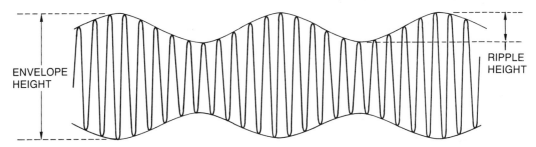

Figure 5-24 Measurements for carrier suppression.

Example 1

For the output signal of the SSB transmitter in Figure 5-25, what is the carrier suppression in dB?

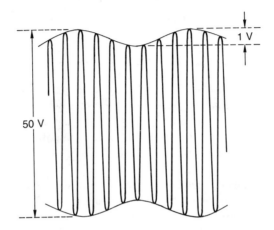

Figure 5-25

Solution

Using the formula for carrier suppression:

$$S_C dB = 20 \log\left(\frac{V_{p\text{-}p}}{V_R}\right)$$ **(Equation 5-7)**

$$S_C dB = 20 \log\left(\frac{50 \text{ V}}{1 \text{ V}}\right) = 20 \log(50) = 20 \times 1.7$$

$$S_C dB = 34 \text{ dB}$$

Conclusion

This section introduced you to one method of transmitter alignment. The purpose of a dummy load was discussed.

5-8 Review Questions

1. Explain the result of having the carrier frequency appear at the output of an SSB transmitter.
2. Describe a dummy load. Why is it needed?
3. Explain a coupling T. Give an example of its use.
4. Describe the oscilloscope pattern of the output wave of an SSB transmitter when no carrier is present.
5. How does the oscilloscope pattern of the output wave of an SSB transmitter appear when the carrier affects the output signal?

MICROCOMPUTER SIMULATION

The fifth troubleshooting simulation on your student disk encourages you to improve your troubleshooting efficiency when using a signal detector for troubleshooting. This program will keep a record of the measurements you have made and then offer suggestions. Having a **troubleshooting history** presented to you can help you improve your troubleshooting efficiency. Your goal here is to make only the actual measurements required for determining which stage, if any, is a fault with the system.

This program prepares you for the next troubleshooting simulation. There you will have the opportunity to use both a signal generator and a signal detector.

CHAPTER PROBLEMS

(Answers to odd-numbered problems appear at the end of the text.)

1. Describe the main features of an AM transmitter.
2. In an AM transmitter, explain the purpose of the RF section and the audio section.
3. What constitutes the input to an AM transmitter? the output?
4. Describe each stage in an AM transmitter, state its purpose, and describe the input and output signals of each stage.
5. State the advantages and disadvantages of high-level transmission.
6. State the advantages and disadvantages of low-level transmission.
7. How many AM stations can be accommodated in a 150-kHz bandwidth if the highest modulating frequency is 10 kHz?
8. For problem 7, how many stations could be accommodated if the highest modulating frequency were 3 kHz?
9. How many AM stations could be accommodated in a 100-kHz bandwidth if the highest modulating frequency were 8 kHz and a 1-kHz guard band were required between each station?
10. For problem 9, how many stations could be accommodated if the guard band were increased to 1.5 kHz?
11. For problem 7, how many stations could be accommodated for (A) DSB? (B) SSB?
12. For problem 9, how many stations could be accommodated for (A) DSB? (B) SSB?
13. What is the power in the carrier and each sideband for an AM signal with 80% modulation and a total power of 2 500 W?
14. Calculate the carrier and total sideband power for an AM signal with 90% modulation and a total power of 1 750 W.
15. If the power content of the carrier of an AM wave is 5 W, determine the power content of each sideband and the total power transmitted when the carrier is modulated 50%.
16. Find the total power transmitted for an AM transmitter and the power in each sideband when the carrier power content is 100 W and the carrier is modulated 75%.
17. For problem 15, what is the power content of the carrier when the modulation is increased to 100%?
18. For problem 16, determine the power content of the carrier when the percent modulation is decreased to 50%.

19. If an AM wave has a power content of 500 W at the carrier frequency, find the power content of one sideband for 85% modulation.

20. Determine the power content of each sideband in an AM transmitter for a carrier power of 12 kW with a modulation index of 0.6.

21. Determine the percent modulation for an AM wave that has a power content of 12 kW in the carrier and 2 kW in each sideband when modulated by a single tone.

22. Find the percent modulation for an AM wave that has a power content of 50 W in the carrier and 15 W in one sideband when modulated with a 5-kHz sine wave.

23. If the total transmitted power of an AM wave is 15 kW, find the percent modulation if the signal in each sideband contains 1 kW.

24. For an AM wave with a total transmitted power of 100 kW, calculate the percent modulation for a sideband power content of 10 kW in each sideband.

25. In a particular AM transmitter, the percent modulation of the AM wave changes from 40% to 60%. If the original power content at the carrier frequency was 800 W, determine the power content of the carrier and each sideband after the percent modulation was increased to 60%.

26. For an AM transmitter, if the original power content of the carrier is 500 W and the percent modulation goes from 30% to 80%, calculate the power content of the carrier and each sideband after the percent modulation has increased to 80%.

27. An SSB signal contains 500 W. State how much power is in the sidebands and how much is at the carrier frequency.

28. For a 750 W SSB signal, how much power is contained in the carrier and the sidebands?

29. If the total transmitted power in a DSB transmitter is 12 kW, determine the power in the carrier and in each sideband. What is the transmitted power when there is no modulating signal?

30. If the total power for a DSB transmitter is 25 kW, calculate the power in each sideband and the carrier. What is the transmitted power when there is no modulating signal?

31. Determine the transmitting power for the SSB transmitter in problem 27 when the modulation index is 0.0.

32. For 0% modulation, what is the total transmitting power for the SSB transmitter in problem 28?

33. An SSB transmitter that contains 8 kW is to be replaced by a standard AM signal of the same power content. Determine the power content of the carrier and each sideband for 75% modulation.

34. An SSB transmitter with 12 kW of transmission power is to be replaced by a standard AM signal of the same power content. Calculate the power content of the carrier and each sideband if the modulation index is 0.6.

35. Describe the circuit in an SSB transmitter that eliminates the carrier.

36. What is the purpose of a balanced modulator?

37. If a diode detector were used to detect a DSB signal, what relation does the detected signal have to the original audio?

38. For a DSB signal detected by a diode detector, what is the original modulating frequency if the detected frequency is 2 kHz?

39. Explain why a carrier must be reinserted for DSB transmission.

40. Is it necessary to reinsert the carrier for SSB reception? Explain.
41. Describe what causes fading.
42. What method of AM transmission reduces the effects of fading? How is the fading reduced?
43. Identify the waveforms in Figure 5-26.

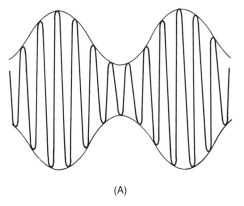
(A)

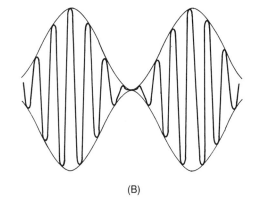
(B)

Figure 5-26

44. Identify the waveforms in Figure 5-27.

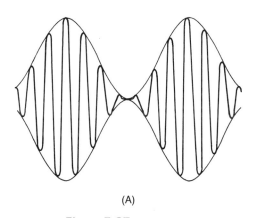
(A)

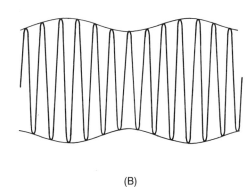
(B)

Figure 5-27

45. Figure 5-28 is a block diagram of an SSB transmitter. Identify each block and state its purpose.
46. If the audio input is a 1-kHz sine wave, what will be the output frequencies at point A in Figure 5-28?
47. What frequency will be at point B of Figure 5-28 for (A) an upper sideband? (B) a lower sideband?
48. What frequency will be at point C of Figure 5-28 for (A) an upper sideband? (B) a lower sideband?

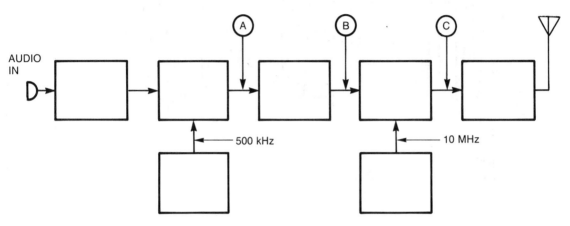

Figure 5-28

49. For the SSB signal in Figure 5-29(A), determine the amount of carrier suppression in dB.
50. For the SSB signal in Figure 5-29(B), what is the amount of carrier suppression in dB?

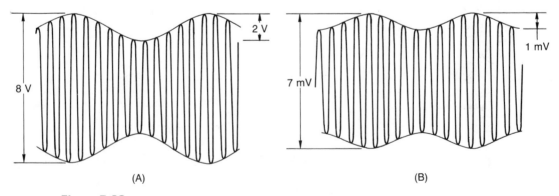

Figure 5-29

CHAPTER 6

Frequency and Phase Modulation Systems

OBJECTIVES

In this chapter, you will study:

- ☐ The similarities of AM and FM transmissions.
- ☐ What frequency modulation is and how it is produced.
- ☐ Details of the frequency spectrum and FM bandwidth.
- ☐ How to analyze FM transmitters.
- ☐ How to analyze FM receivers and what PM is.
- ☐ How stereo FM systems function.
- ☐ Principles of a sweep generator and how it is used.

INTRODUCTION

This chapter introduces you to the details of frequency modulation (FM) and phase modulation (PM) systems. These modulation methods are very similar. There are many advantages to FM and PM. One advantage is a great reduction in noise. You will see how FM stereo functions and how it produces such a realistic sound.

Troubleshooting simulation number six on your student disk gives you the opportunity to use both a signal generator as well as a signal detector. Here you will be able to develop troubleshooting techniques using more than one troubleshooting instrument.

| 6-1 | WHY FM? |

Discussion

Recall that there were three different ways of placing information on a carrier wave:

1. Having the modulating signal change the amplitude of the carrier (amplitude modulation—AM).
2. Having the modulating signal change the frequency of the carrier (frequency modulation—FM).
3. Having the modulating signal change the phase of the carrier (phase modulation—PM).

In this chapter, you will first be introduced to frequency modulation (FM). You will then see that the difference between FM and phase modulation (PM) is very slight.

Advantages of FM

Figure 6-1 shows two transmitted signals. One signal uses AM, the other uses FM. Noise effects the amplitude of radio signals. If the changes in amplitude can be removed inside the receiver, then the noise will not be reproduced in the speaker.

In AM, the changes in amplitude cannot be removed; doing so would also remove the original information. In FM, the changes in amplitude do not represent any intentional information. Thus, in FM receivers, a circuit called a *limiter* is used to remove any changes in amplitude of the received FM signal. This is shown in Figure 6-2. In FM, it is the *change* in *frequency* that represents the original information. The limiter in an FM receiver removes the effects of noise but does not disturb the original modulating information.

A common example of the "noise-free" reception of FM compared to AM is your TV set. Picture information is transmitted to you as AM, but sound information is in FM. So you can observe a "noisy" picture while hearing noise-free sound.

Conclusion

In this section, you saw the main feature of FM compared to AM. As you will see later in this chapter, FM is not completely noise free, but the noise that causes changes in the amplitude of the FM wave can be eliminated. The next section presents a basic way of producing and measuring an FM waveform.

6-1 Review Questions

1. Name three methods of modulating a carrier wave.
2. Where is the modulating information contained in (A) FM? (B) AM?
3. How can noise affect a transmitted signal?
4. How are the effects of noise removed from FM?
5. What is a limiter? Why can't an AM receiver use a limiter?

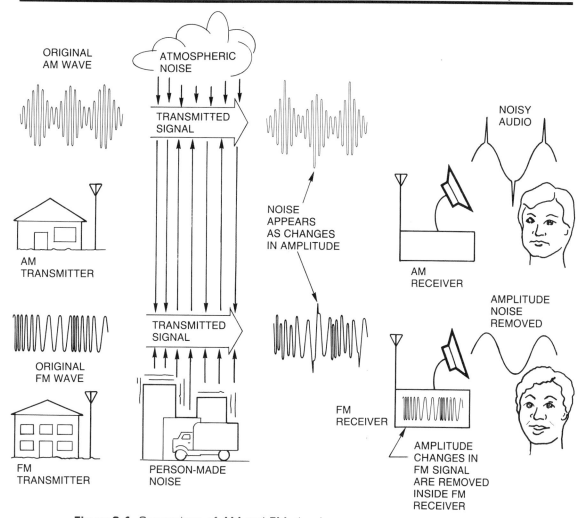

Figure 6-1 Comparison of AM and FM signals.

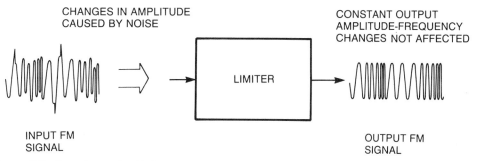

Figure 6-2 Functions of a limiter.

6-2 | FM GENERATION

Discussion

To generate an AM wave, an audio signal had to change the amplitude of the carrier wave. In FM, an audio signal has to change the *frequency* of the carrier wave. There are several techniques for doing this. To present the basic concepts of FM generation, a simplified technique is used in this section. Later in the chapter, a block diagram of a practical FM transmitter will be discussed.

Basic Idea

To generate an FM wave, assume you have an oscillator with an *LC* circuit that controls the frequency of the sine wave produced by the oscillator. See Figure 6-3.

Recall that an oscillator is a circuit that produces its own signal. In this case, the oscillator is producing a high-frequency radio wave. The frequency of the output wave of the oscillator will change if the value of the capacitor or inductor changes. The following presentation uses a microphone that acts as a variable capacitor.

Figure 6-4 shows the construction of a capacitor that is sensitive to sound waves. Such a capacitor is called a *capacitance microphone*. As you learned in AC circuits, a capacitor is two metal plates separated by an insulator. The value of a capacitor is affected by the distance between the plates: the closer the plates, the greater the value of the capacitor.

Figure 6-5 shows the capacitance microphone used as the capacitor in the *LC* circuit of the RF oscillator. The sound received by the capacitor will affect the frequency of the signal produced by the oscillator. Figure 6-5 illustrates one of the most important concepts in the generation of FM. The amount of *frequency change* in an FM signal is determined by the *strength* (amplitude) of the modulating signal.

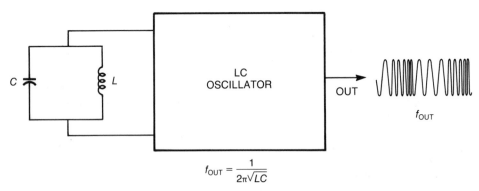

$$f_{OUT} = \frac{1}{2\pi\sqrt{LC}}$$

Figure 6-3 An *LC* oscillator.

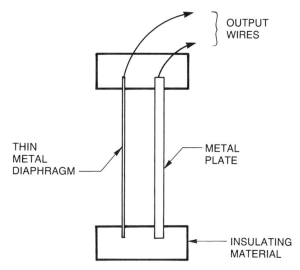

Figure 6-4 Basic construction of capacitance microphone.

Frequency Deviation

Frequency deviation (f_D) means how much the FM carrier frequency *changes* (deviates) from its normal frequency. The *normal* frequency of the FM carrier is the frequency it has without any modulating signal. This normal frequency is also called the *resting frequency* (f_C).

As shown in Figure 6-6, the frequency deviation (f_D) of the FM carrier is caused by the *amplitude* of the modulating signal. Hence, in FM, it is said that the amount of frequency deviation is proportional to the amplitude of the modulating signal.

It is helpful to visualize the relationship between the modulating sine wave and the FM carrier. This relationship is shown in Figure 6-7. The resting frequency is the frequency of the carrier when the amplitude of the modulating signal is zero. This is as it should be, for an amplitude of zero is the same as no modulation at all.

Look closely at Figure 6-7. Note that the full range of change of the FM carrier (from its lowest frequency to its highest) is double its frequency deviation (how much it changes from its resting frequency). For example, if the FM carrier deviated by 5 kHz from its resting frequency, its *total* frequency swing would be 5 kHz above and 5 kHz below the resting frequency for a full range of 10 kHz. This full range of change is called the *carrier swing* and is related to the frequency deviation by

$$f_{CS} = 2f_D$$

(Equation 6-1)

where f_{CS} = Carrier swing in hertz
f_D = Frequency deviation in hertz

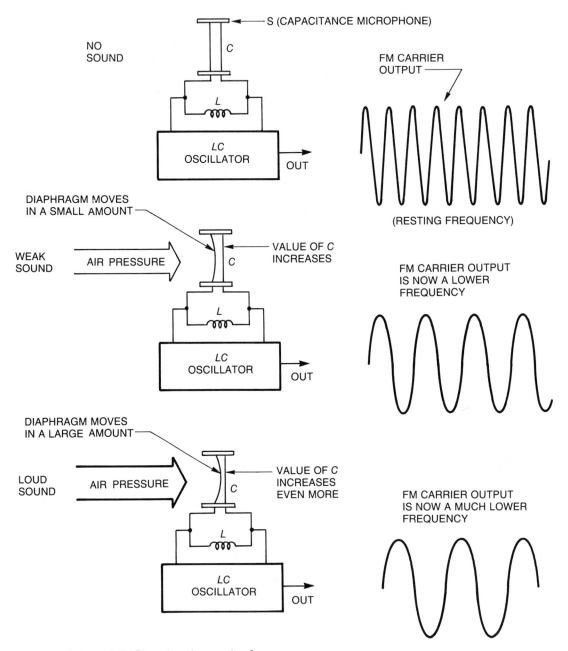

Figure 6-5 Changing the carrier frequency.

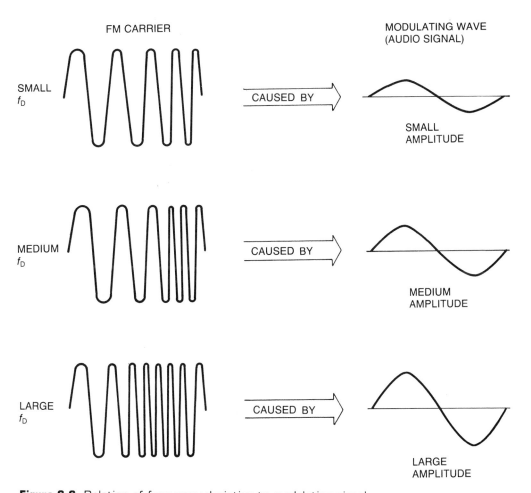

Figure 6-6 Relation of frequency deviation to modulating signal.

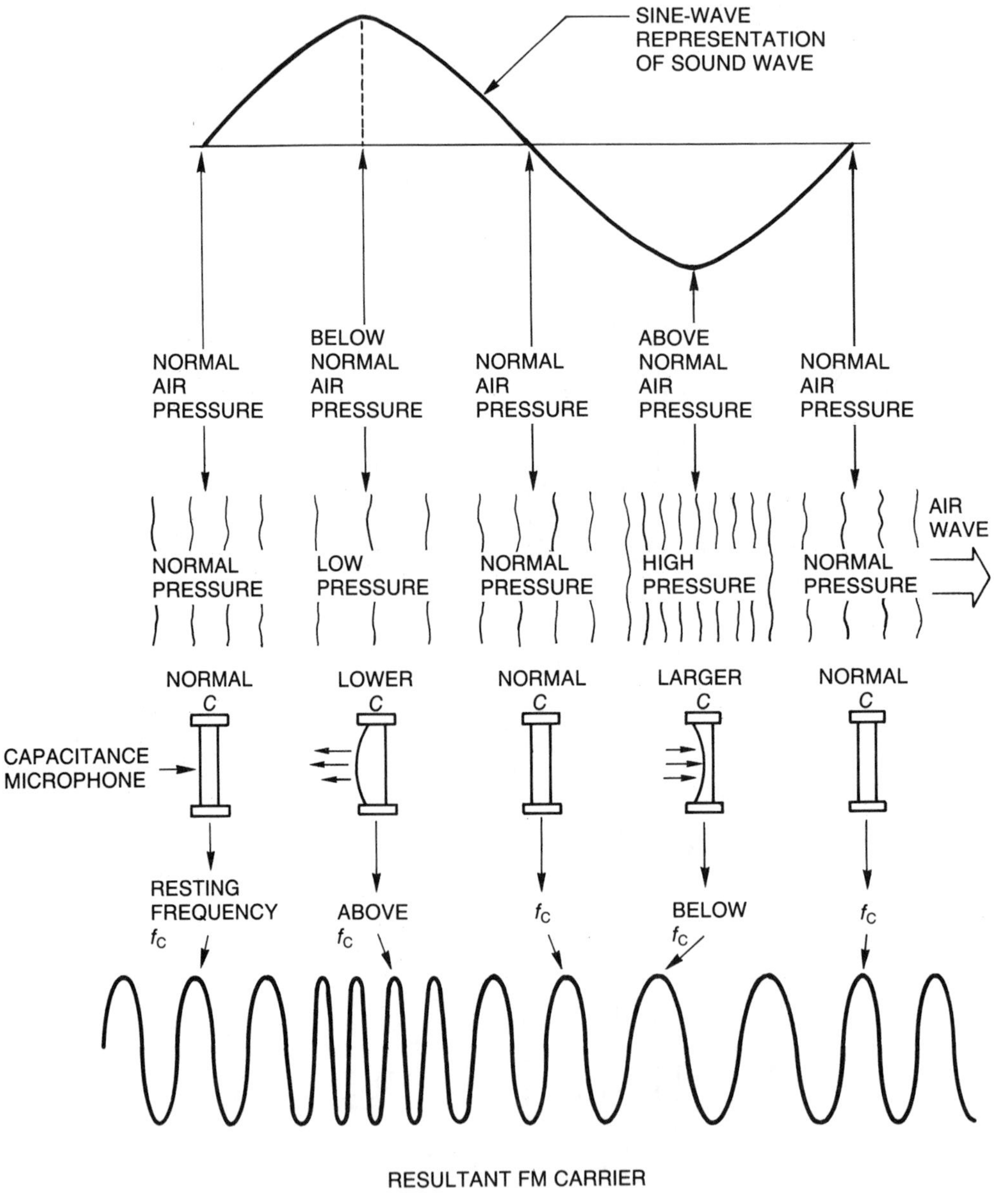

RESULTANT FM CARRIER

Figure 6-7 Relation of audio frequency to FM carrier.

The upper frequency reached by the FM carrier is equal to the resting frequency plus the frequency deviation:

$$f_H = f_C + f_D$$ (**Equation 6-2**)

The lower frequency reached by the carrier is equal to the resting frequency minus the frequency deviation:

$$f_L = f_C - f_D$$ (**Equation 6-3**)

where f_H = Upper frequency reached by FM carrier in hertz
f_L = Lower frequency reached by FM carrier in hertz
f_C = Resting frequency of the FM carrier in hertz
f_D = Frequency deviation of FM carrier in hertz

Example 1

A 100-MHz FM carrier is modulated by an audio tone that causes a frequency deviation of 20 kHz. (A) Determine the carrier swing of the FM signal. (B) Find the lowest and highest frequencies attained by the FM signal.

Solution

(A) Determine the carrier swing:

$$f_{CS} = 2f_D$$ (**Equation 6-1**)
$$f_{CS} = 2 \times 20 \text{ kHz} = 40 \text{ kHz}$$

(B) Find the highest carrier frequency:

$$f_H = f_C + f_D$$ (**Equation 6-2**)
$$f_H = 100 \text{ MHz} + 20 \text{ kHz} = 100.02 \text{ MHz}$$

Find the lowest carrier frequency:

$$f_L = f_C - f_D$$ (**Equation 6-3**)
$$f_L = 100 \text{ MHz} - 20 \text{ kHz} = 99.98 \text{ MHz}$$

Rate of Change

Since the *amplitude* of the modulating wave causes the change in *frequency* of the FM carrier, you may be wondering what does the frequency of the *modulating wave* do to the FM carrier. What happens here is shown in Figure 6-8. The frequency of the modulating signal affects how rapidly (rate of change) the FM wave changes its frequency. For an audio signal of 10 Hz, the FM wave will deviate from its resting frequency 10 times each second. For an audio signal of 100 Hz, the FM wave will deviate from its resting frequency 100 times each second. Putting this another way, how *far* from its resting frequency the

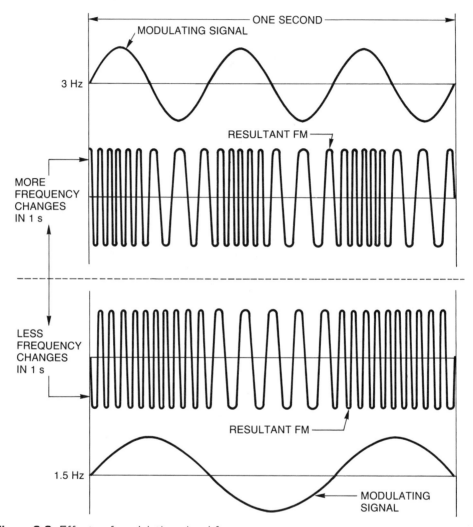

Figure 6-8 Effects of modulating signal frequency.

FM wave deviates is determined by the *amplitude* of the audio signal. How *often* the FM wave deviates from its resting frequency is determined by the *frequency* of the audio signal.

Modulation Index

Since the FM carrier is influenced by both the amplitude and the frequency of the modulating wave, a measurement called the *modulation index* is given as

$$m_I = \frac{f_D}{f_M}$$
(**Equation 6-4**)

where m_I = Modulation index of the FM wave (no units)

f_D = Deviation of the FM wave in hertz

f_M = Frequency of the modulating signal in hertz

Example 2

What is the modulation index of an FM signal with a carrier swing of 10 kHz when the modulating signal is 8 kHz?

Solution

The formula for modulation index uses carrier deviation (f_D), not carrier swing (f_{CS}). Hence, a conversion to frequency deviation must first be made:

$$f_{CS} = 2f_D$$
(**Equation 6-1**)

$$10\text{ kHz} = 2f_D$$

$$f_D = \frac{10\text{ kHz}}{2} = 5\text{ kHz}$$

Now use the definition for modulation index:

$$m_I = \frac{f_D}{f_M}$$
(**Equation 6-4**)

$$m_I = \frac{5\text{ kHz}}{8\text{ kHz}} = 0.625$$

Conclusion

Two important characteristics of an FM carrier were presented. One was the amount of frequency deviation, and the other was the rate of frequency deviation. What caused these changes and how they were measured were also presented.

You will use this information in the next section. There you will learn about the FM spectrum and bandwidth.

6-2 Review Questions

1. Describe the main difference between FM and AM.
2. Explain the basic construction of a capacitance microphone. How does a voice wave cause its capacitance to change?
3. What affect does the modulating signal amplitude have on the resulting FM wave?
4. Define frequency deviation. Explain what is meant by the resting frequency of an FM carrier.
5. Describe the relationship between frequency deviation and carrier swing. How could you determine the highest and lowest carrier frequencies?
6. Explain what effect the frequency of the modulating signal has on the FM carrier.
7. Describe what factors determine the modulation index of an FM carrier.

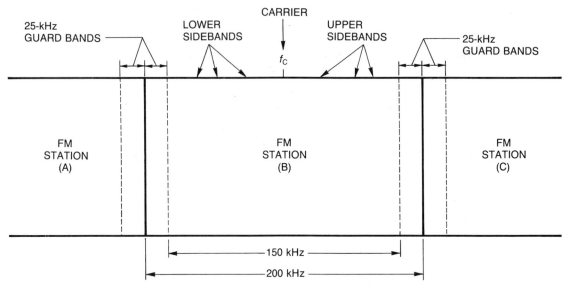

Figure 6-9 Commercial FM frequency allocations.

| 6-3 | FM SPECTRUM AND BANDWIDTH |

Discussion

Each commercial FM broadcast station in the 88–108-MHz band is allocated a 150-kHz channel plus a 25-kHz guard band on either side of the channel. This arrangement is shown in Figure 6-9. You may wonder why such a large bandwidth is required for an FM station, especially since the maximum allowable modulating frequency for commercial FM is 15 kHz. For AM transmission, the required bandwidth is only 10 kHz (2×5 kHz). In this section, you will discover that the frequency modulation of a carrier wave produces many unexpected sidebands.

FM Sidebands

Frequency modulation was first put into practical use in 1936. Before then, some scientist thought that the transmission of FM was not practical. The reason was that FM created an infinite number of sidebands. However the sidebands closest to the carrier contain most of the information. Figure 6-10 gives you an idea of the behavior of the carrier and sidebands of a typical FM signal. Note the changes in the carrier and the surprising creation of new sidebands. This happens even though the modulating frequency is the same; only its amplitude is changing.

Here is a summary of what is taking place.

Figure 6-10A

Here there is no modulating frequency. The frequency spectrum shows only the presence of the FM carrier. This is as you would expect.

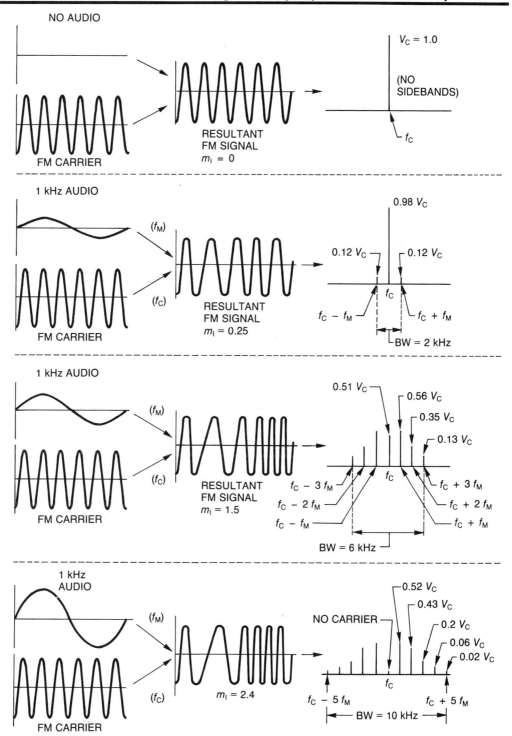

Figure 6-10 Sidebands and bandwidth of FM signal.

Figure 6-10B

Now, a small 1-kHz signal modulates the frequency of the carrier. The resulting spectrum shows two sidebands, each 1 kHz away from the carrier. This is what you get in AM—the only difference in FM is that the amount of carrier is reduced. Other sidebands are being generated (an infinite number), but, since they are so small, they can be ignored.

Figure 6-10C

A real surprise occurs. The only difference between this situation and the last is that the amplitude of the modulating frequency has been increased. Doing this increases the frequency deviation of the carrier. Note the increase in the number of significant sidebands and the corresponding decrease in the carrier. Also note that even though the modulating signal is still 1 kHz, the increase in amplitude has caused the bandwidth to increase to 6 kHz. This is where the FM spectrum and bandwidth differs from the AM spectrum and bandwidth.

Figure 6-10D

Now the carrier has completely disappeared, and more significant sidebands appear. Again, the *frequency* of the modulating signal has not changed; only the amplitude has changed. The bandwidth has increased to 10 kHz in order to accommodate the significant sidebands. The modulating frequency has stayed at 1 kHz; its amplitude is the only quantity that has increased. Again, this is much different from what you'd expect from AM.

FM Bandwidth

Don't be misled by the preceding analysis. The key to the number of significant sidebands of an FM signal is the modulation index. Hence, for the previous example, the amplitude of the modulating signal could have been kept constant and its frequency changed instead. Figure 6-11 shows a graph that relates the bandwidth of the FM signal to the modulation index.

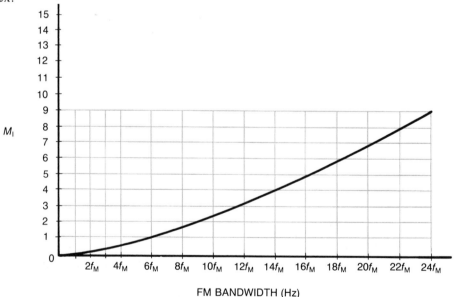

Figure 6-11 Modulation index (M_l) to FM bandwidth relations.

Example 1

What is the bandwidth required to transmit an FM signal with a modulating frequency of 10 kHz and a carrier deviation of 30 kHz.

Solution
First, compute the modulation index:

$$m_I = \frac{f_D}{f_M} \qquad \text{(Equation 6-4)}$$

$$m_I = \frac{30 \text{ kHz}}{10 \text{ kHz}} = 3$$

Use the chart in Figure 6-11 for $m_I = 3$, the bandwidth is close to $12f_M$
Hence,

$$\text{BW} = 12 \times 10 \text{ kHz} = 120 \text{ kHz}$$

Deviation Ratio

Deviation ratio can be defined as the largest modulation index in which the maximum permitted frequency deviation and the maximum modulating frequency are used.

$$R_D = \frac{f_{D \text{ (max)}}}{f_{M \text{ (max)}}} \qquad \text{(Equation 6-5)}$$

where R_D = Deviation ratio (no units)

$\quad f_{D \text{ (max)}}$ = Maximum deviation of FM carrier in hertz

$\quad f_{M \text{ (max)}}$ = Maximum modulating frequency in hertz

For commercial FM, the maximum carrier deviation is 75 kHz, and the maximum modulating frequency is 15 kHz. Hence, the deviation ratio is

$$R_D = \frac{75 \text{ kHz}}{15 \text{ kHz}} = 5$$

For the sound section of commercial television, the maximum carrier deviation is 25 kHz and the maximum modulating frequency is 15 kHz. Thus,

$$R_D = \frac{25 \text{ kHz}}{15 \text{ kHz}} = 1.67$$

Percent Modulation

The percent modulation of an FM carrier can be determined from the formula

$$M_{FM} = \frac{f_{D \text{ (actual)}}}{f_{D \text{ (max)}}} \times 100\% \qquad \text{(Equation 6-6)}$$

where M_{FM} = Percent modulation (no units)

$f_{D \text{ (actual)}}$ = Actual frequency deviation of FM carrier in hertz

$f_{D \text{ (max)}}$ = Maximum allowable frequency deviation of FM carrier in hertz

---

Example 2

For an FM transmission with a frequency deviation of 20 kHz, determine the percent modulation for a commercial FM station.

Solution

The maximum allowable frequency deviation of commercial FM (in the 88–108-MHz band) is 75 kHz.

$$M_{FM} = \frac{f_{D \text{ (actual)}}}{f_{D \text{ (max)}}} \times 100\% \qquad \text{(Equation 6-6)}$$

$$M_{FM} = \frac{20 \text{ kHz}}{75 \text{ kHz}} \times 100\% = 0.266\,7 \times 100\%$$

$$M_{FM} = 26.67\%$$

Narrowband FM

It can be shown that if the modulation index for FM is kept less than $\frac{\pi}{2}$, then the resultant bandwidth depends mainly on the frequency of the modulating signal. Thus, the bandwidth of narrowband FM can be calculated in the same manner as that of AM.

$$BW_{FM \text{ (narrow)}} = 2f_M \qquad \text{(Equation 6-7)}$$

where $BW_{FM \text{ (narrow)}}$ = Bandwidth of FM signal in hertz

f_M = Frequency of modulating signal in hertz

Example 3

What is the bandwidth of a narrowband FM signal generated by a 6-kHz audio that modulates a 125-MHz carrier?

Solution

This is narrowband FM, so

$$BW_{FM \text{ (narrow)}} = 2f_M \qquad \text{(Equation 6-7)}$$

$$BW_{FM \text{ (narrow)}} = 2 \times 6 \text{ kHz} = 12 \text{ kHz}$$

Conclusion

This section presented the complexity of an FM wave. You saw how the sidebands changed with different values of the modulation index. A graph was presented that related the bandwidth of an FM signal to a given modulation index.

The computer applications section for this chapter presents a detailed analysis of the FM spectrum. Using the program presented there gives a valuable insight to FM spectrum analysis that is not normally available.

6-3 Review Questions

1. Define a guard band. How are guard bands allocated in commercial FM stations?
2. If the largest modulating signal allowed for commercial FM is 15 kHz, why is such a large bandwidth required?
3. Describe the major differences between the bandwidths of an AM wave and an FM wave.
4. Explain how the term "deviation ratio" is used in FM.
5. What is the deviation ratio for commercial FM? What is this ratio for the sound section of commercial television?
6. Explain what percent modulation means for FM.
7. What is narrowband FM? How is it different from wideband FM?

6-4 | FM TRANSMITTERS

Discussion

There are many similarities between FM and AM transmitters. This section presents the main parts of an FM transmitter.

The main difference between FM and AM transmitters is that an FM transmitter must cause the carrier *frequency* to represent the modulating signal. You were given a brief introduction to how this could be done in the first section of this chapter.

Basic FM Transmitter

A basic FM transmitter is shown in Figure 6-12. The purpose of each section is as follows:

Audio Amplifier

Amplifies the audio signal to the level needed to operate the FM modulator.

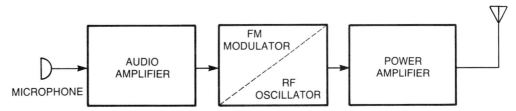

Figure 6-12 Basic FM transmitter.

FM Modulator—RF Oscillator

Causes the frequency of the RF oscillator to change in accordance with the frequency and amplitude of the audio signal.

Power Amplifier

Brings the resultant FM waveform from the previous circuit up to the required power for transmission.

Transmitter Drift

There is a major problem with the basic FM transmitter just presented. At the high frequencies assigned to commercial FM (88–108 MHz), an RF oscillator tends to drift (change its frequency). If a more stable oscillator (such as a crystal oscillator) were used, then there is the problem of making the oscillator frequency deviate the required amount for FM. One solution to these problems is an FM transmitter that uses *frequency multipliers*.

Frequency Multipliers

A *frequency multiplier* is a circuit that causes the output frequency of the circuit to be a multiple of the input frequency. Block diagrams of some frequency multipliers are shown in Figure 6-13.

In practical applications, frequency multipliers that multiply a given frequency by more than 3 are usually not used. If greater frequency multiplications are needed, then the scheme shown in Figure 6-14 is used.

The main features of a frequency multiplier are shown in Figure 6-15.

Recall that *harmonics* are multiples of a *fundamental* frequency. As demonstrated in the computer applications section of Chapter 3, the amplitude of harmonics decreases with increasing frequency of the harmonic. Thus, the fourth harmonic has a smaller amplitude then the third harmonic. This characteristic is one reason why frequency multipliers are cascaded to produce large values of frequency multiplication.

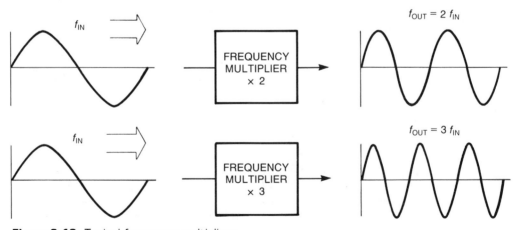

Figure 6-13 Typical frequency multipliers.

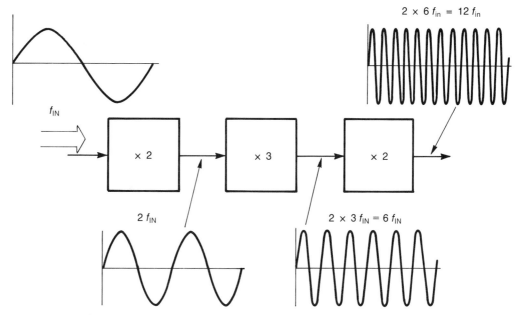

Figure 6-14 Cascading frequency multipliers.

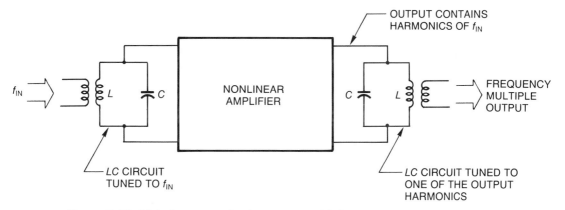

Figure 6-15 Main features of a frequency multiplier.

Typical FM Transmitter

The block diagram of an FM transmitter is shown in Figure 6-16. The transmitter uses a much lower RF oscillator (called the *master oscillator*). Thus, frequency stability of the unmodulated carrier is more easily achieved. In order to increase the frequency of the master oscillator to the required 96 MHz used for transmission, the three frequency multipliers are used.

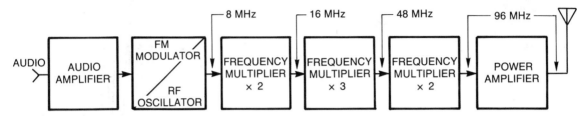

Figure 6-16 Typical FM transmitter.

The maximum carrier deviation on the output of the transmitter is 75 kHz. The frequency multipliers will also increase the amount of deviation of the master oscillator as given by

$$f_{D \text{ (out)}} = N_f f_{D \text{ (MA)}}$$ **(Equation 6-8)**

where $f_{D \text{ (out)}}$ = Frequency deviation of the output in hertz

       N_f = Amount of frequency multiplication (no units)

      $f_{D \text{ (MA)}}$ = Frequency of master oscillator in hertz

---

Example 1

For the FM transmitter in Figure 6-17, what is the final carrier frequency deviation and resting frequency?

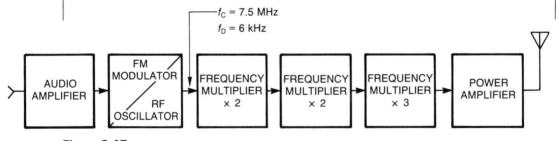

Figure 6-17

Solution

The final carrier frequency is

$$f_c = 12 \times 7.5 \text{ MHz} = 90 \text{ MHz}$$

The final deviation of the carrier is

$$f_{D \text{ (out)}} = N_f f_{D \text{ (MA)}}$$ **(Equation 6-8)**
$$f_{D \text{ (out)}} = 12 \times 6 \text{ kHz} = 72 \text{ kHz}$$

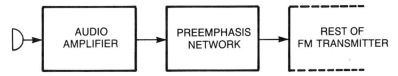

Figure 6-18 FM transmitter with preemphasis network.

FM Noise

It can be shown that for FM noise increases linearly with frequency. This kind of FM noise causes frequency distortion of the FM signal. If a circuit called a *preemphasis network* is used in an FM transmitter, the modulation index for higher frequencies is increased and frequency distortion due to FM noise is greatly reduced. Figure 6-18 is a block diagram of an FM transmitter using a preemphasis network.

Using a preemphasis network at the transmitter end requires a deemphasis network for the FM receiver. This is discussed in the next section.

Conclusion

The transmission of FM is usually accomplished by frequency multiplication. Using this method makes it easier to stabilize the master oscillator, while producing the required frequency deviation of the carrier on the output. FM noise that causes frequency distortion can be greatly reduced by a preemphasis network in the transmitter. The next section presents an introduction to FM receivers.

6-4 Review Questions

1. What is a major problem with a very-high-frequency oscillator?
2. What is the main purpose of an FM transmitter? How does this compare with an AM transmitter?
3. Describe a frequency multiplier. What is the main advantage of using a frequency multiplier in an FM transmitter?
4. Explain what is meant by cascading frequency multipliers.
5. What relationship does the frequency deviation of an FM wave have to the amount of frequency multiplication used in the FM transmitter?
6. Explain why a preemphasis network is used in an FM transmitter. What does it do?

6-5 | FM RECEIVERS AND PHASE MODULATION

Discussion

This section introduces FM receivers. The main difference between an FM receiver and an AM receiver is the method of detection. This should be expected since the two modulating systems differ in how they carry the intelligence.

You will also see the difference between frequency modulation and phase modulation. Phase modulation is used a great deal in space communications. You'll see why PM is the best choice in many applications.

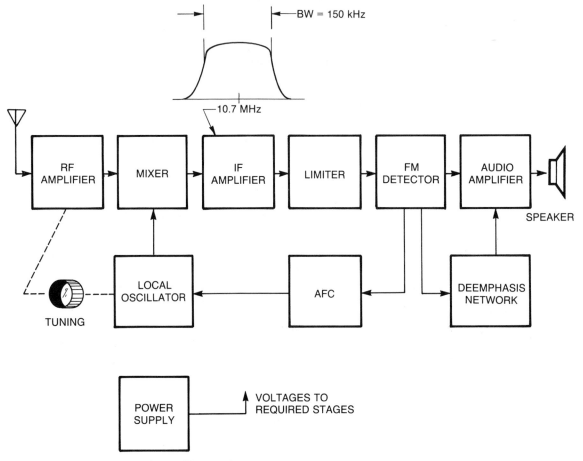

Figure 6-19 Basic FM receiver.

Basic FM Receiver

Figure 6-19 is a block diagram of the basic FM receiver. Notice the similarity between this receiver and an AM receiver. The primary differences are the IF frequency, limiter, deemphasis network, automatic frequency control (AFC) and wider bandwidth requirements.

The purpose of each section follows:

RF Amplifier

Selects the desired FM signal and rejects all others. Amplifies the selected FM signal.

Local Oscillator

Generates a sine wave that is the IF frequency above the incoming signal.

Mixer

Mixes the incoming FM signal with the local oscillator signal. Produces the sum and difference of the two original frequencies.

IF Amplifiers

Intermediate frequency amplifier amplifies the difference frequency. In commercial FM receivers, the IF amplifiers are tuned to 10.7 MHz.

Limiter

Clips the peaks of the FM waveform, thus eliminating the amplitude changes caused by noise.

FM Detector

Converts changes in carrier frequency to changes in amplitude. These amplitude changes now represent the original transmitted audio information.

Deemphasis Network

Compensates for the effect of the preemphasis network at the FM transmitter. Reduces the higher-frequency audio signal.

AFC

Automatic frequency control, circuit used to prevent the local oscillator frequency from drifting. Unlike AM receivers, the local oscillator frequency of an FM receiver is very high (88 MHz + 10.7 MHz = 98.7 MHz at the low end). A drift of just 0.1% would cause a frequency change of at least 98.7 kHz! Hence, the local oscillator frequency is checked by noting the amount of FM detection caused by a small change in the frequency of the local oscillator. This change is then fed back by the AFC circuitry to bring the local oscillator back to the required frequency.

Audio Amplifier(s)

One or more amplifiers that take the reconstructed audio signal and bring it to a level that can operate the receiver speaker system.

Loudspeaker

Converts the changes in electrical energy into changes in air pressure, thus reproducing the original sound.

Power Supply

Supplies voltage and current to the required stages. It may consist of a simple battery or complex circuitry that converts the 120 V AC to different DC voltage levels required by each circuit in the FM receiver.

PM and FM

The main difference between *phase modulation* (PM) and *frequency modulation* (FM) is in how the modulating information affects the carrier. In phase modulation (PM), the only thing required is that the *phase* of the RF carrier be changed with regard to some reference phase. This criteria creates a very narrowband signal. A PM signal with a very narrow bandwidth has the advantage of having its transmission power concentrated where the information is contained. Phase modulation is also more noise free than frequency modulation and requires a much simpler transmitter. These are some reasons why PM is a favorite for satellite communications.

Conclusion

You were introduced to the main parts of an FM receiver. Here you discovered that the FM receiver has more sections than the AM receiver has. In the circuits portion, you will study the details of each of these circuits.

The main differences between FM and PM were presented. You will appreciate these differences more when the circuits that produce these modulation techniques are introduced.

6-5 Review Questions

1. Describe the main differences between FM and AM receivers.
2. What is the purpose of a limiter? What is the purpose of a deemphasis network?
3. Explain what AFC means. What is its function in an FM receiver?
4. What is the main difference between FM and PM?
5. Describe the advantage of PM over FM.

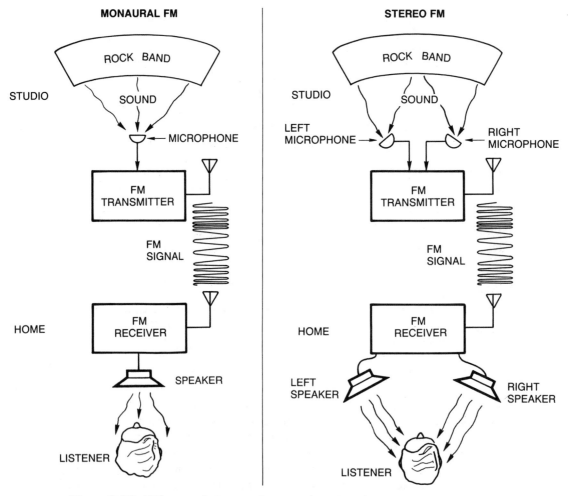

Figure 6-20 Difference between stereo and monaural.

6-6 | FM STEREO

Discussion

Stereo FM transmission was authorized by the FCC in 1961. *Stereo FM,* in contrast to *monaural FM,* has two different audio signals: One signal replicates what the left ear would hear; the other replicates what the right ear would hear if the listener were actually present where the sound was being created. Figure 6-20 illustrates this difference between stereo and monaural sound reproduction.

The difference between the results of stereo and monaural is that stereo gives the listener a sense of direction.

Basic Concept

One very direct method of transmitting stereo is by using two transmitters and two receivers to produce the two different audio signals, one for each ear. Transmitting stereo FM in this manner is illustrated in Figure 6-21.

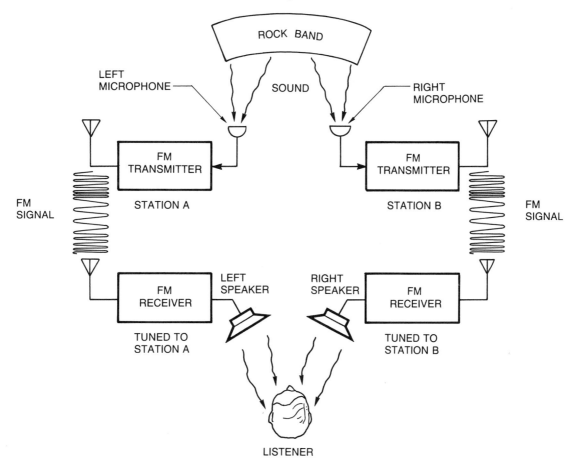

Figure 6-21 One method of creating stereo FM.

This method is not used for several reasons: (1) It requires too much equipment. (2) It requires two separate transmitters, thus using up the limited channel space in the commercial FM band. (3) It requires that the listener tune in two different receivers.

A method had to be found that would allow two different audio signals to be transmitted from one transmitter and received by one receiver. There was also another requirement; whatever system was developed, the transmission of FM stereo had to sound like monaural for those people who already had the old FM monaural receivers. In other words, stereo FM had to be *compatible* with monaural FM. How this was achieved is discussed next.

Adding Audio Signals

First, you need to know what is meant by *signal addition* and *signal subtraction*. This knowledge will help you understand TV stereo as well.

Figure 6-22 shows the concept of signal addition by using square waves. Square waves are used because their addition and subtraction are easy to follow. The same analysis methods apply equally well to complex audio waveforms. This signal addition produces an equivalent monaural signal.

Subtracting Audio Signals

Figure 6-23 shows the same square waves as in Figure 6-22, but now one of them (the right channel) is being sent through an inverter. The other signal is being sent through a delay network so that when the two are again added there is no time delay between them. The resulting signal is not equivalent to a monaural signal. But if it is combined with its equivalent monaural signal, a surprising result occurs.

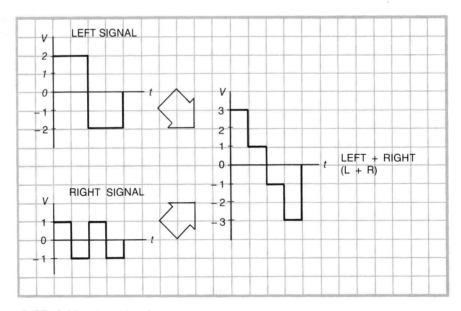

Figure 6-22 Adding two signals.

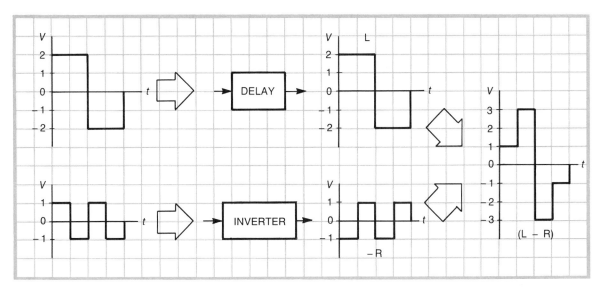

Figure 6-23 Subtracting two signals.

Combining the Two Signals

If the L + R and the L − R signals are transmitted as two different *audio* signals, the L + R signal can be used in monaural receivers. For stereo reception, Figure 6-24 shows how the addition and subtraction of these two signals in a stereo FM receiver can recreate the two original left (L) and right (R) audio signals.

FM Stereo Generation

Figure 6-25 is a block diagram of an FM stereo transmitter. Note the circuits that generate the two audio signals L + R and L − R. The use of a *balanced modulator* and *master oscillator* generate a pilot carrier to produce the frequency spectrum in Figure 6-26.

The FM receiver now need only reverse this process. Figure 6-26 shows that a monaural FM receiver receiving the stereo signal will not reproduce the pilot subcarrier because it is above the response of most audio systems and, certainly, human hearing. Thus the L − R signals are also no problem for a standard FM monaural receiver, so *compatibility* has been achieved.

The pilot subcarrier is put at 19 kHz because it is easier to extract the subcarrier in the receiver. If the pilot subcarrier were at 38 kHz, it is only a few cycles removed from the L − R sidebands, thus making extraction of the carrier difficult.

FM Stereo Reception

The stereo signal-processing circuity of an FM stereo receiver is shown in Figure 6-27. The two signals L + R and L − R are now added and subtracted to produce the original two L and R audio signals.

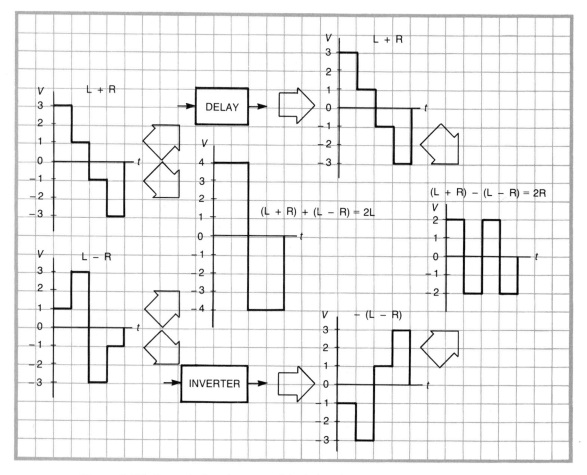

Figure 6-24 Reproducing the two original signals.

Conclusion

The concepts of FM stereo generation and reception were presented here. The process of adding and subtracting two signals to produce compatibility between two different systems is not unique to stereo FM. You know that a black-and-white TV will reproduce a color TV transmission in black and white, while a color TV will reproduce it in color. Again, color TV transmission must be compatible with a black-and-white TV receiver. Chapter 7 uses all of the communication systems you have studied. New systems will also be introduced.

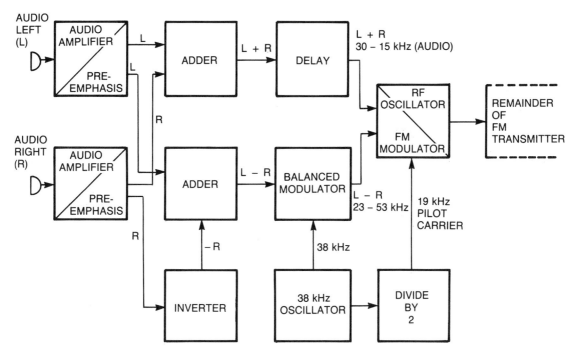

Figure 6-25 Typical FM stereo transmitter.

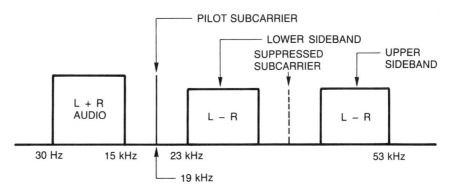

Figure 6-26 FM stereo frequency spectrum.

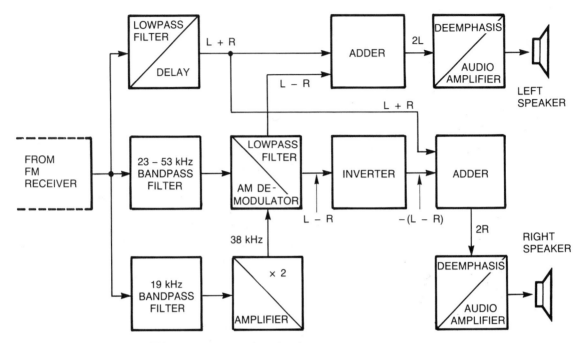

Figure 6-27 FM stereo processing circuits.

6-6 Review Questions

1. Describe the difference between stereo and monaural reception from the standpoint of the listener.
2. Why aren't two separate FM transmitters used to transmit a stereo FM signal?
3. Explain signal addition. Give an example of audio signal subtraction.
4. In terms of stereo FM, explain compatibility.
5. Describe how a single FM station transmits a stereo signal. What portion of this signal is used in a monaural FM receiver?
6. Why is a pilot carrier needed in stereo FM transmission? Explain the reason for the frequency used by the pilot carrier.
7. Describe the basic process of separating L and R audio signals in a stereo FM receiver.

TROUBLESHOOTING AND INSTRUMENTATION

Sweep Generators

Sweep generators are used in testing and aligning communications equipment. Basically, a sweep generator produces a range of radio frequencies over and over again—hence the name *sweep generator*. This frequency repetition allows you to quickly test the frequency response of radio frequency RF and intermediate frequency IF amplifiers.

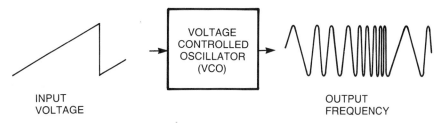

Figure 6-28 Basic construction of a sweep generator.

The construction of sweep generators is based upon a special oscillator circuit that can repeatedly change or sweep its output frequency. One such circuit is called a voltage controlled oscillator (VCO); its operation is presented in this section.

Basic Construction

The basic construction of a sweep generator is shown in Figure 6-28. The VCO creates a sine wave whose frequency is directly proportional to the amplitude of the control voltage. If the control voltage input is a sawtooth waveform, then the output of the VCO will be a sine wave with a sweeping change in frequency that repeats itself according to the frequency of the sawtooth control voltage.

Practical Sweep Generator

It is important that a sweep generator have a constant rate of change of its sweep frequency and a constant output amplitude during the sweep. These qualities are produced by adding the circuits in Figure 6-29.

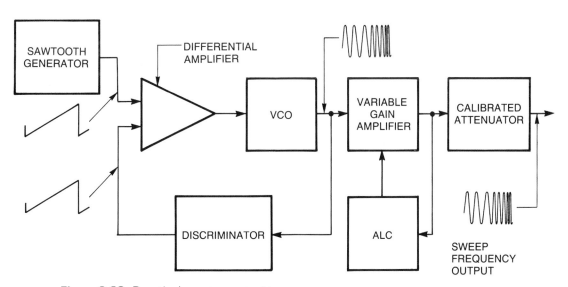

Figure 6-29 Practical sweep generator.

The *discriminator* circuit does exactly the opposite of a VCO. A discriminator gives an output voltage whose amplitude is proportional to the frequency of the input. The *differential amplifier* has an output voltage proportional to the difference between its two input signals. The purpose of using a discriminator with a differential amplifier is to ensure that the output frequency change of the VCO remains linear.

The automatic level control (ALC) is similar to the automatic gain control (AGC) circuit discussed with AM receivers. The ALC makes the variable gain amplifier maintain a constant output amplitude. The *calibrated attenuator* is used to reduce the output of the sweep generator to an amount specified by the user.

Sweep Generator Application

A typical application of a sweep generator is shown in Figure 6-30. If the amplifier under test is an IF amplifier, then the display on the oscilloscope will replicate the frequency response curve of the amplifier.

When alignment of RF and IF amplifiers is required, the sweep generator becomes a valuable tool. Be sure to disable any AGC circuits when doing an alignment. If you don't, you will get an inaccurate alignment because the AGC is trying to keep the amplifier gain constant.

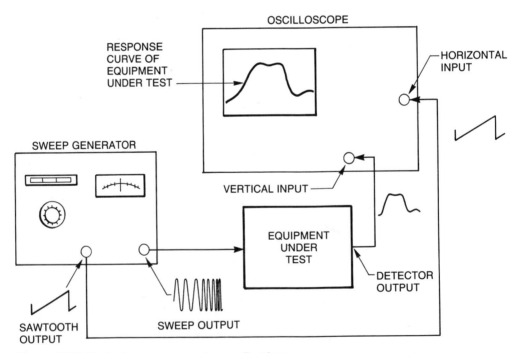

Figure 6-30 Typical sweep generator applications.

Bessel Function

The graph of Figure 6-11 gave an indication of the bandwidth of an FM signal. The graph was derived from a mathematical statement called a *Bessel function of the first kind*. Its form is as follows:

$$J_N(m_I) = \left(\frac{m_I}{2}\right)n\left[\frac{1}{n!} - \frac{(m_I/2)^2}{1(n+1)!} + \frac{(m_I/2)^4}{2!(n+2)!} - \frac{(m_I/2)^6}{3!(n+1)!} + \cdots\right]$$

where m_I = The modulation index
$J_N(m_I)$ = The sideband in question

Conclusion

The sweep generator is used in the alignment of AM and FM radio receivers as well as TV receivers. It is a valuable tool in any communication repair and service area. Sweep generators come in a variety of frequency ranges, depending on the particular equipment to be aligned. Specialized sweep generators, such as those made for commercial TV alignment, have built-in markers and specific key frequencies to mark picture, sound, and color information.

6-7 Review Questions

1. Describe the basic principle of operation of a sweep generator.
2. Why is the term "sweep generator" used to describe the instrument?
3. Explain the operation of a VCO.
4. What effect does an input control voltage of a sawtooth have on the output frequency of a VCO?
5. Describe the purpose of a discriminator.
6. Explain the basic operation of a differential amplifier.
7. Define ALC. How is it used in a sweep generator?
8. Describe a typical application of a sweep generator. How is an oscilloscope used in this kind of application?

MICROCOMPUTER SIMULATION

Be sure to do the **instructions** as well as the **demonstration** for troubleshooting simulation number six on your student disk. For this simulation, you will have the opportunity to use a signal generator as well as a signal detector.

Up to this point, your troubleshooting simulation programs have been a GO/NO-GO situation. This means that if there was a bad stage, then no signal at all could get through. In real troubleshooting situations, this is not always the case, and in the next troubleshooting simulation, you will begin to learn other troubleshooting skills that will help you analyze other types of problems commonly found in communications equipment.

CHAPTER PROBLEMS

(Answers to the odd-numbered problems appear at the end of the text.)

1. State three methods of placing information on a carrier wave.
2. Which of the three methods in problem 1 are similar? What is the major difference?
3. State a major advantage of FM. State a major disadvantage.
4. What circuit inside an FM receiver helps reduce the effects of noise?
5. List the major differences between FM and AM.
6. Give an example of the noise-free operation of FM compared to AM.
7. What is the function of an FM limiter?
8. Why can't an AM receiver use a limiter?
9. Explain the basic construction of a capacitance microphone.
10. What effect does the amplitude of the modulating signal have on the FM carrier?
11. Define resting frequency and frequency deviation.
12. Explain the relationship between carrier swing and frequency deviation.
13. An 88-MHz FM carrier is modulated by an audio tone that caused a frequency deviation of 15 kHz. Determine (A) the carrier swing of the FM signal and (B) the lowest and highest frequencies obtained by the FM signal.
14. Repeat problem 13 for a carrier frequency of 105 MHz and a frequency deviation of 10 kHz.
15. Explain the effect of the modulating signal frequency on the FM carrier.
16. What happens to the FM carrier if the modulating frequency doubles, while its amplitude stays the same.
17. What factors determine the modulation index of an FM wave?
18. What would happen to the modulation index of an FM wave if the modulating frequency increased? if the amplitude of the modulating frequency increased?
19. Compute the modulation index of an FM signal with a carrier swing of 10 kHz when the modulating signal is 5 kHz.
20. Solve problem 19 if the carrier swing is 15 kHz for a modulating signal of 8 kHz.
21. The maximum frequency of an FM signal modulated by a 3-kHz sine wave is 100.02 MHz, and the minimum frequency is 99.98 MHz. Determine (A) the carrier swing, (B) the carrier frequency, (C) the frequency deviation of the FM signal, and (D) the modulation index.
22. Solve problem 21 if the modulating sine wave is 5 kHz and the maximum and minimum frequencies of the carrier are 100.04 MHz and 99.96 MHz, respectively.
23. For an FM signal in the 88–108-MHz broadcast band with a frequency deviation of 15 kHz, determine the percent modulation.
24. Determine the percent modulation of an FM signal in the audio portion of a TV broadcast if its frequency deviation is 20 kHz.
25. Compute the amount of carrier swing necessary to produce an 80% modulation for the audio portion of the TV band.
26. Determine the amount of carrier swing required to produce a 75% modulation for the 88–108-MHz FM broadcast band.

27. Find the percent modulation of an FM signal in the TV audio band that has a carrier swing of 50 kHz.
28. Determine the percent modulation of a commercial 88–108-MHz FM signal with a carrier swing of 110 kHz.
29. For a 75-MHz FM carrier modulated by a 5-kHz audio tone that causes a frequency deviation of 15 kHz, determine (A) the modulation index and (B) the FM signal bandwidth.
30. Calculate the modulation index and the FM signal bandwidth of a 50-MHz FM carrier modulated by a 3-kHz audio tone that causes a frequency deviation of 7 kHz.
31. Calculate the modulating frequency that causes an FM signal to have a bandwidth of 50 kHz when its frequency deviation is 15 kHz.
32. An FM signal with a frequency deviation of 20 kHz has a bandwidth of 50 kHz. Determine the frequency of the modulating signal.
33. If a 100-MHz FM carrier is modulated by a 10-kHz sine wave, with a carrier swing of 50 kHz, determine (A) the modulation index and (B) the bandwidth.
34. Redo problem 33 if the FM carrier is modulated by a 5-kHz sine wave.
35. What is the bandwidth of a narrowband FM signal generated by a 2-kHz audio signal and a 110-MHz FM carrier?
36. Determine the bandwidth of a narrowband FM signal caused by a 6-kHz sine wave and a 150-MHz FM carrier.
37. Determine the bandwidth of an FM signal when a 3-kHz audio signal modulates a 75-MHz FM carrier causing a frequency deviation of 1.5 kHz.
38. Find the bandwidth of an 80-MHz FM carrier when modulated by a 5-kHz audio signal that causes a frequency deviation of 3.4 kHz.
39. For an FM transmitter, the initial carrier frequency is 8 MHz with a frequency deviation of 1 kHz. If the total frequency multiplication of the transmitter is 12, determine the final FM carrier frequency and the final frequency deviation.
40. Repeat problem 39 for an initial frequency deviation of 2.5 kHz and a carrier frequency of 6 MHz.
41. What is a preemphasis network?
42. Describe the use of a deemphasis network. Why is it needed?
43. What does an FM detector do?
44. What is the difference between an AM detector and an FM detector?
45. What is the purpose of AFC in an FM receiver?
46. Draw the block diagram of an FM receiver and explain the function of each block.
47. Describe the difference between FM and PM.
48. How is information contained in PM? What are some of the advantages?
49. From the standpoint of the listener, what is the difference between AM and FM?
50. Describe the basic requirements of a stereo system.
51. Give an example of the addition of two signals and an example of the subtraction of two signals.
52. Show graphically how two signals are added and subtracted.
53. How are the L + R and L − R signals combined in FM stereo to produce the original left and right audio signals?

54. Graphically demonstrate how the addition and subtraction of the L + R and L − R stereo FM signals produce the original left and right audio signals.
55. What is the frequency for the pilot subcarrier for commercial stereo FM?
56. Explain why the pilot subcarrier uses a different frequency and not its original frequency for stereo FM.
57. Describe a sweep generator.
58. Explain what a VCO does.
59. Draw the block diagram of a sweep generator.
60. Show how a sweep generator could be used to check the response characteristics of an FM IF amplifier.

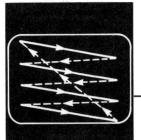

CHAPTER 7

Television Systems

OBJECTIVES

In this chapter, you will study:

- [ ] The transmission requirements of TV in the United States.
- [ ] How the TV picture is created and its relation to the transmitted TV signal.
- [ ] Basic operation of a black-and-white TV receiver.
- [ ] Basic operation of a color TV receiver.
- [ ] The principles of stereo TV and its other possibilities.
- [ ] Principles of video cassette recorders.
- [ ] Basic principles of computer-aided troubleshooting.

INTRODUCTION

This chapter covers material that will introduce you to the fascinating world of television systems. Television is a unique area of electronic communication that, like the invention of the telephone and radio, has revolutionized the way we live, work, and play. This important part of electronic communication is covered from a system point of view. The circuit details appear in the circuit section of this text.

7-1   TELEVISION SYSTEM REQUIREMENTS

Discussion

Television allows you to experience events from places you may never visit. Television, as a mode of communication, has had such an impact on our lives that it controls our attitudes, opinions, and buying habits.

177

Television is the primary mode of communication sought by politicians, advertisers, and many nonprofit organizations. It is so readily accepted within the privacy of the home that much of family life is centered around TV program scheduling.

The basic requirements of this most amazing invention are presented in this chapter.

Required Information

A TV station must be able to transmit all of the information shown in Figure 7-1. The complete (composite) TV signal consists of

- Monaural or stereo sound
- Light and dark (intensity) picture information
- Color information
- Synchronizing information

All of this information must be transmitted so that it can be received by any TV receiver (black and white or color).

Frequency Allocations

Television stations are separated by frequency-division multiplexing (FDM). Each TV station is assigned a transmitting frequency called a *channel*. This channel has an assigned bandwidth of 6 MHz. The reason for such a large bandwidth requirement is that much information must be contained in the TV signal. The frequency allocations for commercial TV in North America are:

- Channels 2–6   54–88 MHz
- Channels 7–13   174–216 MHz
- Channels 14–83   470–890 MHz

Channels 2–6 are just below the commercial FM band, which makes it easier for manufacturers to combine an FM radio with a TV receiver.

Components of a TV Signal

Figure 7-2 graphically shows the major components of a TV signal and how they are distributed within the TV receiver. There are key circuits that separate individual elements of the composite TV signal into individual components. These signals are then routed to specific parts of the TV receiver. A knowledge of this routing process is essential to troubleshoot and repair TV receivers.

TV Picture Tube

The basic construction of a TV picture tube is shown in Figure 7-3. The basic operation of the TV picture tube is similar to that of the cathode ray tube (CRT) in your laboratory oscilloscope. Electrons are "pulled" from the electron gun by a very large positive voltage applied to the picture tube. The tube in Figure 7-3 is a black-and-white tube because there is only one electron gun and the phosphor coating consists of a single type of "white" phosphor. Unlike the CRT of your laboratory oscilloscope, the TV picture tube uses *magnetic deflection* instead of electrostatic deflection.

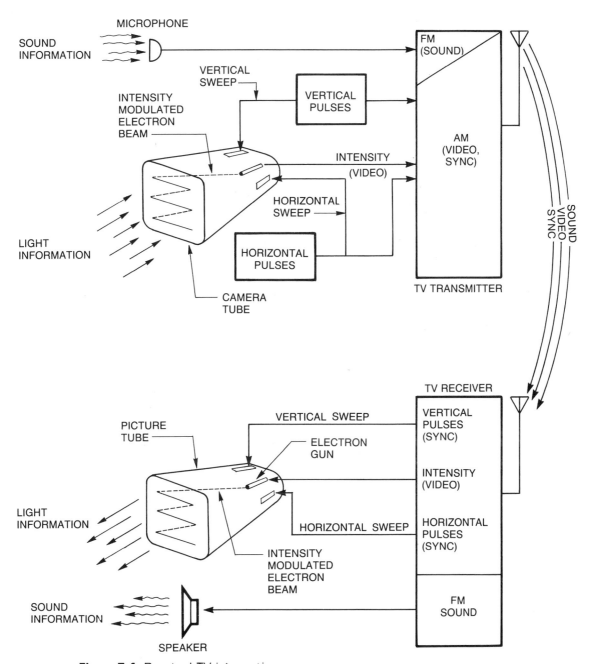

Figure 7-1 Required TV information.

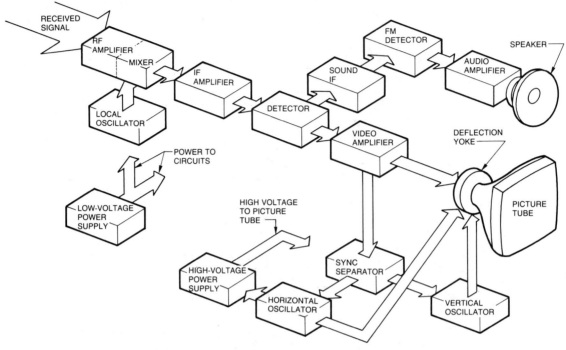

Figure 7-2 Components of a TV receiver.

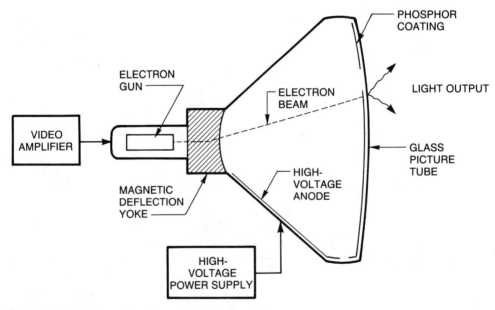

Figure 7-3 Basic construction of TV picture tube.

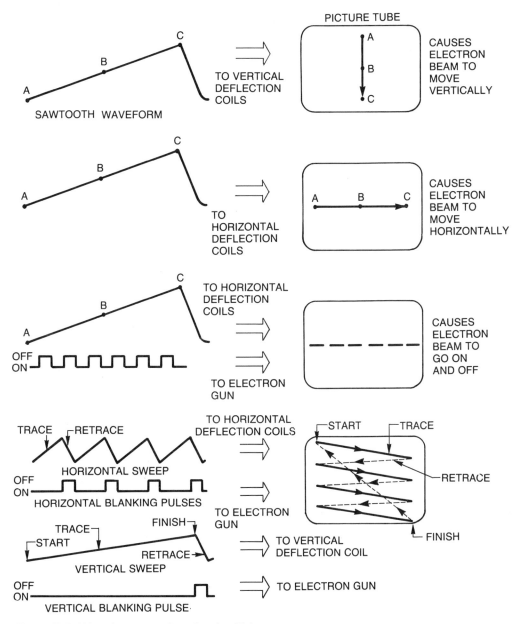

Figure 7-4 Waveforms used to develop TV raster.

Raster Development

Figure 7-4 shows the development of a TV *raster,* which is used to present a TV image. Note that only four lines are developed across the face of the TV screen. Actually, there are 525 lines across the picture face that form the complete TV raster. These lines appear to be present all at the same time because of the persistence of the phosphor. *Persistence* is the ability of the phosphor to glow for a short time after it has been struck by an electron.

The sawtooth waveforms that make up the TV raster originate from oscillators within the TV. The waveforms produced by these oscillators must be synchronized with the picture information arriving from the TV station. This synchronization is accomplished by *sync* (synchronizing) pulses.

Composite TV Signal

The complete (composite) TV signal is shown in Figure 7-5. You can see the location of the horizontal sync and blanking pulses along with the corresponding picture information contained between the horizontal blanking pulses.

Figure 7-6 shows the relation of the TV signal to picture development on the TV screen. As the figure shows, a large amplitude will decrease the intensity of the electron beam. If the signal is large enough (as for the blanking pulses), the electron beam can be completely cut off. The *horizontal blanking* pulses cut off the beam so that the *horizontal retrace* lines do not appear across the picture screen.

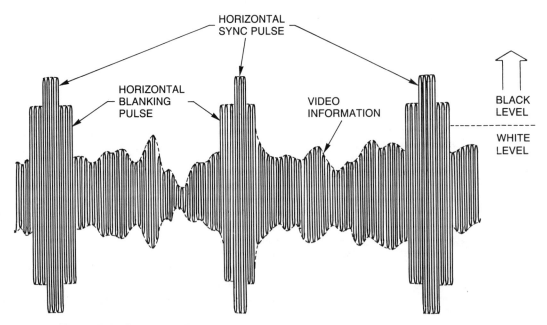

Figure 7-5 Composite TV signal.

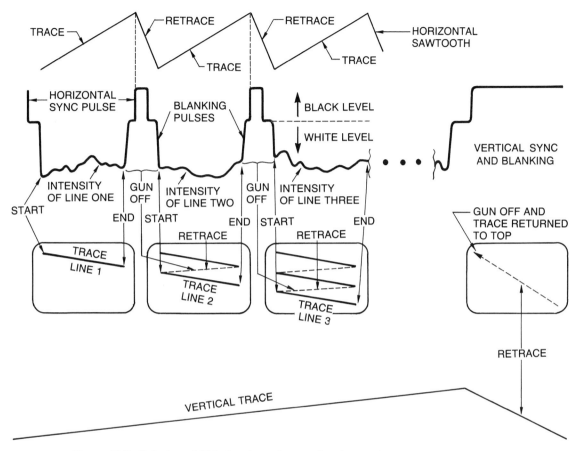

Figure 7-6 Relation of TV signal to picture development.

Conclusion

This section gave you an overview of the major requirements of a TV system. Almost every major electrical communication circuit can be found in a TV receiver. The real engineering feat is the production of such a complex piece of equipment at such a relatively low cost to the consumer.

7-1 Review Questions

1. Is the information used to represent a TV picture transmitted all at the same time? Explain.
2. What information must be contained in a TV signal?
3. What is a TV channel? What are the bandwidth requirements? Why is such a large bandwidth required?
4. What kind of modulation is used to transmit picture information? sound information?
5. Name the major components of a TV receiver.
6. Explain the basic operation of a TV picture tube.

7. Describe how a raster is developed. How many oscillators are required to do this?
8. What is a sync pulse? What is a blanking pulse?
9. Describe a composite video signal.

7-2 | TELEVISION SYSTEM SPECIFICATIONS

Discussion

The previous section gave an overview of the requirements of a TV system. This section gives the specific requirements of a TV system. Television works for the same reason that motion pictures work. The human eye has a certain persistence that ignores the flickering between motion picture frames, provided the frames are presented fast enough. The same principle holds for television; even though the electron beam is scanning one line at a time, you won't see any "flicker" if the lines making up the frames are presented fast enough.

Horizontal Deflection

How fast should the horizontal line scan the TV screen? How many lines are necessary to make a complete picture? Two major factors must be considered:

1. The information must be presented rapidly enough so that persistence of vision does not result in flicker.
2. The bandwidth requirements of the system must be reasonable.

It was decided to use AM to transmit the *video* (picture) information in order to hold down the bandwidth requirements. However, the more rapidly picture information is sent, the larger the bandwidth. The final result was a compromise between presenting information fast enough to prevent flicker and keeping the bandwidth a reasonable size.

Interlaced Scanning

It was decided to present the viewer with one complete TV picture consisting of 525 lines every $1/30$ of a second, called a *frame*. This decision kept down the video bandwidth requirements but would produce flicker. To overcome the flicker, scientists sent two TV *fields* consisting of 262.5 lines each that were *interlaced* with each other. Each field was presented every $1/60$ of a second. Having two interlaced fields produced a single television frame at a rate of $2/60$ or $1/30$ of a second. This produced a flashing rate of $1/60$ of a second, which exceeds the flicker rate (about 30 Hz) of the average eye.

The interlaced scanning method is shown in Figure 7-7. Note that there is a total of 525 lines, all of which are not seen by the viewer.

Scanning Frequencies

The interlaced scanning method determined the frequencies of the horizontal sawtooth. The total number of lines to be scanned each second is given by

$$f_S = L_T F_T$$

(Equation 7-1)

where f_S = Scanning frequency in hertz
L_T = Total lines per frame
F_T = Total frames per second

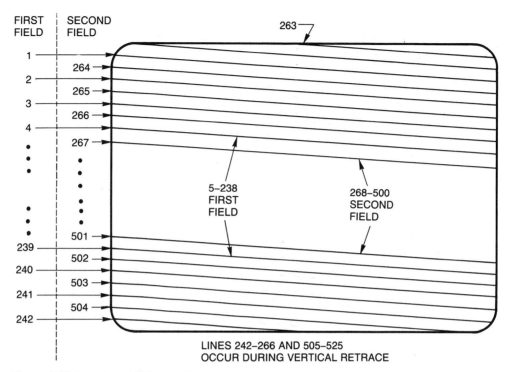

Figure 7-7 Interlaced TV scanning.

Example 1

What must be the horizontal scanning frequency of a commercial TV receiver?

Solution
Using the relationship

$$f_s = L_T F_T$$ **(Equation 7-1)**
$$f_s = 525 \times 30 \text{ Hz} = 15\ 750 \text{ Hz}$$

Thus, the horizontal scanning frequency must be 15.75 kHz.

Vertical Blanking Signal

The composite video signal for commercial TV is shown in Figure 7-8. The purpose of the equalizing pulses before and after the vertical sync pulse is to keep the horizontal oscillator in sync during vertical blanking. The pulses present during vertical sync are used for the same thing.

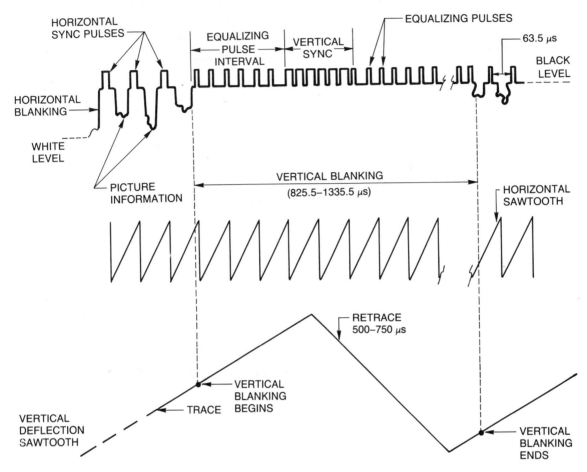

Figure 7-8 Vertical blanking signal.

Teletext

In 1983, the FCC authorized the use of the vertical blanking interval for the transmission of digital data. This digital data can appear on your TV screen as text and graphics.

Closed-caption information, automatic color balance information, or other test signals can also be sent at this time.

To get information from the teletext transmission, you need to buy a teletext *decoder* or buy a newer TV with this capability built in. Teletext services bring stock market reports as well as weather and sports reports. Teletext is usually accompanied by advertising, where the advertiser picks up the costs of the transmission.

To see if teletext is available on any of your TV channels, simply adjust your TV's vertical hold until you can see the vertical blanking (this will appear as a thick horizontal bar across the screen). If the bar appears to have randomly moving white dots rather than being almost pure black, then teletext is being transmitted. The moving white dots are the digital data being transmitted during the vertical blanking.

Bandwidth Requirements

To provide enough picture detail, the video signal must produce small picture *elements*. A picture element is the smallest bit of information that can be displayed on the screen (about 0.05 inch). You need 525 vertical picture elements because there are 525 lines on the screen. The ratio of the screen width to its height (called the *aspect ratio*) for commercial TV is 4 to 3. Hence, there must be $^4/_3$ picture elements for each horizontal line. This works out to

$$525 \times \frac{4}{3} = 700 \text{ horizontal picture elements}$$

Since there are 525 vertical picture elements, the total number of picture elements for one picture must be

$$525 \times 700 = 367\ 500 \text{ picture elements}$$

Since these picture elements are presented 30 times each second, the total number of picture elements scanned each second is

$$30 \times 367\ 500 = 11\ 025\ 000 \text{ picture elements per second}$$

This is quite a lot of information, but, unlike our ears, the human eye requires great quantities of information!

Consider each peak of a sine wave (the positive and the negative) as capable of producing a picture element. Then converting the required picture elements per second to sine wave frequency:

$$\frac{1}{2} \times 11\ 025\ 000 = 5\ 512\ 500 \text{ Hz}$$

Since TV sets are mass produced, there will be some interlacing problems and the shape of the electron beam will not be perfect. The FCC assumes that these imperfections will reduce the required video frequency by about 25%. This factor is called the *utilization ratio*. Hence, the highest required video frequency would be

$$(0.75)(5\ 512\ 500 \text{ Hz}) = 4\ 134\ 375 \text{ Hz}$$

The FCC has set 4.2 MHz as the maximum video signal that commercial TV may transmit. Hence, an AM bandwidth of 8.4 MHz would be needed. To reduce this bandwidth requirement and maintain picture detail, vestigial sideband transmission is used. Recall that vestigial sideband maintains its own carrier, so the receiver doesn't need to be as complex as an SSB receiver. See Figure 7-9. Note the location of the sound. As stated in the previous chapter, TV sound is transmitted as FM with a 25-kHz maximum frequency deviation.

Conclusion

The specifications of a commercial TV system were developed to meet specific requirements. There is the requirement of conveying a large amount of information for the video signal while maintaining a reasonable bandwidth to allow as many TV stations as possible.

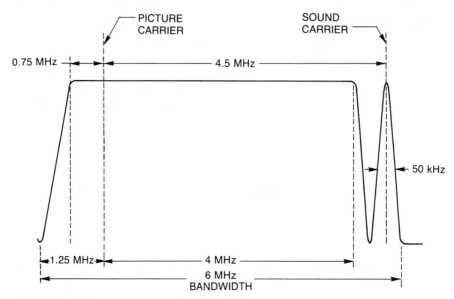

Figure 7-9 Television spectrum requirements.

The frequency allocations of commercial TV stations are listed in Table 7-1. They are divided into two major categories: VHF (*very high frequency*) and UHF (*ultrahigh frequency*).

The channels in the UHF band are continuous from channel 14 through and including channel 83. Table 7-1 lists only seven of these UHF channels to indicate their frequency range.

Table 7-1	TV FREQUENCY ALLOCATIONS				
VHF Band (Lower)		**VHF Band (Upper)**		**UHF Band**	
Channel No.	**Frequency Range (MHz)**	**Channel No.**	**Frequency Range (MHz)**	**Channel No.**	**Frequency Range (MHz)**
2	54–60	7	174–180	14	470–476
3	60–66	8	180–186	24	530–536
4	66–72	9	186–192	34	590–596
	(4 MHz not in TV band)	10	192–198	44	650–656
5	76–82	11	198–204	54	710–716
6	82–88	12	204–210	64	770–776
FM	88–108	13	210–216	83	884–890

7-2 Review Questions

1. What are the two major factors that determine how fast the picture information of a TV receiver is presented?
2. Explain why video information is presented as AM.
3. Describe interlaced scanning. Why is it used?
4. What is the difference between a *frame* and a *field* in a TV picture? How often is a frame presented? How often is a field presented?
5. What two factors determine the horizontal scanning frequency?
6. Describe a picture element. What determines the maximum number of vertical picture elements?
7. What do the terms "aspect ratio" and "utilization ratio" mean?
8. Briefly describe the spectrum requirements of a TV signal.
9. How is the transmitted TV signal bandwidth kept at 6 MHz?

7-3 | BLACK-AND-WHITE TELEVISION

Discussion

This section presents a detailed block diagram of a standard black-and-white TV receiver. The circuits contained in each block will be presented in detail in the circuits section of this text. For now, it's important to understand a TV receiver system. In the next section, you will be introduced to a color TV receiver.

Television Receiver

The block diagram of a standard black-and-white TV receiver is shown in Figure 7-10. The function of each stage follows:

UHF Tuner

Selects the desired UHF station. Consists of a UHF mixer and oscillator that change the incoming selected UHF signal to 45 MHz, which is amplified by the RF amplifier of the VHF tuner.

VHF Tuner

Contains an RF amplifier that selects the desired VHF station. It also prevents the local oscillator frequency from radiating back through the receiving antenna and improves the signal-to-noise ratio.

Fine-Tuning Controls

Adjust the local oscillator frequency. The fine-tuning control for the UHF tuner consists of a variable capacitor; for the VHF tuner, it consists of a variable inductor.

Video IF

Usually consists of two or more stages of amplification. IF amplifiers provide most of the signal gain and produce required selectivity.

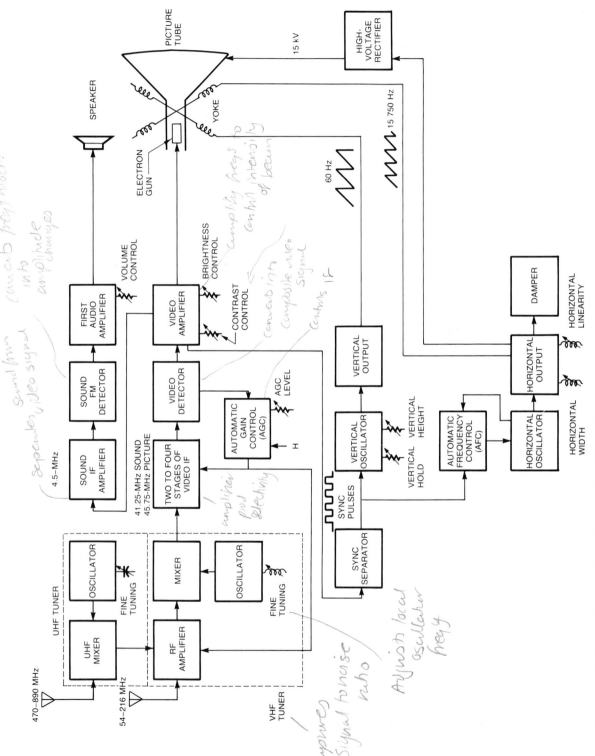

Figure 7-10 Block diagram of black-and-white TV receiver.

Video Detector

Converts video IF into the composite video signal (ideally consisting of 0 MHz to 4.2 MHz) and a lower-frequency sound IF of 4.5 MHz.

Automatic Gain Control

Controls gain of IF amplifiers to maintain a constant IF gain as the strength of the incoming signal varies due to atmospheric conditions or changing from station to station. Note the AGC level control adjustment.

Video Amplifier

Wideband amplifiers that amplify frequencies up to 4.2 MHz, which are used to control the intensity of the electron beam as it sweeps across the TV screen.

Contrast Control

Basically a video amplifier gain control that controls the output signal amplitude of the video amplifier, thus affecting the amount of difference between black and white.

Brightness Control

Controls the strength of the electron beam: the stronger the beam, the brighter the picture displayed.

Sound IF

Amplifier that separates the 4.5-MHz sound IF from the video signal. The FM sound IF signal is amplified and sent to the sound detector.

Sound Detector

Converts the frequency modulation into amplitude changes that represent the original sound. The output of this stage is audio signals.

Audio Amplifier

Amplifies the detected audio signal. Note the variable-resistor volume control. Causes the speaker to vibrate at the reconstructed audio rates, thus reproducing the original sound.

Sync Separator

Separates the vertical and horizontal sync pulses from the composite video signal.

Vertical Oscillator

Produces the 60-Hz sawtooth required for vertical deflection.

Vertical Height Control

Controls the amplitude of the vertical oscillator output, which controls the height of the displayed raster.

Vertical Hold Control

Helps control the frequency of the vertical oscillator. Adjusting this properly prevents the picture from "rolling" vertically.

Vertical Amplifier

Amplifies the output sawtooth of the vertical oscillator to provide sufficient signal to operate the vertical deflection coil around the CRT.

Vertical Linearity Control

Helps maintain the results of the vertical sawtooth so that the picture is displayed in a uniform vertical field.

Horizontal Oscillator

Creates the sawtooth used for horizontal deflection of the electron beam.

Automatic Frequency Control

Helps control the frequency of the horizontal oscillator by comparing the received horizontal sync pulse to the output of the horizontal oscillator.

Horizontal Amplifier

- Provides the drive for the horizontal deflection coil.
- Produces high-voltage pulses used by the high-voltage power supply.
- Creates pulses used by AFC, AGC, and horizontal blanking circuits.
- When a high-voltage tube is used, it provides the filament voltage for the tube.

Horizontal Width Control

Controls the picture width by adjusting the amplitude of the horizontal sawtooth.

Horizontal Hold

Helps control the frequency of the horizontal oscillator, which prevents the picture from "tearing" horizontally when adjusted properly.

Horizontal Linearity Control

Helps maintain a uniform display of the picture along the horizontal.

Damper

Helps develop the complete picture and helps to create the high voltage for the picture tube.

High-Voltage Rectifier

Causes high-voltage pulses to appear as a steady DC voltage for the TV screen.

Low-Voltage Power Supply

Produces the voltages required by the various stages in the TV receiver.

Conclusion

A TV receiver is a very complex system. It consists of more than 25 separate and unique circuits, ranging from RF amplifiers to complex pulse circuits.

The next section introduces a color TV receiver. As you might expect, even more circuits are required for this system.

7-3 Review Questions

1. Name the major circuits contained in a TV tuner.
2. What TV signals are amplified by the video IF amplifier?
3. What is the purpose of the sync separator?

4. How many oscillators does a black-and-white TV receiver contain? What are their frequencies?
5. Describe how the output of the video amplifier affects the picture seen on the picture tube.

7-4 | COLOR TELEVISION

Discussion

A color TV signal must meet many different requirements:

- ■ It must be compatible with black-and-white TV receivers.
- ■ It must stay within the 6-MHz bandwidth already allocated to black-and-white TV.
- ■ It must be able to simulate the wide variety of colors seen by the human eye.

If it were not for the compatibility requirements, the color TV system could be much simpler.

Colors

Colors may be formed from primary colors by *subtractive* or *additive* color mixing. Subtractive mixing uses the filtering action of dyes, inks, and pigments on *reflective* light. This kind of color mixing is most familiar to artists, printers, and painters. Additive mixing uses the color mixing from light *sources*. The three primary colors for additive mixing are *red, green,* and *blue.* All of the colors that you perceive on your color TV use additive mixing. An example of additive mixing is shown in Figure 7-11.

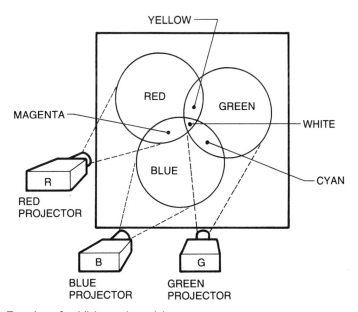

Figure 7-11 Results of additive color mixing.

The important thing to note is that white is made up of all three of the primary additive colors. White is formed when the intensity of each primary color is carefully controlled. If the intensity of red is 30%, green 59%, and blue 11%, then the color additive mixing process will produce the brightest white possible.

A Color Transmission System

One method of transmitting a color TV signal is shown in Figure 7-12, but the system is impractical for the following reasons:

- The viewer must buy at least three TV receivers to view the color and one more to view a black-and-white transmission.
- Since the bandwidth requirements for video information are 0 to 4.2 MHz, it would require a total bandwidth of 3×4.2 MHz = 12.6 MHz just for the video information of one color TV transmission.

Obviously, the color TV system you use doesn't require three or four sets. The method used requires a very special color TV picture tube.

Color TV Picture Tube

The basic construction of a color TV picture tube is shown in Figure 7-13. Note that the color phosphor dots are arranged in triangles. A single triangular arrangement of the three color dots is called a *triad*.

Each electron gun always excites the same color phosphor. Thus the electron gun that excites the green phosphor is called the *green gun;* the others are called the *red* and *blue* guns.

These electron guns sweep the screen in unison (as if they were a single gun) in the same manner as black-and-white TV. By controlling the intensity of the electron beam from each electron gun, the color phosphor triads can simulate all the colors you see on your color set. When white is to be reproduced, the color intensities are 30% red, 59% green, and 11% blue.

The Color Signal

The color signal has to be added to the existing black-and-white TV signal within existing bandwidth requirements, and it must be compatible with a black-and-white TV receiver. This addition is carried out like the addition in FM stereo, where left and right signals were added and subtracted.

The production of a color TV signal at the studio is illustrated in Figure 7-14. The figure shows that the black-and-white signal (called the Y signal) is transmitted along with the difference between the red and white (R − Y) and the difference between the blue and white (B − Y). To allow compatibility with black-and-white TV, the R − Y and B − Y signals are placed at a higher frequency by using suppressed-carrier techniques. See Figure 7-15.

The *composite color signal* is the vector sum of the R − Y and B − Y signals. This is so because the two color signals are 90° out of phase with each other. Note the use of balanced modulators to produce the color signals in order to suppress the color carrier. If

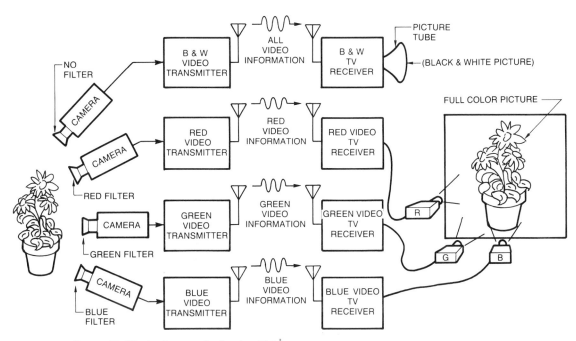

Figure 7-12 An impractical color TV system.

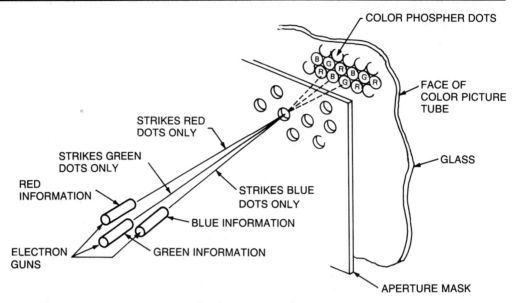

Figure 7-13 Basic construction of color picture tube.

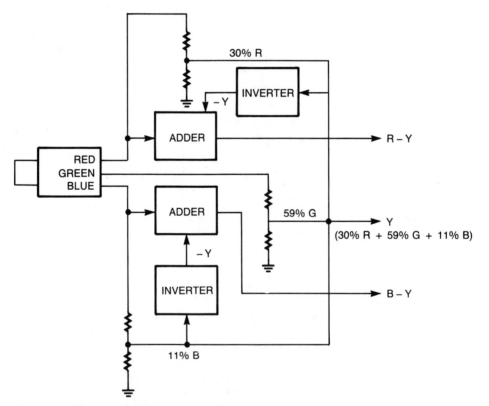

Figure 7-14 Production of color TV signals.

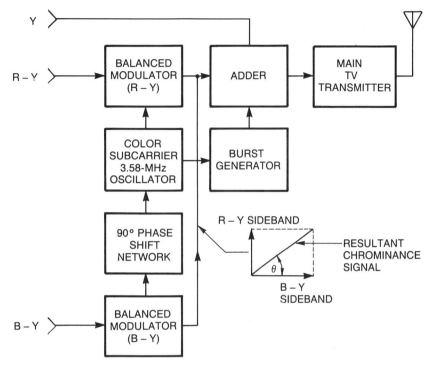

Figure 7-15 Creation of composite color signal.

the color carrier were not suppressed, a fixed pattern of 920 kHz would appear on your TV screen during color transmission. This frequency is the difference between the color carrier and sound carrier frequencies. The *burst generator* ensures that the color sync pulse is transmitted only during horizontal blanking.

A subcarrier frequency of 3.579 545 MHz was chosen in order to reduce interference with the rest of the composite video signal. The color subcarrier frequency produces an interleaving process of the color signal (called the chrominance) with the rest of the composite video signal. The interleaving is shown in Figure 7-16.

Because a suppressed carrier is used, the carrier must be reinserted by the color TV receiver. Signal reinsertion is done by a crystal-controlled oscillator contained in each color TV receiver. To keep the phase relationship of the color oscillator correct for carrier insertion, the transmitted color TV signal contains a *color sync burst*. This burst is located on the "back porch" of each horizontal blanking pulse, as shown in Figure 7-17.

The resulting color signals are called *chrominance signals*. They are phase modulated with reference to the phase of the transmitted color sync burst. The phase relationships and corresponding colors are shown in Figure 7-18.

The transmitted color TV signal now contains a phase modulated component that represents color information. Such a composite signal is shown in Figure 7-19.

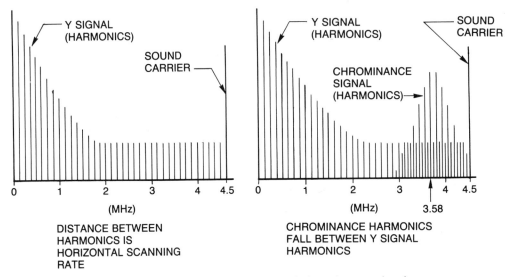

Figure 7-16 Harmonic distribution of luminance and chrominance signals.

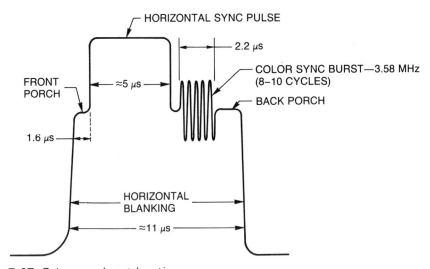

Figure 7-17 Color sync burst location.

Color TV Receiver

Combining the R − Y and B − Y signals with the Y signal gives the original red and blue signals. If the reconstructed red and blue signals are then subtracted from the Y signal, then the green signal is left (because the Y signal is made up of red, blue, and green). Therefore, in a color TV receiver, all three colors can be restored to operate each color gun. Using this method makes it possible to squeeze all of the color information into the required 6-MHz commercial TV bandwidth.

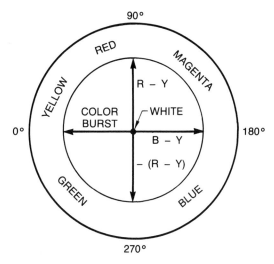

Figure 7-18 Phase relationships and corresponding colors.

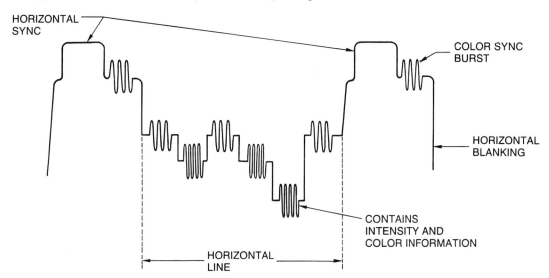

Figure 7-19 Composite color TV signal.

A block diagram of the color portion of a color TV receiver is shown in Figure 7-20. The function of each stage follows:

Chroma Bandpass Amplifier

Separates the high-frequency color (chrominance) signal from the rest of the composite video signal received from the video amplifier and amplifies it.

Color Control

Controls the amplitude of the chrominance signal from the bandpass amplifier. Determines the amount of color intensity (saturation) on the screen.

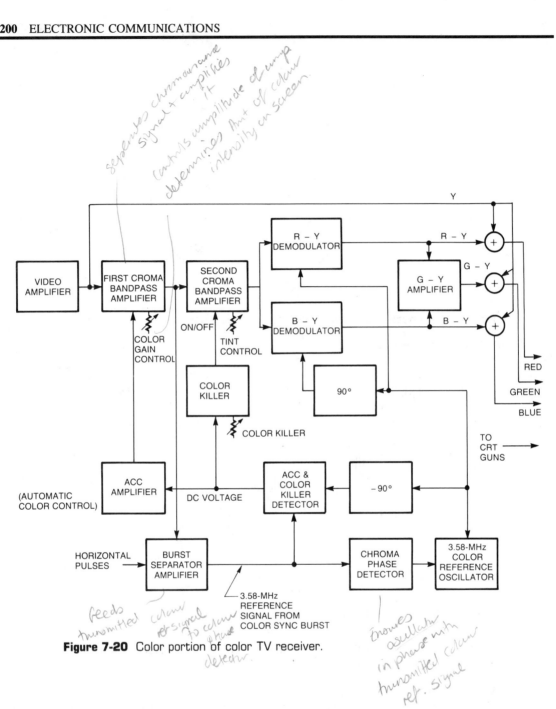

separates chromourama
signal & amplifies
it
controls amplitude of amp
determines Amt of colour
intensity on screen

Feeds
transmitted colour
ref signal
to colour phase
detector.

Ensures
oscillator
in phase with
transmitted colour
ref. signal

Figure 7-20 Color portion of color TV receiver.

Color Killer

Ensures that no color appears on the screen while viewing a black-and-white signal (when color dots do appear on the screen during a black-and-white reception, this problem is called "confetti"). The color killer turns off the bandpass amplifier when there is no color signal.

Color Killer Control

Adjusts the level required to activate the color killer control.

Color Burst Separator and Amplifier

Amplifier is turned on when the color burst appears on the back porch of the horizontal blanking pulse. It feeds the transmitted color reference signal to the color phase detector.

Color Phase Detector (*chroma phase detector*)

Ensures that the 3.58-MHz color oscillator is in phase with the transmitted color reference signal.

Tint Control

Adjusts the phase of the 3.58-MHz color oscillator relative to the phase of the transmitted color signals. Since the phase difference between these signals controls the color, adjusting the tint control affects the resulting picture color.

Chroma Oscillator (color reference oscillator)

The crystal-controlled 3.58-MHz oscillator used as the subcarrier that was suppressed at the transmitter. Its phase acts as the reference for the incoming color signals.

R − Y and B − Y Demodulators

Convert high-frequency chrominance signal into color difference low-frequency video. Have the same chrominance signal from the chroma bandpass amplifiers and the 3.58-MHz signal from the color reference oscillator. The chroma signals fed into each of these are 90° out of phase with each other.

G − Y Amplifier

Matrix that converts the incoming R − Y and B − Y signals into a G − Y output. Amplification of this signal also takes place here.

Conclusion

This section gave a brief introduction to the principles and major circuits of color TV. Keep in mind that there are textbooks available on just color TV. You should have an understanding of the major requirements of a commercial TV system.

7-4 Review Questions

1. Describe some major requirements for the transmission of commercial color TV.
2. Explain the difference between additive and subtractive color mixing.
3. Give an example of an impractical color TV system that requires three different cameras and transmitters. Explain how the system would work and why it is impractical to implement.
4. Describe the basic construction of a color TV picture tube.

5. Why are the electron guns given the names red, green, and blue? Are the electrons colored?
6. What factor determined the selected frequency of the color carrier? Why isn't the color carrier transmitted?
7. Explain the purpose of the color sync burst. When is it transmitted?
8. Name the major components of a color TV receiver.

7-5 STEREO TELEVISION

Discussion

In 1984, the Federal Communications Commission (FCC) authorized *multichannel TV sound* (MTS). This system includes the option of transmitting a TV signal with a stereo sound signal. The first commercially televised program to use MTS was the 1984 Olympics in Los Angeles, California.

As you will see, multichannel TV sound (MTS) is much more than just the addition of stereo sound transmission. It is a very flexible arrangement that allows many different transmission options.

BTSC System

The *Broadcast Television Systems Committee* (BTSC), appointed by the FCC and consisting of representatives from TV broadcasting and equipment manufacturers, recommended the presently accepted MTS system to the FCC. Therefore, the MTS system is called the BTSC system.

A "fully loaded" BTSC system is shown in Figure 7-21. Observe the similarity in the use of L + R and L − R signals, as was done for stereo FM. Hence, the transmission of stereo TV sound is compatible with TV receivers that do not have stereo capabilities.

The L − R signal is a DSB-SC (double-sideband suppressed-carrier). The pilot carrier is a multiple of the suppressed carrier and thus easily reinserted in the TV receiver equipped to receive it.

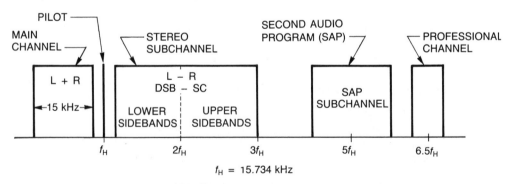

Figure 7-21 A fully loaded BTSC system.

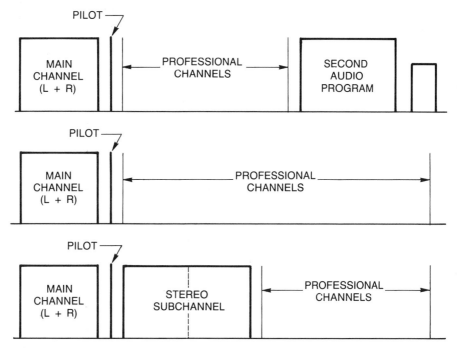

Figure 7-22 MTS options.

A second audio channel can be used to transmit the audio in another language. This channel is referred to as the *second audio program* (SAP). An optional professional channel is also available, which can be used for communications between TV stations or between a TV station and its remote units.

The actual configuration of the MTS audio baseband leaves many options. As shown in Figure 7-22, the only requirements by the FCC are the main audio channel and the frequency of the pilot carrier.

Conclusion

Multichannel TV sound is a new and exciting addition to TV communications. There is great flexibility available in this system. It is conceivable (though not likely) that the professional channel could be used to turn off your VCR during commercials.

7-5 Review Questions

1. Explain how the standards for stereo sound broadcasting on commercial TV were established.
2. Define MTS as it applies to commercial TV.
3. Describe what other options are available with MTS.
4. What is a fully loaded BTSC system?
5. State one use of an SAP channel.
6. What are some uses of professional channels?

| 7-6 | VIDEO CASSETTE RECORDERS |

Discussion

In the last few years, the most popular consumer item in the electronics marketplace has been the video cassette recorder (VCR). This instrument allows the recording, duplication, and playback of TV signals. This section introduces the operation of the VCR and the differences between the two most popular recording techniques: Beta and VHS.

Magnetic Recording

In a VCR, three items must be recorded:

1. Video information (including color)
2. Sound information
3. Control information (to ensure proper tape speed)

This information is recorded by placing it on a $1/2$-inch polyester-base recording tape using recording/playback heads, as shown in Figure 7-23. As the tape moves under the recording head, the frequency and strength of electrical signals can produce a permanent magnetic pattern on the tape. This magnetic tape pattern can be used to reproduce the same electrical currents when the tape is moved under a playback head. See Figure 7-24.

The relationship of the wavelength of the recorded signal and the tape speed is

$$\text{wavelength} = \frac{\text{tape speed}}{\text{input signal frequency}}$$

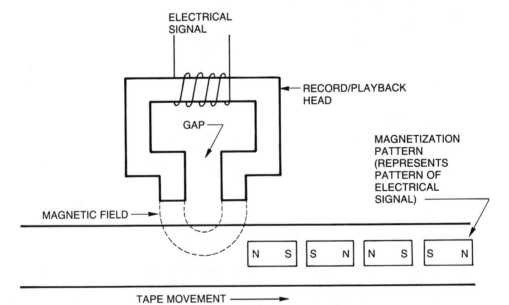

Figure 7-23 Basic elements of video tape recording.

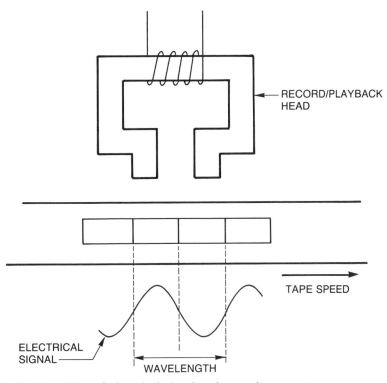

Figure 7-24 Relationships of electrical signal and record tape patterns.

For audio signals where the frequency range is from 20 Hz to about 20 kHz, a tape speed of about 19 cm/s is sufficient for good-quality recordings. However, for video signals, where the frequency range is from 0 Hz to over 4 MHz, a tape speed of 2 280 meters per minute would be required. This would make a 1-hour recording tape larger then the average home TV set!

Rotating Tape Heads

The high-frequency video recording requirements are satisfied by having the video recording and playback heads rotate on a drum while the tape moves across the drum. This increases the relative speed of the tape to the record/playback head. Thus less recording tape is used. See Figure 7-25. The head and track arrangement is the same for VHS and Beta. The difference between the two systems is their relative speed. Beta is 6.9 m/s, and VHS is 5.8 m/s. The speed of the drum cylinder is 1 800 rpm for VHS and Beta.

VCR Recording

The VCR servo system ensures precise tracking of the rotating heads. Beta and VHS both use an automatic servo control system.

The vertical sync pulses of the TV signal are used to synchronize the rotating heads with the tape movement. This is done by having the heads rotate just slightly faster than 1 800 rpm (30 Hz). The vertical sync pulses are sent to a divide-by-2 circuit, which produces the control signal. This control signal is recorded on the tape by a separate

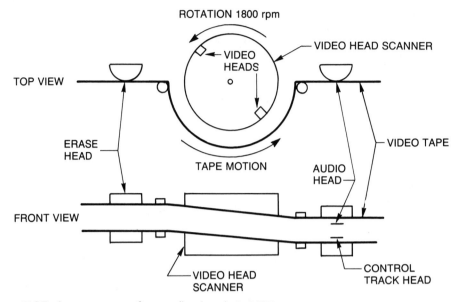

Figure 7-25 Arrangement of recording heads in VCR.

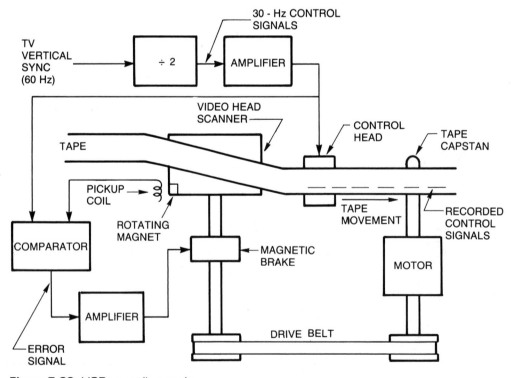

Figure 7-26 VCR recording mode.

stationary control track head located adjacent to the audio head. This arrangement is shown in Figure 7-26.

The figure shows that the rotating magnet on the video head scanner is fed into a comparator circuit, which compares the actual rotating frequency to the vertical scanning frequency of the TV signal. This method ensures that the control signal is in step with the received video signal.

VCR Playback

Figure 7-27 shows the VCR servo system in the playback mode. The comparator now compares the head scanner speed to that of the recorded 30-Hz control signal. Thus synchronization of the tape in playback mode is achieved.

The recording pattern of a video cassette tape is shown in Figure 7-28.

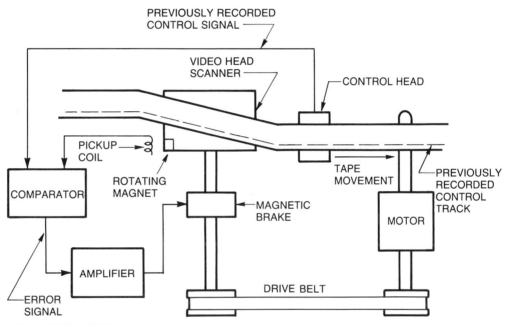

Figure 7-27 VCR in playback mode.

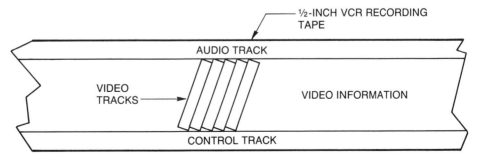

Figure 7-28 Recording pattern of video cassette tape.

Conclusion

This section briefly described video cassette recording and playback. You were introduced to the requirements of this system and how these requirements were achieved. Video cassette recorders have opened new opportunities in home entertainment and educational systems. The maintenance and repair of these instruments also mean new and exciting job opportunities for the service technician.

7-6 Review Questions

1. State the three kinds of information that a VCR must record.
2. Name the two major VCR recording methods.
3. Explain why the VCR record/playback heads must rotate in relation to the moving recording tape.
4. Describe what portion of the TV signal is used to synchronize the video information on the recording tape.
5. What is the frequency of the VCR control signal? How is it used in the record mode? in the playback mode?

TROUBLESHOOTING AND INSTRUMENTATION

Computer-Aided Troubleshooting

Computer-aided troubleshooting (CAT) consists of using a computer when troubleshooting and repairing electronic equipment. Any personal computer can be used that can accommodate an interface to allow signals to be analyzed. Figure 7-29 shows the basic requirements for CAT.

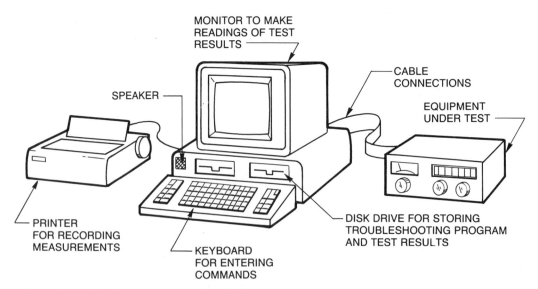

Figure 7-29 Basic requirements for CAT.

CAT Example

Suppose a technician wants to use CAT to measure the gain of an amplifier. The amplifier under test is shown in Figure 7-30. The system works as follows:

1. The computer interface takes the input and output signals of the amplifier and converts them to digital information.
2. The *peak detector* outputs the peak reading of its input signal. The magnitude of the peak detector output is then sent to an *analog-to-digital converter* (A/D converter).
3. The A/D converter inputs digital information into the computer.
4. A program (software) inside the computer records the two readings.

Levels of CAT

You may think of CAT as being divided into different levels. Theoretically, a computer could not only diagnose the problem but also cause a robot arm to replace the defective part.

Level 1 CAT

The technician gives all instructions to the computer, and the computer simply displays the values of the measurements. The technician decides what to do with the data.

Level 2 CAT

The technician gives all instructions to the computer, and the computer displays both measured and calculated values.

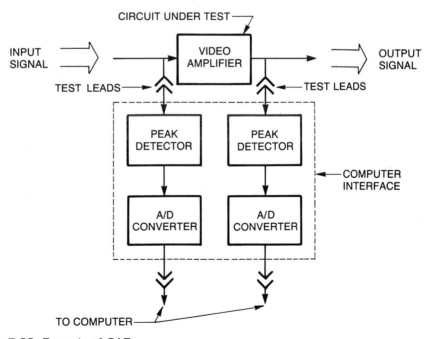

Figure 7-30 Example of CAT.

Level 3 CAT

The computer recommends troubleshooting procedures to the technician. These procedures are based on previous readings made by the computer. The technician still decides what measurements are to be made.

Level 4 CAT

The computer automatically makes tests and displays results for observation or technical documentation. Voice synthesizers inform technician of possible difficulties. The technician may override the computer by voice or keyboard commands.

Level 5 CAT

Same as level 4, except the computer directs a robot arm or similar device to change suspected parts.

Level 6 CAT

Same as level 5, but the computer is integrated into parts inventory control. Parts are automatically ordered as needed to maintain proper inventory level.

You can speculate on higher levels of CAT. CAT is used where it is more economical than a technician with hand tools and equipment. Large organizations that frequently service the same type of equipment depend heavily on CAT.

Future of CAT

For the communication service technician, CAT represents exciting new opportunities. Technicians familiar with software and hardware have new employment potentials. The field of electronics has never been stagnant. Its future and the future of all technicians depends on the dynamic changes inherent in the science and applications of electronics.

Conclusion

The microcomputer simulation section that follows presents a sample CAT program for TV troubleshooting. Many technicians are concerned about the growing use of CAT. Remember, it is technicians who must understand and even repair the CAT system itself! A CAT system that troubleshoots communication equipment requires technicians who understand communications as well as computer science.

7-7 Review Questions

1. List the basic components needed for CAT. What is the purpose of each?
2. Discuss a typical application for CAT. Why is a computer interface circuit needed?
3. List some advantages and disadvantages of CAT.
4. Explain the purpose of an A/D converter.
5. Describe different levels of CAT. What is the highest level you can envision?

 ## *MICROCOMPUTER SIMULATION*

The seventh troubleshooting simulation on your student disk presents the troubleshooting of systems using gain measurements. In this simulation an RF type voltmeter is used to measure the input and then the output of an amplifier. From this, a calculation of gain may be made.

This is then compared to the required specification for the system being analyzed. From this, you can then determine if the system is within acceptable standards.

CHAPTER PROBLEMS

(Answers to odd-numbered problems appear at the end of the text.)

1. Describe the information that a TV station must transmit.
2. Explain the purpose of (A) horizontal sync pulses and (B) vertical sync pulses.
3. Describe the use of FM and AM in TV transmission.
4. Explain how one TV station is separated from another.
5. What is the relation of commercial FM frequency allocations to commercial TV allocations?
6. Does the relationship of the commercial FM band to commercial TV band have any benefit? Explain.
7. Describe how the major components of a TV signal are distributed inside a TV receiver.
8. State the major sections of a black-and-white TV and explain the purpose of each.
9. In a TV picture tube, what is the purpose of the (A) electron gun and (B) the large picture tube voltage?
10. State the kind of deflection used in a TV picture tube. What causes the electron beam to be seen on the screen?
11. What determines the "color" of a black-and-white picture tube?
12. Explain the development of a TV raster. Why is a raster necessary?
13. How many horizontal lines are used to develop a TV raster?
14. What two circuits inside a TV receiver develop the raster? How are these circuits timed to ensure that they are in step with the same circuits at the transmitting station?
15. List the main parts of the composite video signal.
16. What is the relationship of the horizontal blanking pulse to the horizontal sync pulse? Why is this relationship necessary?
17. In a composite video signal, what is the relationship between the amplitude of the signal and the intensity of the electron beam in the receiver picture tube?
18. Is all the information seen on a TV screen presented at the *same* time? Explain. How does this relate to motion pictures?
19. What two principles prevent the TV viewer from observing flicker?
20. Why was AM chosen to transmit TV picture information? What is the relationship between the amount of picture information sent per unit time and the bandwidth used to transmit this information?
21. Explain interlaced scanning. Why is it used?
22. State the difference between a TV field and a TV frame as it relates to picture information. Is there a difference between the scanning frequencies of a frame and a field? Explain.
23. What is the relationship between the scanning frequency and the total lines per frame?
24. If the number of lines per frame were to double over that presently used in North America, what would be the new scanning frequency required?

25. How much time is allocated for horizontal retrace? for vertical retrace?
26. Define equalizing pulses. Explain their purpose.
27. What is a picture element? How many picture elements are required in the vertical direction?
28. What determines the number of vertical picture elements?
29. Define aspect ratio. What is the aspect ratio in commercial TV receivers?
30. How does the aspect ratio determine the number of picture elements in the horizontal direction? For an aspect ratio of 1:1, what is the relationship of the number of picture elements in the vertical direction to the number in the horizontal direction?
31. Define utilization ratio. How is this applied to commercial TV?
32. What factors determine the highest required video frequency to be transmitted?
33. What is the highest video frequency set by the FCC for commercial TV?
34. Explain how the transmitted bandwidth of commercial TV is reduced to 6 MHz. Explain why this reduction in bandwidth was not taken further.
35. What is the relationship of the sound carrier to the picture carrier in commercial TV? What is the maximum deviation allowed for the sound portion of commercial TV?
36. Name the two major categories of TV frequency transmission. Which category has the most available TV stations?
37. What is the total number of available TV stations?
38. Sketch the frequency spectrum for channel 5. Give each significant frequency.
39. Sketch the frequency spectrum for channel 9. Give each significant frequency.
40. Draw the block diagram of a black-and-white TV receiver and describe the purpose of each block.
41. The RF amplifier in the tuner serves two purposes; what are they?
42. Describe the differences between UHF and VHF tuners. What are their similarities?
43. What section of a TV receiver determines the bandwidth and produces the most signal gain?
44. Explain the purpose of the video detector. What two signals does it produce?
45. What is AGC? What does it do? What sections of a TV receiver are controlled by AGC?
46. Describe the difference between a video amplifier and an audio amplifier. What are their similarities?
47. Explain how the contrast control affects the TV picture. What circuit does the contrast control affect?
48. Explain the difference between the contrast control and the brightness control.
49. What is the function of the sync separator? If this section of the TV receiver experienced problems, what would be the most likely result on the picture?
50. Draw the block diagram of the vertical oscillator. Show what controls are available with the vertical oscillator and describe their functions.
51. Draw the block diagram of the horizontal oscillator. Show what controls are available with the horizontal oscillator and describe their functions.
52. If the horizontal oscillator stopped oscillating, what would happen to the output of the high-voltage power supply? Why would this occur?
53. List the requirements for a color TV signal.
54. Explain the difference between additive and subtractive color mixing. Which method is used in color TV?

55. What percentage of the primary colors used in color TV are needed to produce the brightest white?
56. List the main differences between a color picture tube and a black-and-white picture tube.
57. Why are the electron guns in a color TV picture tube called the green, red, and blue guns?
58. Describe how the color TV signal is produced in order to provide compatibility with black-and-white TV receivers.
59. Describe the components of a composite color TV signal.
60. What is the subcarrier frequency for color transmission? Why was this frequency selected?
61. What circuit in a color TV receiver reinserts the suppressed carrier of the color signal? What part of the color TV signal ensures that the phase of this circuit is correct?
62. Sketch the color wheel and explain how the phase relations develop different colors in the color TV receiver.
63. What is a chrominance signal? How is the color information contained in such a signal?
64. What modulation process is used for color transmission?
65. Explain how the primary colors are reconstructed in a color TV receiver.
66. Sketch the block diagram of the color circuits used in a color TV receiver. Describe the purpose of each block.
67. Explain MTS. What options are provided for this kind of transmission?
68. Sketch the frequency spectrum of a fully loaded BTSC system. Briefly describe how each section may be used.
69. Explain the method used to transmit stereo sound. Describe how this method is similar to commercial FM stereo transmission.
70. Describe some of the other options available with MTS.
71. Describe the basic requirements for a VCR system.
72. Sketch the recording heads used in a VCR and show their relationship to the recording tape.
73. Explain what portion of the TV signal is used to control the speed of the video tape.
74. Describe a basic system used for CAT.
75. What is meant by CAT *levels?*

CHAPTER 8

Pulse Communication Techniques

OBJECTIVES

In this chapter, you will study:

- ☐ The reasons why pulse modulation is superior to other forms of electronic communications.

- ☐ The basic concepts of signal sampling.

- ☐ The different methods of pulse modulation, including pulse amplitude modulation, pulse duration modulation, and pulse position modulation.

- ☐ What constitutes pulse code modulation and delta modulation.

- ☐ The systems used to develop and receive these pulse communication techniques.

- ☐ Frequency-division multiplexing as used in telephone systems.

INTRODUCTION

Pulse communications is one of the fastest-growing areas of electronic communications. Its concepts go back as far as 1812, but its growth into consumer electronics has exploded in the last decade. Part of this was brought about because of the availability of inexpensive integrated circuits (ICs). Communication technicians must understand the basic principles of pulse communications. This chapter offers the essential introduction.

8-1 BASIC IDEA

Discussion

There are some basic principles of pulse communications that make it more desirable than other forms of electronic communications. This section introduces you to some of the fundamental techniques of pulse communications.

Sampling

The reason for sampling can be demonstrated by the following example. Suppose a factory contains several processing vats. Each vat has a thermometer that must be carefully monitored. The temperatures of these vats can be observed in two main ways. See Figure 8-1.

Observe from the figure that to receive *continuous* data, four workers are required. This may not be necessary, since any change in the temperature of the vats will be gradual and such a system is not very economical when you consider the salaries and fringe benefits of four people.

The other method shown in Figure 8-1, *sampling* data, is a more efficient system. Only one worker must monitor the temperatures of all four vats. If the worker makes the samples faster than the thermometers can change, then the same effect as *continuous* sampling is achieved at a reduced cost (one fourth less in this example).

This situation is similar to transmitting and receiving information in electronic communications. Obviously, *sampling* data rather than *continuous* monitoring produces greater system efficiency and allows many different kinds of information to be transmitted on one carrier. Thus, only one transmitter and one receiver are needed.

Sampling Electronic Signals

Suppose it was necessary to transmit three different electrical signals over a single wire. By sampling techniques, this transmission could be carried out as in Figure 8-2. Both switches are changing in such a manner that when the transmitting switch is at position A so is the receiving switch. This kind of electronic sampling is called *time-division multiplexing* (TDM). This name was applied because multiple signals are made by sampling them at different times. The chart recorders are assumed to have a slow response time that is much less than the sampling rate. Thus the graphs of each waveform look smooth and continuous.

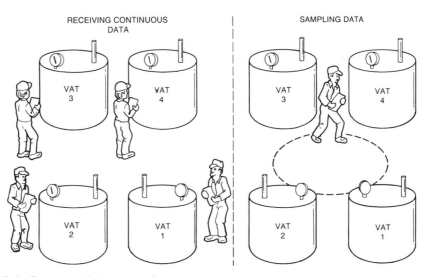

Figure 8-1 Concept of data sampling.

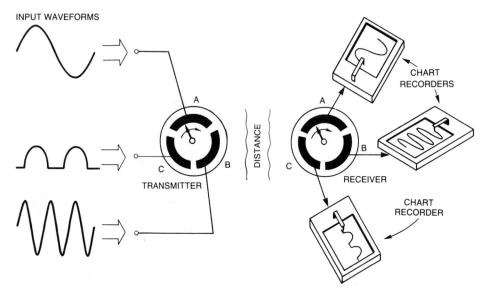

Figure 8-2 Electronic sampling.

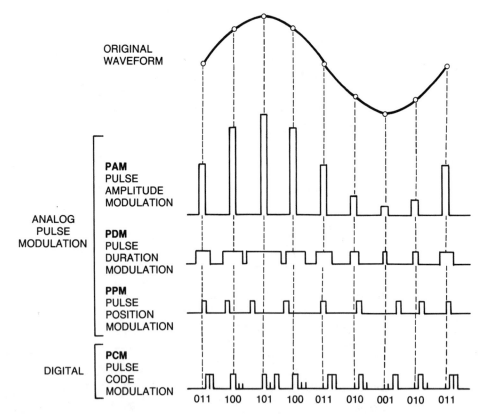

Figure 8-3 Different pulse techniques.

Pulse Modulation Techniques

Think of these electronic sampling schemes as using information samples called *pulses*. For example, consider the worker sampling the temperature of all four vats. Every "peek" at a particular thermometer could be considered as a "pulse" of information received by the worker's eyes. Similarly, each time a signal was received by the chart recorders, it remained for a short time, and thus could be called a "pulse."

In each case, amplitude information (amount of temperature, amplitude of each waveform) of each signal was sent for only a short time. This technique is called *pulse amplitude modulation* (PAM). It is called this because the amplitude of the resulting pulse changes in step with the modulating signal.

Several types of pulse modulation techniques are used in electronic communications. See Figure 8-3. The first three techniques are called *analog pulse modulation*. They are called this because some characteristic of the pulse (amplitude, duration, or position) is continuously varied.

The fourth method is called *digital modulation*. It is called this because the modulating signal is sampled and this sample is then converted into a digital code. This digital code now represents the amplitude of the modulating signal. You will discover how electronic communication systems produce and demodulate all four of the pulse modulation techniques just presented.

Conclusion

This section presented the basic concepts of information sampling. These concepts will be used in the rest of this chapter.

8-1 Review Questions

1. Describe the difference between continuous monitoring of data and sampling of data.
2. When is the data sampling more efficient than continuous monitoring? When is sampling not as accurate as continuous monitoring?
3. For the electronic sampling scheme in Figure 8-2, what are some of the factors that would cause the waveforms on the chart recorders to not look continuous?
4. What is time-division multiplexing? How is this different from frequency-division multiplexing?
5. Describe the four methods of converting a sine wave into pulse modulation.
6. What is the difference between analog pulse information and digital pulse information?
7. Which form of pulse modulation is similar to AM?

8-2   SAMPLING THEOREM

Discussion

In electronic communications, *sampling* is the process of taking a periodic sample of the waveform to be transmitted and transmitting the samples. If enough samples are sent, the waveform can be reconstructed at the receiving end. One method of sampling a waveform is shown in Figure 8-4.

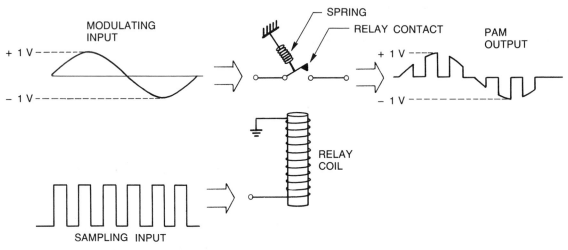

Figure 8-4 Waveform sampling.

Using a mechanical device such as a relay is not a practical method of sampling, since electronic means are available, but it does serve to illustrate the concept of waveform sampling.

Sampling Frequency

The more samples that are taken, the more the final outcome looks like the original wave. However, if fewer samples of one wave could be taken, then other kinds of information could be transmitted. This is analogous to having one person reading the temperatures of several vats. The less time spent reading the temperature of any single vat, the more time is left to read the temperatures of other vats (or to get other kinds of information).

The question is, what is the minimum sampling rate that can be used for any signal so that the signal will be correctly restored at the receiver? The answer is given by the *sampling theorem* (called the Nyquist sampling theorem): The *sampling frequency* of a pulse modulated system must be equal to or greater than twice the highest signal frequency in order to convey all the information of the original signal.

Mathematically this theorem says

$$f_s = 2f_{N(\max)}$$ **(Equation 8-1)**

where f_s = Minimum sampling frequency to
   ensure that the samples contain
   all of the information of the
   original signal

$f_{N(\max)}$ = Maximum frequency of the modulating signal

Example 1

A pure 1-kHz sine wave is to be sampled at the lowest possible rate for transmission as pulses. What is the minimum sampling frequency required to ensure that all components of the wave are restored at the receiver?

Solution
Using the sampling theorem—

$$f_s = 2f_{N(max)}$$ (Equation 8-1)
$$f_s = 2 \times 1 \text{ kHz} = 2 \text{ kHz}$$

Hence, the minimum sampling frequency must be 2 kHz.

Aliasing Distortion

Aliasing distortion occurs when the sampling rate is less than the sampling theorem allows. An example of this kind of distortion is in old western movies where the wheels of wagons seem to be turning the wrong way. This effect is shown in Figure 8-5.

The reason for the "wagon wheel" distortion is that the sampling rate of the movie camera was less than twice the radian frequency of the rotation of the wheel. This kind of distortion with pulse sampling can be illustrated by using a frequency spectrum analysis. This is explained in the following subsection.

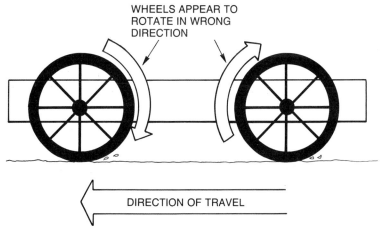

WHEELS APPEAR TO
ROTATE IN WRONG
DIRECTION

DIRECTION OF TRAVEL

Figure 8-5 Example of aliasing distortion.

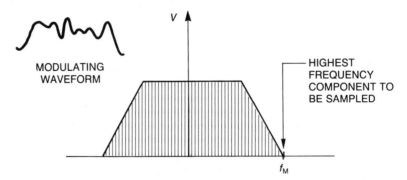

MODULATING WAVEFORM

HIGHEST FREQUENCY COMPONENT TO BE SAMPLED

f_M

Figure 8-6 Frequency spectrum of a modulating waveform.

Effects of Sampling on Frequency Spectrum

Assume that the frequency spectrum of the modulating wave is as in Figure 8-6. Note that this represents some kind of complex wave (such as voice) and that the higher-order harmonics are smaller in amplitude.

Figure 8-7 shows what happens when the modulating wave is sampled at twice its maximum frequency. Theoretically, the harmonics of the sampling extend to infinity. But, in practice, the resulting spectrum need only be passed through a lowpass filter to be restored. So that a practical lowpass filter can be used to pass only the highest modulating frequency (f_N) and not any components of the higher sampling harmonics, the sampling rate is made slightly larger than $2f_N$. This higher sampling rate (f_s) creates a *guard band* between f_N and the lowest-frequency component ($f_s - f_N$) of the sampling harmonics. Therefore a practical lowpass filter can be used to restore the original modulating signal. See Figure 8-8.

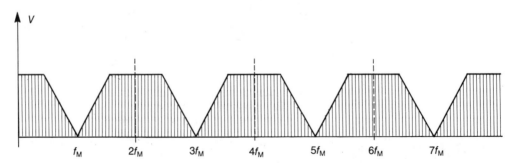

f_M $2f_M$ $3f_M$ $4f_M$ $5f_M$ $6f_M$ $7f_M$

Figure 8-7 Results of sampling at $2f_N$.

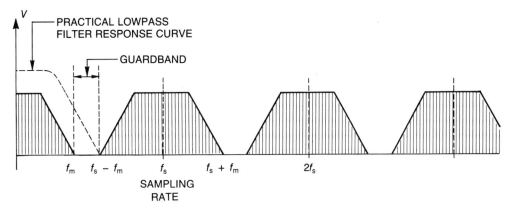

Figure 8-8 A more practical sampling rate.

The restoration of the modulating signal from the PAM waveform by a lowpass filter is shown in Figure 8-9. This is similar to the action of the filter capacitor in an AM detector.

Aliasing distortion occurs when the sampling rate is lower than the sampling theorem allows. The effect on the resulting spectrum is shown in Figure 8-10.

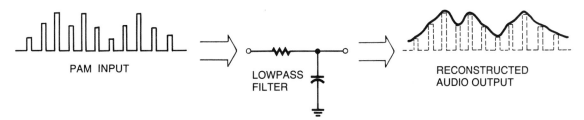

Figure 8-9 Restoring the sampled waveform.

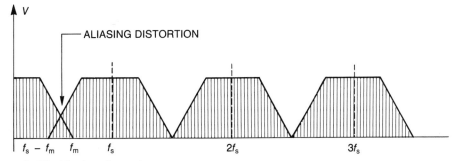

Figure 8-10 Aliasing distortion spectrum.

Conclusion

The sampling theorem demonstrated the minimum requirements for transmitting sampled information. This theorem is contained in the subject of *information theory*. Information theory is in itself a complete field of knowledge with many useful applications.

8-2 Review Questions

1. Describe the process of sampling as used in electronic communications.
2. What is an advantage of taking as few samples as possible of a modulating wave without distortion?
3. State the sampling theorem.
4. Explain aliasing distortion. Give an example.
5. State why a practical sampling rate is usually slightly more than the sampling theorem requires.
6. When sampling, what is a guard band? Why is it used?
7. Describe the action of a lowpass filter in regard to restoring the modulating frequency of a PAM signal.

8-3 PULSE AMPLITUDE MODULATION

Discussion

Pulse amplitude modulation (PAM) is the most direct form of pulse modulation. In this sampling process, the amplitude of the pulse varies in proportion to the amplitude of the modulating signal.

Types of PAM

The two general types of pulse amplitude modulation are *dual polarity* and *single polarity*. Both are shown in Figure 8-11.

The *analog bilateral switch* is an IC circuit that uses electronic (nonmechanical) switches. Practical methods of generating PAM can use this type of switch. An example is a 4016 IC. This chip contains 14 pins and is shown in Figure 8-12. Note that there are four separate electronic switches. Each switch has its own separate control. For example, the switch between pins 1 and 2 is controlled by an input voltage on pin 13.

Using the analog switch to create PAM is shown in Figure 8-13.

Demodulation

To demodulate PAM, a lowpass filter is used. The slope of the filter must be steep enough to pass the highest modulating frequency and to eliminate the lowest sampling frequency component. This means that the filter's cutoff must fall well within the guard band of the particular PAM system.

Pulse amplitude modulation is transmitted over wire, or it can modulate a carrier wave. When carrier wave modulation is used, PAM is very susceptible to noise. Thus PAM is not the most popular form of pulse modulation.

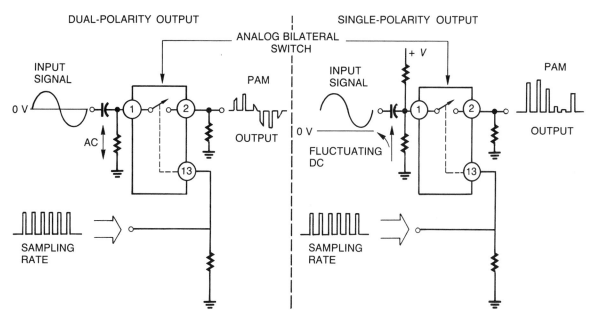

Figure 8-11 Generation of PAM waveform.

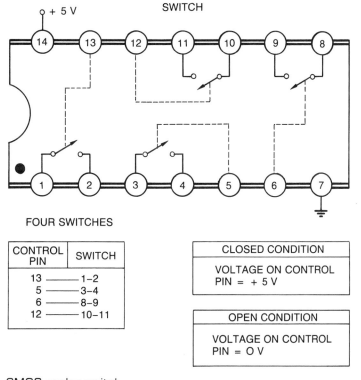

Figure 8-12 CMOS analog switch.

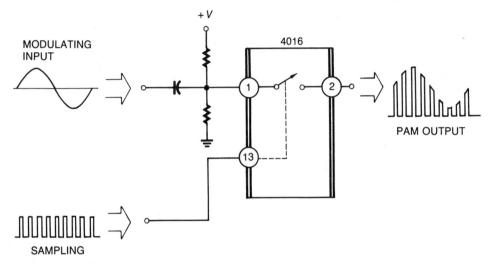

Figure 8-13 PAM using analog switch.

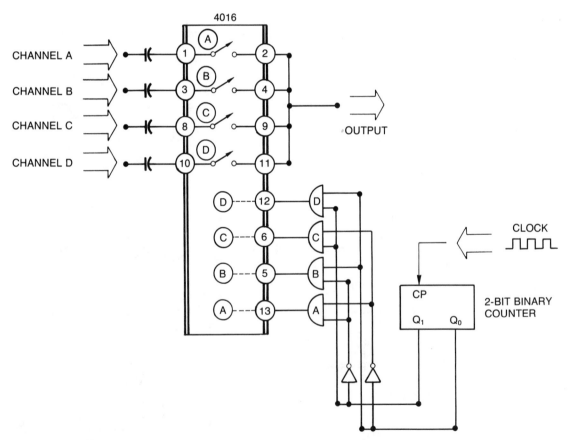

Figure 8-14 Sampling system transmitter.

Time-Division Multiplexing

Pulse amplitude modulation allows more than one signal to be transmitted at a time. This concept is called *time-division multiplexing* (TDM). Figure 8-14 shows a possible sampling system using the 4016 analog switch. The system allows four separate channels of information to be multiplexed and transmitted over a single output. The system consists of four AND gates controlled by a 2-bit binary counter. When the counter is at 00_2, AND gate A is activated, thus closing switch A. When the counter is at 01_2, the only AND gate activated is B, which closes switch B. Thus only one switch at a time is activated and a time-division multiplexed signal appears at the output.

How fast the sampling takes place is determined by the frequency of the clock that controls the binary counter.

A sampling system receiver is shown in Figure 8-15. The receiving time-division multiplexing system is similar to the transmitter. Here, the received signal is switched between the proper channels by another 2-bit binary counter, which is controlled by the AND gates in the same manner as the TDM transmitter. For this system to function properly, the transmitter clock must be synchronized with the receiver clock.

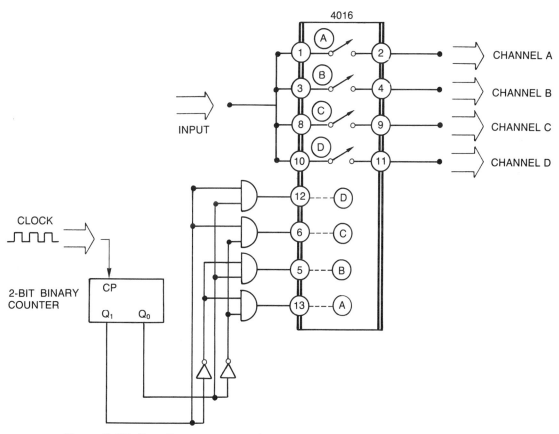

Figure 8-15 Sampling system receiver.

Conclusion

This section showed practical circuits that could be used to generate PAM. You also learned some of the limitations of pulse amplitude modulation (PAM) and some more information about time-division multiplexing (TDM).

8-3 Review Questions

1. Describe PAM.
2. Explain the difference between dual-polarity PAM and single-polarity PAM. Describe a generalized circuit that can be used to generate each kind.
3. Discuss the operation of an IC switch.
4. Describe the process of TDM. Why is it used?
5. What is the main disadvantage of PAM?
6. Explain how an integrated switch can be used to multiplex signals.
7. Describe a generalized circuit for a four-channel TDM receiver.

8-4 | PULSE DURATION MODULATION

Discussion

This section introduces you to the concepts of *pulse duration modulation* (PDM). Pulse duration modulation is in the class of *pulse time modulation* (PTM), where some time variation of the pulse rather than the amplitude is affected by the modulating signal. In the next section, you will see the other method of PTM called pulse position modulation (PPM). PAM, PDM, and PPM are equivalent to AM, FM, and PM (phase modulation).

General Idea

In pulse duration modulation (PDM), some aspect of the duration of the pulse represents the information about the modulating signal. Figure 8-16 shows the three types of pulse duration modulation.

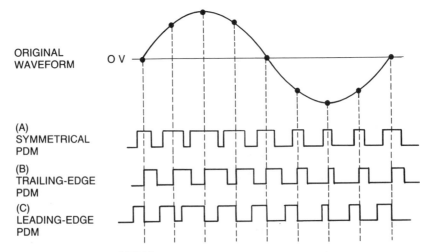

Figure 8-16 Three types of PDM.

Figure 8-16(A) shows a symmetrical PDM waveform. In this case, both the leading and trailing edges of the pulses are varied according to the amplitude of the modulating signal. Note that when the modulating signal is at zero, the width of the pulse is at a "reference" width. Thus, as the amplitude of the modulating signal goes positive, the pulse width increases; and as the amplitude of the modulating signal goes negative, the pulse width decreases. It is important to note that the spacing between the centers of the pulses remains constant.

Trailing-Edge PDM

Figure 8-16(B) shows trailing-edge PDM. In this kind of sampling, the trailing edge of the pulse varies in accordance with the amplitude of the modulating signal. Note that the leading edges of the pulses remain at a fixed rate relative to each other. Therefore the timing between each leading pulse edge is constant.

Leading-Edge PDM

Figure 8-16(C) shows leading-edge PDM. Here the leading edge changes in accordance with the amplitude of the modulating signal. Note that the trailing edge of each pulse is fixed, and the timing between the trailing edges is constant.

Generating PDM

Figure 8-17 shows a method of generating PDM. There are two signals: a sawtooth wave and the modulating signal. Both are combined in a linear amplifier (one that does not produce distortion). Recall from Chapter 3 that when two signals are added this way the result is the waveform in Figure 8-17.

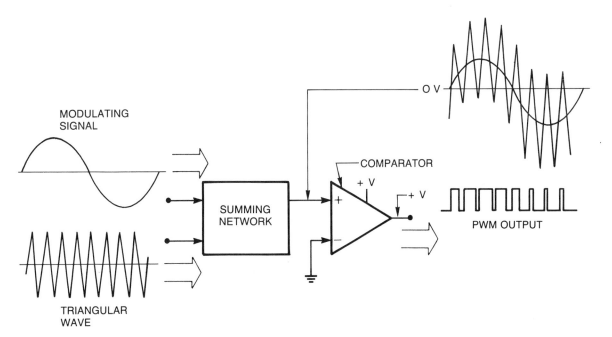

Figure 8-17 Method of generating PDM.

The next circuit in the PDM generator is a *comparator*. This circuit will produce an output voltage of either $+V$ or zero volts and nothing in between. That is, if the voltage at its $(+)$ terminal is more than the voltage at its $(-)$ terminal, then the output will be $+V$. This condition is called *saturation* because the output signal cannot get any larger than $+V$. If the voltage at its $(+)$ terminal is less than the voltage at its $(-)$ terminal, then the output will be zero volts. This action is shown in Figure 8-18.

If the reference voltage at the $(-)$ terminal of the comparator is increased to $+1$ V, then the output of the comparator will not saturate until the input signal is greater than $+1$ V. At values less than $+1$ V, the input signal keeps the comparator output at zero volts.

To produce symmetrical PDM, a triangular wave was used as the sampling waveform. Figure 8-19 shows how leading-edge and trailing-edge PDM are produced.

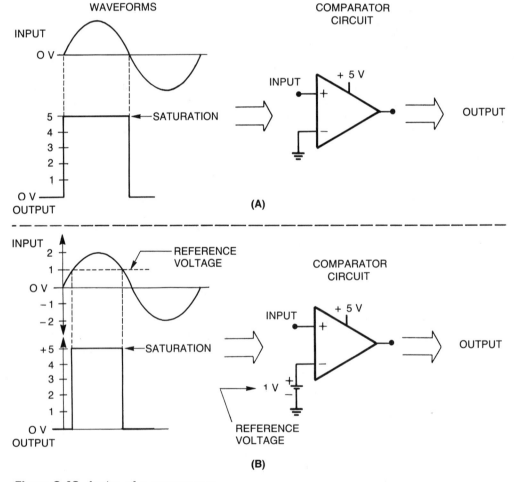

Figure 8-18 Action of a comparator.

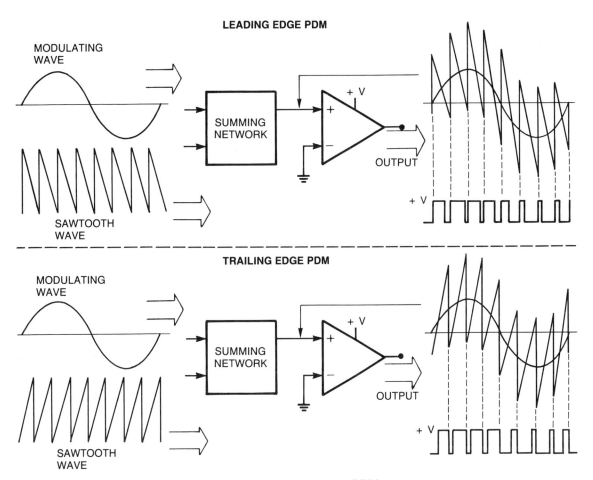

Figure 8-19 Producing leading- and trailing-edge PDM.

PDM Demodulation

A simplified method of demodulating a PDM signal is shown in Figure 8-20. The input PDM signal is differentiated and fed into a *positive-peak clipper* and a *negative-peak clipper*. The resulting two waveforms operate a *ramp generator,* which produces an ideal sawtooth output while it is on. When the off input of the ramp generator is activated, its output returns to zero. This is similar to the action of an RS flip-flop. The difference is that the output is a sawtooth whose amplitude is proportional to the amount of time between an on input condition and an off input condition.

The resulting sawtooth is effectively an AM wave that is fed into a lowpass filter or *integrator*. The resulting waveform is a close reconstruction of the original modulating signal.

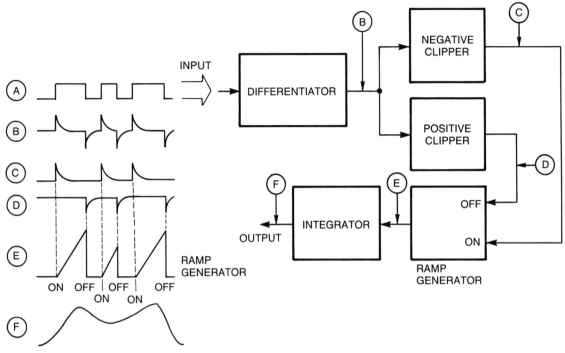

Figure 8-20 Demodulating PDM.

Conclusion

In this section, you saw how a PDM signal was created and demodulated. The advantage of PDM is similar to that of FM—a more noise-free reception. In the next section, you will learn the other aspect of time-division multiplexing (TDM) called pulse position modulation (PPM).

8-4 Review Questions

1. Explain the difference between PAM and PDM.
2. What is the relationship between PTM and PDM?
3. Name the three types of PDM.
4. Describe the action of a comparator.
5. Describe a method of producing symmetrical PDM.
6. Explain the difference between producing leading-edge PDM and trailing-edge PDM.
7. Describe the results of differentiating a square wave.
8. What is a clipper? What is the difference between a positive-peak clipper and a negative-peak clipper?
9. Explain the operation of a ramp generator.
10. Describe a simple PDM demodulator.

8-5 | PULSE POSITION MODULATION

Discussion

Pulse position modulation (PPM) is another form of pulse time modulation (PTM). In PPM, the position of the pulse relative to a reference is changed in accordance with the amplitude of the modulating signal. Pulse position modulation has less noise due to amplitude changes, because the received pulses may be clipped at the receiver, thus removing amplitude changes caused by noise.

Typical PPM Waveform

Figure 8-21 shows a typical pulse position modulation (PPM) waveform. Note the reference train of pulses. The PPM pulses are measured with respect to these reference pulses. Observe that when the signal goes positive, the modulated pulses fall behind the reference pulses. When the signal goes negative, the modulated pulses start before the reference pulse. As you will see in this section, this is not the only form of PPM. The criterion for PPM is that the position of the pulse be changed by the amplitude of the modulating wave.

Producing PPM

A method of producing PPM is shown in Figure 8-22. The modulating signal is first converted into PDM. Then the PDM signal is fed into a differentiator. The time constant of the differentiator is such that positive and negative "spikes" are produced on the output. This is then set to a positive-peak *clipper*. Recall from the last section that a positive-peak clipper is a circuit that effectively removes the positive parts of the differentiated waveform. The resulting waveform triggers a one-shot flip-flop. This circuit goes "true" (+5 V) for a predetermined amount of time and then returns to its "false" (0 V) state. This circuit is also referred to as a *monostable multivibrator*.

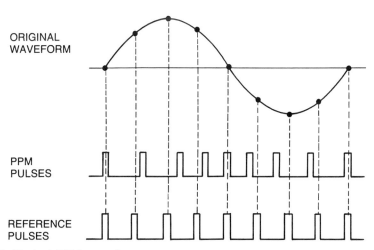

Figure 8-21 Typical PPM waveform.

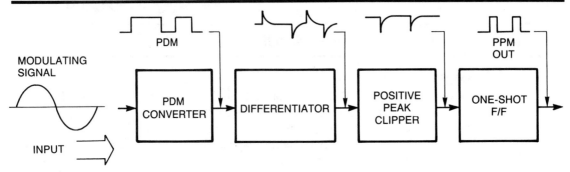

Figure 8-22 A method of producing PPM.

Figure 8-23 shows the relative position of the various signals in the development of PPM. Note that the PDM is trailing-edge modulation. This makes the leading edge fixed, which in turn makes the positive part of the differentiated waveform unchanged by the modulating signal. Thus, the positive portion of the differentiated wave is clipped. The resulting negative portion of the differentiated wave contains the information about the modulating signal in terms of its relative position. The one-shot causes pulses of equal duration to appear on the output. The *position* of the resulting pulse now carries the information about the modulating signal.

PPM Demodulation

Figure 8-24 shows a typical PPM demodulator. The received PPM signal is first sent to a limiter, which removes any amplitude variations due to noise. Reference pulses along with the PPM pulses are used to control the output of a flip-flop. The flip-flop output is

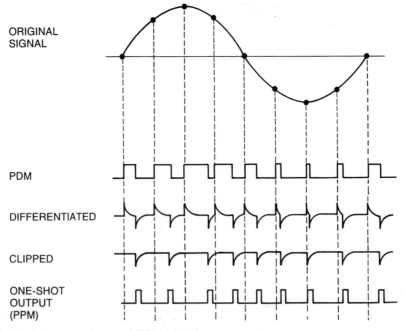

Figure 8-23 Relative position of PPM signals.

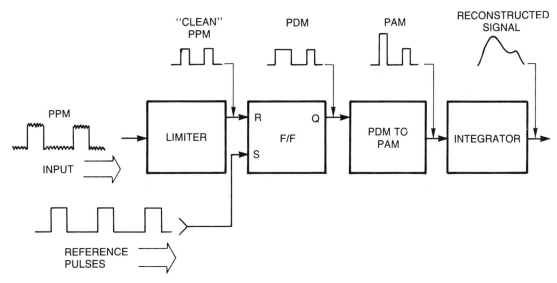

Figure 8-24 Typical PPM demodulator.

now proportional to the amount of time between the reference pulses and the PPM pulses. The resulting signal is now PDM. This signal is sent to a PDM-to-PAM converter (this circuit was explained in the last section). A lowpass filter is then used to reconstruct the original modulating signal. Figure 8-25 shows the time relationships of the resulting waveforms.

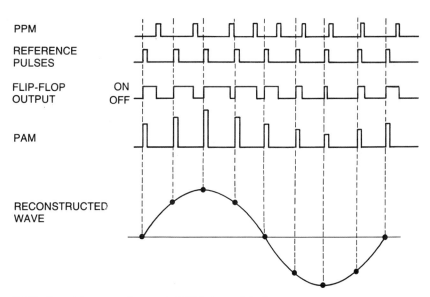

Figure 8-25 Resulting waveforms in PPM demodulation.

Conclusion

This section concluded the major ways of producing pulse modulation. In the next section, you will be introduced to digital modulation.

8-5 Review Questions

1. Explain how information is transmitted using PPM.
2. What is the relationship between PPM and PTM?
3. Define reference pulses. How are they used in PPM?
4. Describe the process of producing PPM.
5. What is a monostable multivibrator?
6. Describe the process of PPM demodulating.

8-6 PULSE CODE MODULATION

Discussion

Pulse code modulation (PCM) is a form of digital communications. The amplitude of the modulating signal is converted into a digital code (it could be a binary number that represents the signal amplitude). This process is similar to the action of an analog-to-digital converter (A/D converter) where the amplitude of the analog signal is converted into a digital code. Analog-to-digital converters are discussed in Chapter 9.

On the receiving end, PCM can now be reconstructed into the original modulating waveform. But it doesn't have to be. Since the transmission is a digital code, this digital code can be processed at the receiving end by a computer, thus producing a wide variation of signals based upon the original modulating signal. This process is the essence of digital sound recordings, where noise can be removed by computer software rather than by hardware. The difference is that ideal hardware can be represented by software when it is not possible or economically feasible to obtain such hardware.

Quantized Pulses

In PCM, the pulses result from sampling the modulating waveform. In essence, the modulating waveform is "sliced" into small units or *quantized*. These *quantum* points are then converted into a binary code that represents the amplitude of the waveform at that point. Figure 8-26 illustrates the process.

In the figure, the sine-wave amplitude is 7 volts. Each sampling point is taken at equal intervals. These points represent the quantizing points of the waveform. Observe that these quantizing points are similar to PAM, but these amplitudes are not continuous they are quantized. Note from the figure that the first quantizing point falls exactly at 3.5 volts. Since the quantizing levels are expressed in whole numbers, the value of 3 (011_2) is given to this point. The next sampling point falls exactly at 5 volts; hence, the quantizing point has a value of 5 (101_2). The important point is that each quantizing level is equal to or less than the actual value of the modulating wave at that sampling point.

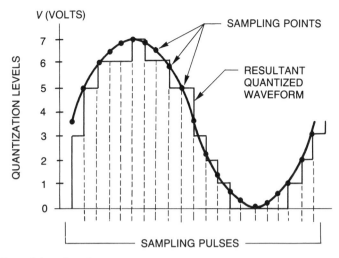

Figure 8-26 Quantizing the sine wave.

Quantizing Noise

Quantizing noise is the error or distortion introduced by PCM when the modulating signal is not an exact value of the resulting binary code. Quantizing noise occurs when the modulating wave lies somewhere between two quantizing points in PCM. This problem can be reduced by increasing the sampling points, but that would also increase the required bandwidth of the system and the number of pulses required for each code. In some digital sound recording systems, there are 65 536 quantizing points, which require 16 binary places ($1111\ 1111\ 1111\ 1111_2 = 65\ 535_{10}$). Thus, in these systems, great accuracy can be obtained, and the effect of quantizing noise minimized. Compare this with the example here of using only eight quantizing points (0 through 7).

PCM Coding

Figure 8-27 shows the resultant binary code of the quantized wave of Figure 8-26. Each portion of the wave is converted to a binary number that represents the final wave. Since there are eight levels of the wave, 3 bits are needed, 000_2 to 111_2, giving a range from 0 to 7 volts. The resulting pulse code along with the quantized wave is shown in Figure 8-28.

In practical systems, an 8-bit code is usually used for PCM along with a sync or reference pulse. Using an 8-bit code reduces quantizing noise since it allows 256 quantizing steps.

PCM Transmitter

A PCM transmitter is an encoder. An *encoder* is a circuit that converts a signal into a specified code. For the circuit in Figure 8-29, an analog-to-digital converter converts the incoming analog (continuously changing) modulating signal into a digital (discrete) signal.

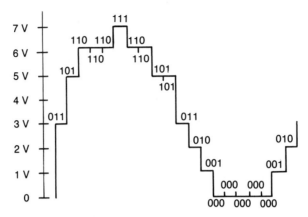

Figure 8-27 Quantized wave coding.

QUANTIZING LEVEL (V)	BINARY NUMBER	PULSE CODE
0	000	
1	001	
2	010	
3	011	
4	100	
5	101	
6	110	
7	111	

Figure 8-28 Resulting PCM waveforms.

This PCM encoder circuit is sometimes called a *flash converter,* because the input signal is easily converted into a binary number code. The circuit consists of three differential amplifiers that have either a 0-V (false) or a +5-V (true) output, depending on the level of the input signal (which varies between 0 and +3 V). The outputs of these differential amplifiers operate the logic circuity consisting of an *exclusive NOR* gate and two *AND* gates. The resulting logic output is a binary number representation of the input signal. Thus, the modulating signal is quantized, and PCM is achieved.

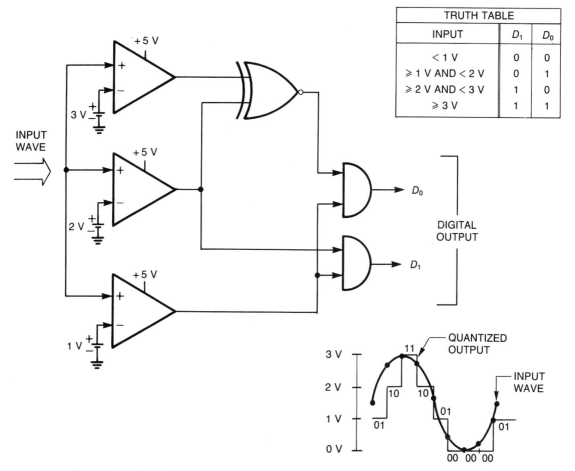

TRUTH TABLE		
INPUT	D_1	D_0
< 1 V	0	0
≥ 1 V AND < 2 V	0	1
≥ 2 V AND < 3 V	1	0
≥ 3 V	1	1

Figure 8-29 PCM encoder.

PCM Demodulator

A PCM demodulator is shown in Figure 8-30. The PCM demodulator uses the 4016 analog switch, AND gate decoders, and a serial-in–parallel-out shift register. The closure of each electrical switch produces a discrete output voltage determined by the values of each of the individual voltage sources (these voltages could also have been achieved by a voltage divider).

The 2-bit PCM signal is fed into the shift register, and the 2-bit ring counter activates the parallel output so that the complete 2-bit PCM code activates the proper AND gate.

The resulting waveform is then passed through a *lowpass* filter where the original modulating wave is restored. In this simplified system, quantizing noise is quite large. The next chapter will present practical A/D (analog-to-digital) and D/A (digital-to-analog) converters along with a practical serial-to-parallel register called the *UART* (universal asynchronous receiver transmitter). The UART can be used to transmit and receive digital data.

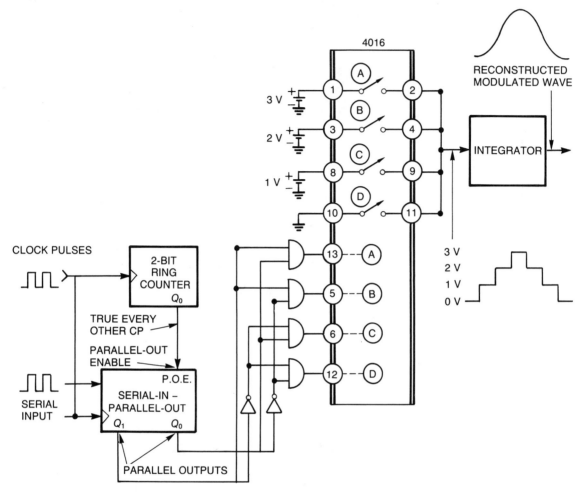

Figure 8-30 PCM decoder.

Companding

Companding is a process used to overcome two difficulties of PCM:

1. The uniform step size in quantization means that weak signals will have more quantizing noise.
2. Signals with a large dynamic range (from very small to very large values) require many encoding bits, which may not be practical for a particular system.

Both problems are illustrated in Figure 8-31.

Companding increases the number of quantization steps for small signals and decreases the number of steps for large signals. The method of companding is shown in Figure 8-32. On the receiving end, a *complementry expander* is used to restore the companded signal.

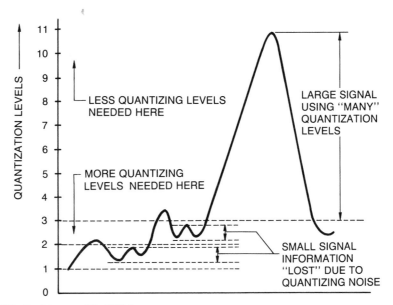

Figure 8-31 Problems with PCM.

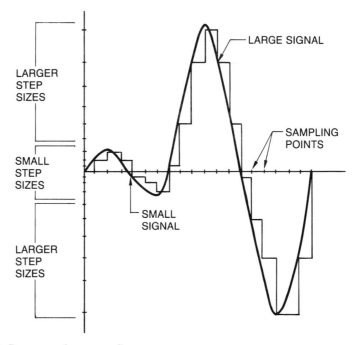

Figure 8-32 Process of companding.

Conclusion

This section explained the concepts of digital modulation. Digital modulation techniques can produce many unique waveforms based upon the original modulating signal. For PCM, the modulating signals can be video or audio. Video and sound synthesizers can be produced, which can take pictures and sounds of "real" objects and transform them into creations limited only by the imagination of the user and the system software.

8-6 Review Questions

1. Describe the main difference between PCM and other forms of pulse modulation.
2. Explain the basic function of a D/A converter.
3. Describe the essence of digital recordings.
4. What does quantizing a waveform mean?
5. Give an example of quantizing a sine wave.
6. What are some advantages of converting a signal into PCM?
7. Describe the action of a flash converter.
8. What is quantizing noise? How can it be reduced?
9. Describe the action of a PCM decoder.
10. What is the purpose of companding? Briefly describe the companding process.

8-7 | DELTA MODULATION

Discussion

Delta modulation, in its simplest form, transmits just 1 bit per sample, and the polarity of the bit indicates if the signal is larger or smaller than the previous sample. See Figure 8-33.

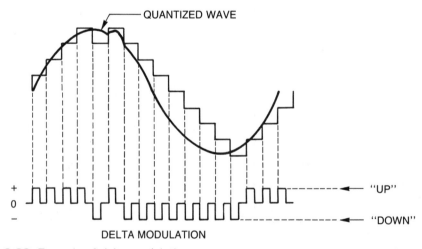

Figure 8-33 Example of delta modulation.

Advantages of Delta Modulation

Delta modulation is a form of PCM. Its advantages are

- Simplified encoding
- Simplified decoding
- Reduction of bits to be transmitted
- Responds more rapidly to fast-changing signals (music and video)

Delta Modulation Transmitter

A simplified delta modulation transmitter is shown in Figure 8-34. It consists of an *up-down* binary counter. When the up input is true, the counter counts up to large numbers during each clock pulse, thus making the output voltage of the digital-to-analog converter (D/A converter) larger. When the down input of the counter is true, it counts toward zero with each clock pulse, thus making the output of the D/A converter smaller. The resulting output of the D/A converter is a *staircase* waveform that follows the modulating signal. The output of the comparator goes to a type D flip-flop. Every time the output of the comparator is zero (meaning that the modulating signal is decreasing), the Q output goes to 0 at the next clock. Every time the output of the comparator is +5 V (meaning that the modulating signal is increasing), the Q output of the flip-flop goes to a +5 (or 1). The output code is shown in Figure 8-34.

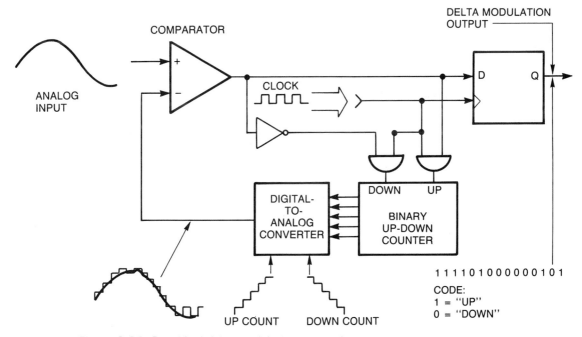

Figure 8-34 Simplified delta modulation transmitter.

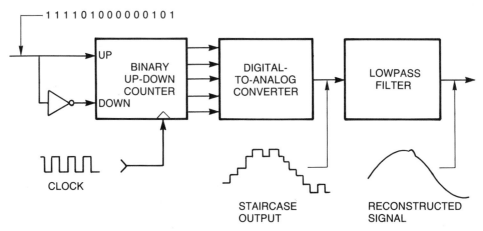

Figure 8-35 Delta modulation decoder.

Delta Modulation Decoding

A simplified delta modulation decoder is shown in Figure 8-35. Note that the incoming signal operates an up-down binary counter. The output of the delta modulation decoder is a pulsed reconstruction of the original modulating signal. This signal is then sent through a lowpass filter to restore the modulating signal.

Conclusion

Delta modulation is easier to develop and decode. This concludes the different kinds of pulse modulating systems.

8-7 Review Questions

1. Explain delta modulation.
2. What are some advantages of delta modulation?
3. Describe an up-down counter.
4. How does a staircase generator work? What causes the staircase waveform to change direction?
5. Explain how a delta modulation transmitter functions.
6. Describe the action of a delta modulation decoder.

8-8 FREQUENCY-DIVISION MULTIPLEXING

Discussion

Another form of multiplexing is *frequency-division multiplexing* (FDM), which, like time-division multiplexing (TDM) is used to transmit information from several different sources. The main difference between frequency-division multiplexing (FDM) and time-division multiplexing (TDM) is that in FDM information about each modulating signal is always being transmitted (just at a different frequency). In TDM, the same transmission frequency is used, but the information about each modulating signal is not continuously transmitted.

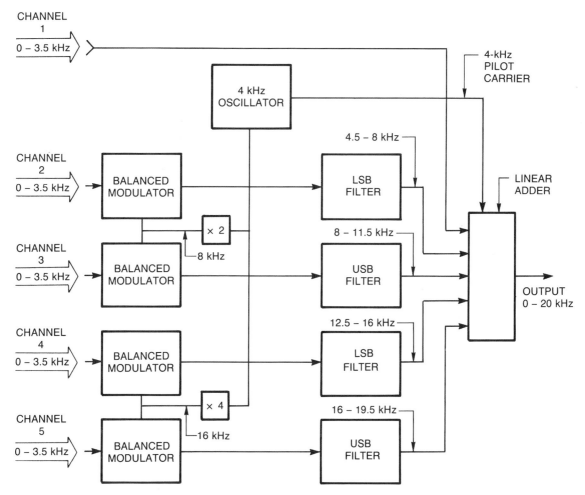

Figure 8-36 A simplified FDM system.

Simplified FDM System

Figure 8-36 shows a simplified frequency-division multiplexing (FDM) system with a five-channel FDM transmitter. Each channel input represents a telephone, and the output is a single cable. The purpose of this system is to allow five separate telephones to transmit voice over the same cable. Here is how it's done.

1. Channel 1 is transmitted as is. Its frequency spectrum is from 0 to 3.5 kHz.
2. A pilot carrier oscillator of 4 kHz drives a ×2 and a ×4 frequency multiplier. This produces carrier frequencies of 8 kHz and 16 kHz.
3. Channel 2 is fed into a balanced modulator along with the 8-kHz carrier. The output is fed into a *lower sideband* (LSB) filter, resulting in an output frequency range of 4.5 kHz to 8 kHz for channel 2.

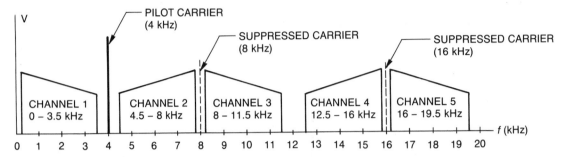

Figure 8-37 Resulting frequency spectrum of five-channel FDM transmitter.

4. Channel 3 is also fed into a balanced modulator with the same 8-kHz carrier as channel 2. However, this time the output signal goes to an *upper sideband* (USB) filter, and the output frequency range for channel 3 becomes 8 kHz to 11.5 kHz.

5. The same process is repeated with channels 4 and 5, except that the suppressed-carrier signal is now 16 kHz instead of 8 kHz. The use of a lower sideband filter for channel 4 and an upper sideband filter for channel 5 produces the two distinct ranges of frequencies for each channel.

6. All of the resulting frequencies are combined in a *linear adder,* which combines the signals without distortion. The resulting bandwidth is from 0 to 20 kHz.

The goal has now been reached. Five different telephone voice channels can now be carried on a single cable! The resulting frequency spectrum is shown in Figure 8-37. Note the location of the pilot carrier. Also note that each channel has its own separate range of frequencies and ideally will not interfere with the adjacent channel.

FDM Receiver

The five-channel receiver for the FDM signal is shown in Figure 8-38. The operation of the receiver is as follows:

1. First, each channel is separated by bandpass filters; this includes the separation of the pilot carrier.

2. The separate pilot subcarrier is used to synchronize a 4-kHz oscillator, which in turn is fed to frequency multipliers to produce the 8-kHz and 16-kHz carriers.

3. Channel 1 needs no further frequency reduction.

4. Channel 2 is fed into a balanced modulator. The output is filtered to produce the original audio signal. The same modulating is done with channel 3.

5. Channels 4 and 5 are reconstructed in the same manner using the 16-MHz signal in their respective balanced modulators. Thus, the original audio frequencies of these two channels are also restored.

This system has completely restored the original voice modulation and separated it into five distinct receivers. The important point in FDM is that all five signals were continuously sent and received.

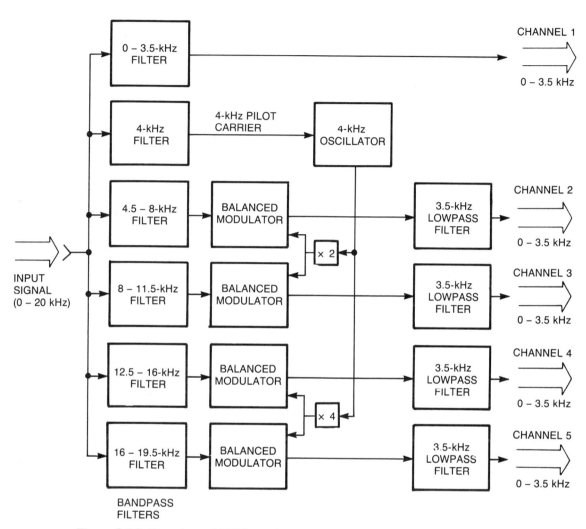

Figure 8-38 Five-channel FDM receiver.

Disadvantage of FDM

Frequency-division multiplexing requires very linear amplifiers (do not introduce any distortion). The reason is that distortion generates harmonics, and these higher frequencies will interfere with adjacent voice channels.

If harmonics are produced by channel 1, they could interfere with the other channels. This kind of interference is not a factor with time-division multiplexing (TDM) because only one channel at a time is transmitted.

Conclusion

Frequency-division multiplexing was first introduced in Chapter 2. It was used to explain how one radio station could be distinguished from another. In this section, you saw how FDM could be applied to telephone communications. You will come across this technique again in fiber optics and laser transmission.

8-8 Review Questions

1. Explain the difference between FDM and TDM.
2. Define channel as it applies to FDM.
3. What factors must be considered if the number of channels in an FDM system is increased?
4. What is the function of the pilot carrier in the FDM scheme explained in this chapter?
5. State an advantage of using balanced modulators in FDM.
6. In the FDM system in this section, how is the carrier reinserted in the receiver?

TROUBLESHOOTING AND INSTRUMENTATION

Pulse Measurements

There is a difference between an *ideal* pulse and a real pulse. The pulses presented in this chapter were assumed to be ideal; that is, they looked like perfect square waves. In communications, it is often necessary to make measurements with real pulses that will be seen on an oscilloscope. These measurements will tell you how close the measured pulse comes to the ideal.

Real-Pulse Measurements

Figure 8-39 shows a real pulse along with measurements of various properties of that pulse. The definition of each pulse measurement is as follows:

1. *Preshoot*
   A change of amplitude of the opposite polarity that precedes the pulse.
2. *Rise Time*
   The amount of time it takes for the pulse to go from 10% of its maximum value to 90% of its maximum value.
3. *Nonlinearity*
   Any variation from a straight line drawn from the 10% to 90% amplitude points.
4. *Overshoot and Rounding*
   Changes that occur in the pulse after the initial transition.
5. *Settling Time*
   The time it takes for the pulse to reach its maximum amplitude (not counting overshoot).
6. *Ringing*
   Amplitude changes that follow overshooting. They are usually a series of dampened sine waves.

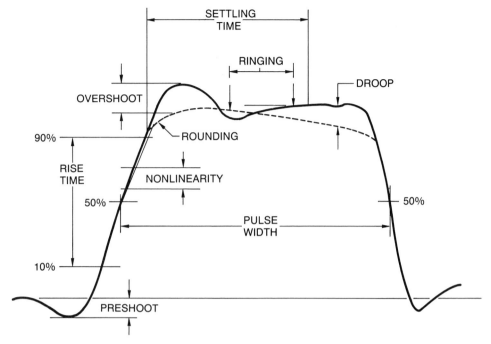

Figure 8-39 Pulse measurements.

7. *Droop*

A decrease in the pulse amplitude with time.

8. *Pulse Width*

The amount of time between the 50% amplitude points of the pulse.

These pulse measurements are important in data communications devices.

Pulse Spectrum

The frequency spectrum of a pulse changes dramatically as the pulse width or pulse repetition rate changes. See Figure 8-40. Observe that as the width of the pulses decreases, the spectrum changes. In the limit, when the width of the pulse is zero, it is no longer a pulse but an impulse. This condition is impossible, since it implies zero rise time. Note that the spectrum of an impulse consists of harmonics of equal amplitude and infinite bandwidth, which implies that an impulse needs infinite power—another reason why a pulse of zero width is impossible.

From the figure, observe what happens as the duration between pulses increases. In the limit, when the period approaches infinity, the spectral components become indistinguishable from one another and the amplitudes merge into an almost continuous wave.

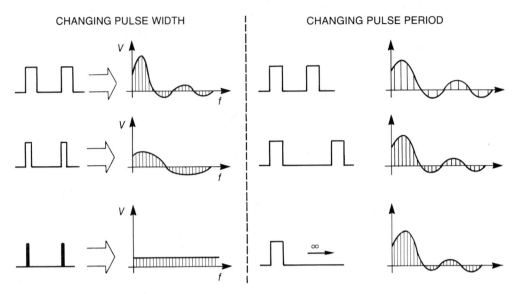

CHANGING PULSE WIDTH   CHANGING PULSE PERIOD

Figure 8-40 Spectrum changes due to pulse variations.

Conclusion

The actual shape of a pulse and its spectral components are important considerations in pulse communications. In this section, you were introduced to actual pulse measurement considerations as well as bandwidth requirements.

8-9 Review Questions

1. What is an ideal pulse? Does it exist?
2. Name some factors that characterise a real pulse.
3. How is the rise time of a pulse measured?
4. Explain how pulse width is measured.
5. Describe the changes in the pulse spectrum as the pulse duration is decreased. What happens in the limit (when the pulse becomes an impulse)?
6. Describe the changes in the pulse spectrum as the period between pulses is increased. What happens in the limit (pulse period approaches infinity)?

MICROCOMPUTER SIMULATION

Troubleshooting simulation number eight on your student disk helps you develop skills in performing gain measurements of multi-stage amplifiers. In this simulation, you are required to first measure the overall system gain using an RF type voltmeter. From this measurement, you can calculate the overall system gain. If this is not within specification, then you must determine which stage is causing the problem.

CHAPTER PROBLEMS

(Answers to odd-numbered problems appear at the end of the text.)

1. Explain how pulse communications is different from other forms of communications presented thus far.
2. Define *sampling* as it applies to getting information. Give an example.
3. Name some advantages and disadvantages of sampling.
4. What pulse parameters can be modulated by a modulating signal?
5. Explain the difference between analog and digital pulse modulation.
6. What are the three methods of analog pulse modulation?
7. Sketch a sawtooth wave and show how this waveform can be represented using (A) PAM, (B) PDM, (C) PPM.
8. Sketch a square wave and show how this waveform can be represented using (A) PAM, (B) PDM, (C) PPM.
9. Explain the difference between TDM and FDM.
10. What are some of the advantages of sampling a modulating waveform? What must be considered about the sampling frequency?
11. State the sampling theorem. What will happen if the sampling rate is less than that given by this theorem.
12. Describe aliasing distortion. Give an example.
13. What is the minimum sampling frequency for a modulating signal of 10 kHz?
14. Derive the minimum sampling frequency for a modulating signal of 15 kHz.
15. Describe the two different types of PAM.
16. Draw the schematic of a simple circuit that would produce the two different kinds of PAM.
17. Explain PTM.
18. What are the three basic forms of PTM?
19. Why is a practical sampling rate greater than the minimum allowed value?
20. Sketch the resulting spectrum of a sampling rate that is
    (A) equal to the minimum sampling rate.
    (B) greater than the minimum sampling rate.
    (C) less than the minimum sampling rate.
21. Sketch the schematic of a simple PAM demodulator.
22. Name some disadvantages of PAM.
23. Give an example of an analog bilateral switch.
24. How many equivalent switches are in the 4016 analog switch?
25. What controls the switches in the 4016 analog switch?
26. Sketch the schematic of a simple PAM modulator using an analog switch.
27. Sketch the schematic of a sampling system transmitter that samples two channels. Be sure to show all of the required decoding circuits.
28. Sketch the schematic of a sampling system receiver for two channels. Show all of the required decoding circuits.
29. What is PDM? How is it different from PAM?
30. What are the three different types of PDM?
31. What is an advantage of PDM over PAM?

32. What kind of pulse modulation compares to (A) AM? (B) FM? (C) PM?
33. Draw the schematic of a comparator. Briefly explain its operation.
34. Sketch the output waveform of the comparator in Figure 8-41(A).
35. Sketch the output waveform of the comparator in Figure 8-41(B).

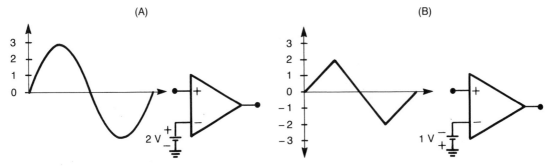

Figure 8-41

36. What are the two output voltage values of a comparator?
37. Define saturation.
38. Draw the schematic of a circuit used to generate PDM.
39. State how leading-edge and trailing-edge PDM are produced.
40. Explain the results of (A) positive clipping and (B) negative clipping.
41. Describe the operation of a ramp generator.
42. How is a ramp generator used in demodulating PDM?
43. Describe the main difference between PDM and PPM.
44. What is a differentiator? Sketch the resulting waveform when a square wave is applied to the input of a differentiator.
45. What role does a differentiator play in the production of PPM?
46. Sketch the schematic of a PPM demodulator. What is the purpose of the reference pulses in PPM?
47. Describe the difference between a quantized sine wave and the original sine wave.
48. For the sine wave in Figure 8-42, sketch the resulting quantized waveform for the indicated sampling points.
49. Sketch the resulting quantized waveform for the sine wave in Figure 8-42 if the number of sampling points is doubled.
50. Sketch the resulting quantized waveform for the sine wave in Figure 8-42 if the number of quantization levels is doubled.
51. List the quantized wave coding for the first half-cycle of the quantized wave in problem 48.
52. List the quantized wave coding for the first half-cycle of the quantized wave in problem 50.

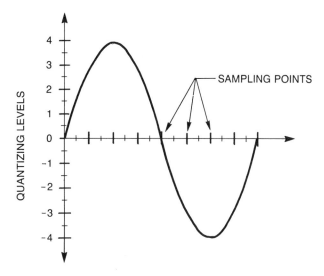

Figure 8-42

53. Describe the action of a D/A converter.
54. What is a flash converter? Sketch the schematic and briefly describe its operation.
55. Explain quantizing noise. How can it be reduced?
56. Explain the companding process. What is its advantage?
57. What is delta modulation? What is its advantage?
58. Sketch a sawtooth wave and show the corresponding delta modulating signal.
59. Define a staircase waveform. Where is it used?
60. Sketch the schematic of a simple delta modulation encoder. Describe the function of each section.
61. Sketch the schematic of a simple delta modulation decoder. Describe the function of each section.
62. What are the two basic types of multiplexing?
63. What is TDM? FDM?
64. What are some advantages of FDM over TDM? What are some of the disadvantages?
65. Sketch the schematic of an FDM system that multiplexes three channels. Each channel has a frequency range of 20 Hz to 2 kHz.
66. Sketch the schematic of an FDM receiver for the multiplexer in problem 65.
67. Sketch the frequency spectrum of the system in problems 65 and 66.
68. Define the following terms as they apply to pulse measurements: (A) pulse width, (B) nonlinearity, (C) settling time, and (D) rise time.
69. Describe what happens to the spectrum of a repetitive pulse as the pulse width decreases.
70. Describe what happens to the spectrum of a repetitive pulse as the pulse repetition rate decreases.

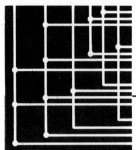

CHAPTER 9

Digital Communication Techniques

OBJECTIVES

In this chapter, you will study:

- [ ] How digital signals are converted into analog signals.
- [ ] The process of synthesizing a specified waveform.
- [ ] How analog signals are converted into digital signals.
- [ ] Different methods of transferring digital information.
- [ ] Standard systems used for interfacing one digital system to another.
- [ ] Basic techniques used in the process of data modulation.
- [ ] Fundamental principles of digital meters.

INTRODUCTION

The revolution in electronic communications is taking place in "digitech." This is the area where digital technologies are transforming telephones, radios, stereos, and almost every electrically operated device we use. This chapter introduces digital communications and its many applications. It is recommended that you refer to Appendix A before beginning this chapter.

9-1 DIGITAL-TO-ANALOG CONVERSION

Discussion

Digital-to-analog (D/A) converters are widely used with computers to digitize an analog signal. With the increasing availability of digital circuits for communication systems, D/A converters are now used in many communication applications. This section will introduce to you a method of converting a digital signal to an analog signal.

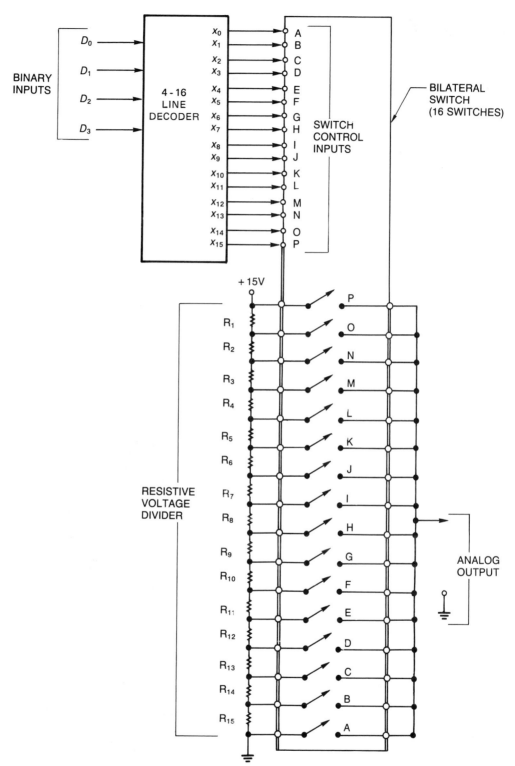

Figure 9-1 Digital-to-analog converter.

Digital Conversion

One method of converting a digital signal into an analog signal is shown in Figure 9-1. The D/A converter consists of the following major sections:

- A 4–16-line decoder.
- A resistive voltage divider. Each resistor drops 1 volt.
- A bilateral switch consisting of 16 switches and 16 control lines.

Figure 9-2 shows the logic circuits contained inside the 4–16-line converter (sometimes called a 1-of-16 decoder). The logic consists of 16 four-input AND gates and four

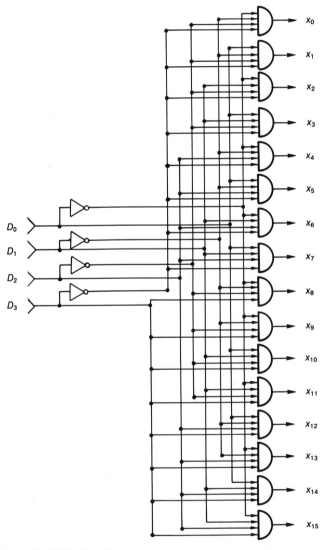

Figure 9-2 Logic for 1-of-16 decoder.

inverters. The conditions of the digital inputs (D_0, D_1, D_2, and D_3) determine which of the 16 outputs (x_0–x_{15}) will be true. Only one output at a time will be true; all others will be false. This relationship is shown in Table 9-1. From the truth table, you can see that the output that is true corresponds to the binary number represented by the digital input. This is used in the D/A converter of Figure 9-1, resulting in one switch closed at a time.

Since the resistive voltage divider is such that the voltage drop across each resistor is 1 volt, the amount of voltage that appears on the *analog output* of the D/A converter is equal to the binary value of the *digital input*. For example, when the binary input is 0101_2 (representing 5_{10}), the sixth line from the top of the 4–16-line decoder x_5 will be true and cause the sixth switch from the bottom (F) of the bilateral switch to close. Thus the output voltage will be 5 VDC.

Table 9-1				1-OF-16 DECODER TRUTH TABLE															
Inputs				**Outputs**															
D_3	D_2	D_1	D_0	x_0	x_1	x_2	x_3	x_4	x_5	x_6	x_7	x_8	x_9	x_{10}	x_{11}	x_{12}	x_{13}	x_{14}	x_{15}
0	0	0	0	1	0	0	0	0	0	0	0	0	0	0	0	0	0	0	0
0	0	0	1	0	1	0	0	0	0	0	0	0	0	0	0	0	0	0	0
0	0	1	0	0	0	1	0	0	0	0	0	0	0	0	0	0	0	0	0
0	0	1	1	0	0	0	1	0	0	0	0	0	0	0	0	0	0	0	0
0	1	0	0	0	0	0	0	1	0	0	0	0	0	0	0	0	0	0	0
0	1	0	1	0	0	0	0	0	1	0	0	0	0	0	0	0	0	0	0
0	1	1	0	0	0	0	0	0	0	1	0	0	0	0	0	0	0	0	0
0	1	1	1	0	0	0	0	0	0	0	1	0	0	0	0	0	0	0	0
1	0	0	0	0	0	0	0	0	0	0	0	1	0	0	0	0	0	0	0
1	0	0	1	0	0	0	0	0	0	0	0	0	1	0	0	0	0	0	0
1	0	1	0	0	0	0	0	0	0	0	0	0	0	1	0	0	0	0	0
1	0	1	1	0	0	0	0	0	0	0	0	0	0	0	1	0	0	0	0
1	1	0	0	0	0	0	0	0	0	0	0	0	0	0	0	1	0	0	0
1	1	0	1	0	0	0	0	0	0	0	0	0	0	0	0	0	1	0	0
1	1	1	0	0	0	0	0	0	0	0	0	0	0	0	0	0	0	1	0
1	1	1	1	0	0	0	0	0	0	0	0	0	0	0	0	0	0	0	1

Practical Decoders

A practical 4–16-line decoder, the 74154, is shown in Figure 9-3. The four binary inputs of the decoder are pins 23 through 20 (internally connected to A, B, C, and D, which represent D_0, D_1, D_2, and D_3, respectively). Pins 18 and 19 are two control lines called G_1 and G_2. These lines must be connected to 0 V before the chip becomes active. These lines are used in systems where more than one decoder will be used and only one line at a time is to be activated. The output lines are pins 1 through 17. For this chip, the selected output line is a logical 0 while all the other output lines are a logical 1. This method was chosen because it works well with practical computer circuits.

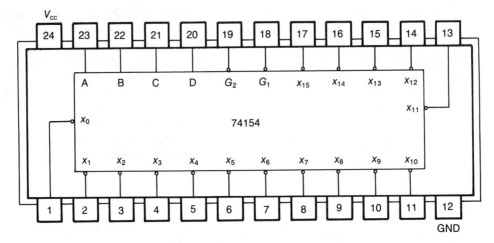

Figure 9-3 74154 4-line to 16-line decoder.

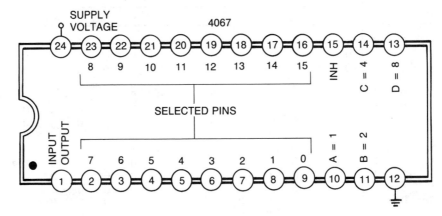

Figure 9-4 4067 1-of-16 analog switch.

Practical Analog Switch

A practical analog switch consisting of 16 switches is shown in Figure 9-4. The 4067 is a 24-pin bilateral switch that can be used in an A/D converter. As shown in Figure 9-4, all of the switches have one end tied together and connected to pin 1. This configuration reduces the number of pins required for the chip while maximizing the number of switches available in the chip. This works out well for the A/D converter, since one end of all the switches must be tied together, Figure 9-1. The chip has a built-in 1-of-16 decoder, and the digital control pins are 10 through 14 (excluding pin 12), labeled A, B, C, and D.

Example 1

If a 4-bit binary counter were connected to the digital inputs of the D/A converter of Figure 9-1, what would the resulting output waveform look like for 2 cycles?

Solution

Connecting the digital inputs to a 4-bit binary counter would cause the following to happen:

D_3 D_2 D_1 D_0	Switch Closed	Output Voltage (Volts)
0 0 0 0	A	0
0 0 0 1	B	1
0 0 1 0	C	2
0 0 1 1	D	3
0 1 0 0	E	4
0 1 0 1	F	5
0 1 1 0	G	6
0 1 1 1	H	7
1 0 0 0	I	8
1 0 0 1	J	9
1 0 1 0	K	10
1 0 1 1	L	11
1 1 0 0	M	12
1 1 0 1	N	13
1 1 1 0	O	14
1 1 1 1	P	15

The output waveform would appear as shown in Figure 9-5. Observe that the wave-form is a *staircase* with each step equal to 1 volt.

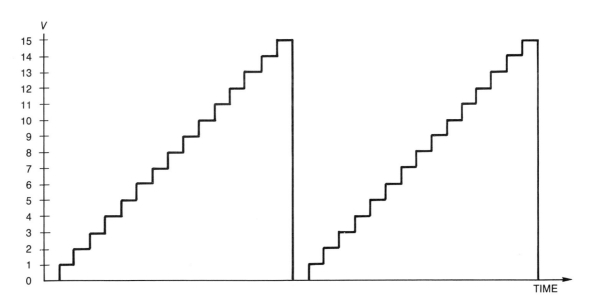

Figure 9-5

Conclusion

You have been introduced to D/A converters. In the next section, you will see a practical application.

9-1 Review Questions

1. State the purpose of a D/A converter.
2. Describe the basic sections of a practical D/A converter.
3. Explain the operation of a 4 to 16-line converter.
4. What is the purpose of the resistive voltage divider used in the D/A converter?
5. Is it possible for more than one switch to be closed in the D/A converter described here? Explain.

9-2 | WAVEFORM SYNTHESIS

Discussion

Waveform synthesis is the process of creating a specified waveform. This process is used in music synthesizers and other applications. Different musical instruments can be distinguished from one another by their inherent waveforms. Even though each instrument may be playing the same "note," the resulting waveform will be different; that is why different musical instruments, such as an oboe or a piano, do not sound alike.

Creating a Complex Waveform

Figure 9-6 shows a circuit that generates a complex waveform. Assume that the purpose of the circuit is to reproduce the sound of a certain note from a guitar. The desired output signal is from the output of the lowpass filter, as shown.

The waveform synthesizer consists of the following circuits:

- A 3-bit binary counter
- A 3 to 8-line decoder
- A diode control *matrix* (this circuit actually determines the shape of the output waveform)
- A bilateral switch consisting of four switches
- A resistive voltage divider, where each resistor drops 1 volt

The action of this circuit is similar to that of the D/A converter presented in the last section. The diode control matrix has been added.

Diode Control Matrix

The diode control matrix determines which of the four bilateral switches will be closed when an output line of the 3 to 8-line decoder is active. In Figure 9-6, when the binary count is 000_2, the x_0 line is +5 V and the diode will conduct, causing a voltage drop across R_1. Note that the only output lines from the decoder connected to the resistors are those connected through a diode. No other lines are connected.

The reason for using the diodes to connect the output of the decoder is shown in Figure 9-7. When x_1 is active, diode D_1 will conduct as shown and is therefore considered to be

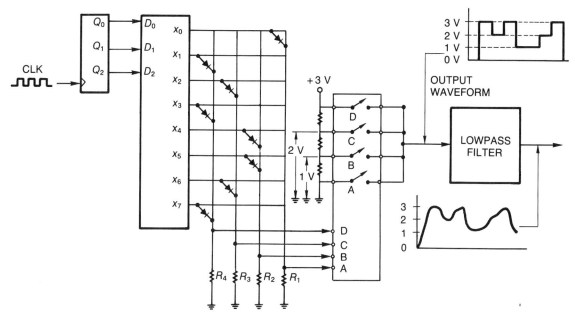

Figure 9-6 Waveform synthesizer.

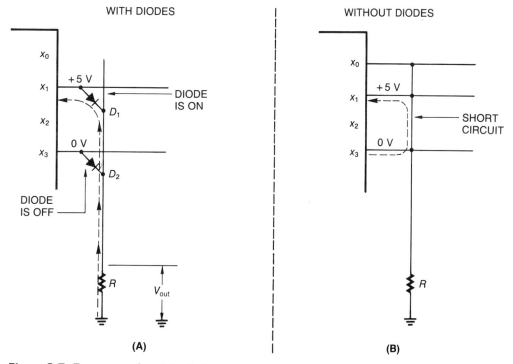

Figure 9-7 Response of matrix diodes.

on. At the same time, diode D_2 will not conduct because there are 0 volts on both sides of it. The reason for the diodes is illustrated in Figure 9-7(B). Without the diodes, the output lines of the decoder would become shorted.

Example 1

Design a waveform synthesizer that will produce the waveform in Figure 9-8.

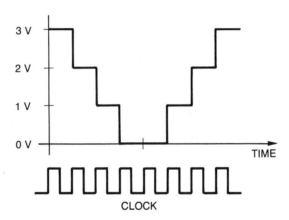

Figure 9-8

Solution

The waveform synthesizer need only have a different arrangement of diodes in its diode matrix. This arrangement with the corresponding output signal is shown in Figure 9-9.

ROM

The diode matrix used in both of the waveform synthesizers is called a *read only memory* (ROM). This name was adopted because the matrix contains a specified "pattern" that will cause a predictable sequence of events to occur. In this case, the predictable sequence was the desired waveform.

When using a *music synthesizer,* the ROM (the diode matrix) that represents the desired signal is selected by the operator. Thus, many different musical instruments may be replicated. Since the controlling signals are digital, a computer can be used to make any conceivable kind of desired waveform. In this manner, musical "instruments" that do not even exist can have their sounds created.

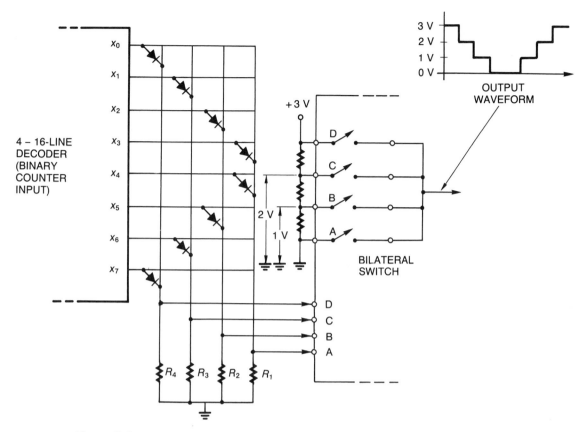

Figure 9-9

Conclusion

This section presented an application of a D/A converter. The use of a waveform synthesizer was presented here to unfold an important concept in communications. The next section introduces analog-to-digital (A/D) converters.

9-2 Review Questions

1. Define waveform synthesis.
2. Explain why one musical instrument sounds different from another even when they are playing the same note.
3. Name the major sections of a waveform synthesizer of the type described in this chapter.
4. Describe the action of a diode control matrix. Why are diodes used?
5. What is a ROM? Describe how they are used in waveform synthesizers.

9-3 | ANALOG-TO-DIGITAL CONVERSION

Discussion

Analog-to-digital (A/D) conversion is the process of converting an analog quantity into a digital quantity. It is necessary when information, such as sound or video, must be processed in digital form. Analog-to-digital converters are used in the creation of digital sound recordings. The music is converted into digital form, then recorded on a laser disk as digital information. In the recording process, the digitized sound can be processed in a computer to remove noise or otherwise enhance or modify the original musical piece. Recorded digital information is reconstructed during playback by D/A converters.

Simultaneous A/D Converter

In the last chapter, you were introduced to a flash converter which converted the analog signal into a 2-bit binary number. This system is limited to only four levels of binary representation ($2^2 = 4$).

A more elaborate simultaneous A/D converter is shown in Figure 9-10. This system uses seven differential amplifiers, a resistive voltage divider, and a priority encoder. A

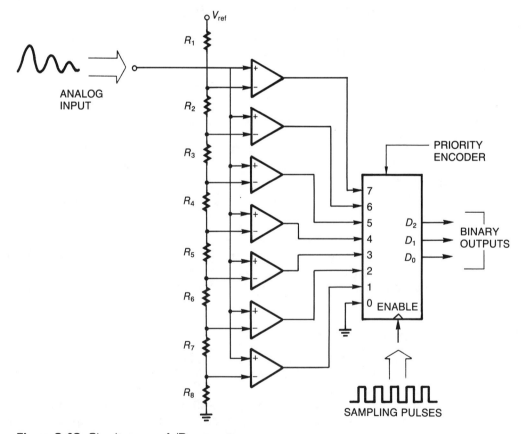

Figure 9-10 Simultaneous A/D converter.

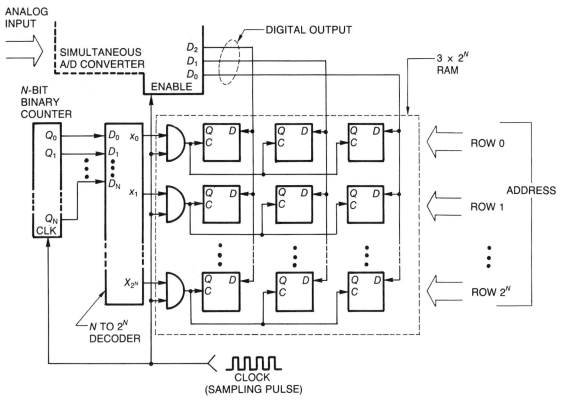

Figure 9-11 Storing a digitized signal in RAM.

priority encoder will have a binary output equal to the largest input value that is true during a *sampling pulse*. The three binary outputs can be transferred to a storage device such as a digital memory.

A/D Converter with RAM

RAM has come to mean *randomly accessible read-write memory*. It consists of groups of flip-flops that can store the result of some digital process. Figure 9-11 shows a 3-bit simultaneous A/D converter digitizing an analog signal and storing the results in RAM. For each sampling pulse, the binary counter is incremented, causing a new row of flip-flops to store the resulting digital output of the A/D converter. The rows of flip-flops are called *addresses,* and the information stored in each row (address) is called *data*. This is collectively referred to as *memory,* and more memory can be used to store the resulting digital code from more signals. After the digital information is stored in memory, it can be transferred to a computer to be processed for various purposes.

Stairstep A/D Converter

Another common method of converting an analog signal to a digital signal is by a *stairstep A/D converter*. See Figure 9-12. This system is sometimes referred to as a *ramp* A/D converter because of the shape of the stairstep waveform.

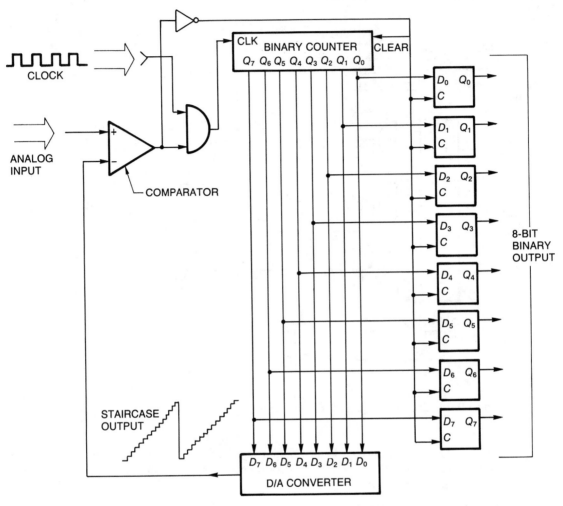

Figure 9-12 Stairstep A/D converter.

The operation of the stairstep A/D converter can be explained as follows:

1. Assume that the binary counter starts at zero ($0000\ 0000_2$), and the output of the comparator is zero.
2. An analog signal voltage is applied to the *analog input* of the comparator. If this voltage is greater than zero, it will cause the output of the comparator to switch to a High output, which in turn enables the AND gate and allows the clock pulses to cause the binary counter to begin counting.
3. The binary counter count causes a stairstep output from the D/A converter. This is fed to the ($-$) input of the comparator.
4. The counter continues its count until the stairstep voltage is larger than the input analog signal. As soon as this happens, the output of the comparator goes to zero.

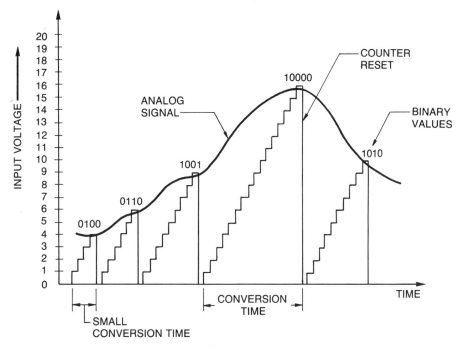

Figure 9-13 Example of stairstep A/D converter operation.

5. When the output of the comparator goes to zero, the AND gate inhibits any more clock pulses from incrementing the binary counter. At the same time, the count of the binary counter is stored in the data flip-flops.
6. The value stored in the data flip-flops now represents the binary value of the analog signal at the time the signal was sampled.

   This process is illustrated in Figure 9-13, which shows that this method is slower than the simultaneous A/D converter, because the larger the value of the input voltage, the longer the binary counter must count. In the worst case, the counter would have to go through 256 counts. For each sample of the input signal, the counter must start from zero up to the point where the stairstep reference voltage reaches a value larger than the input analog signal. Hence, the *conversion time* depends on the value of the analog signal.

Conclusion

This section introduced the basic concepts of analog-to-digital conversion. Some very important applications were mentioned, such as digital sound recordings and digital signal processing. In the next section, you will be introduced to methods used in transferring digital information from one device to another.

9-3 Review Questions

1. Briefly describe the process of A/D conversion. Give an application.
2. Describe the operation of a simultaneous A/D converter.
3. How is RAM used with an A/D converter?

4. Describe the operation of a stairstep A/D converter.

5. What is meant by conversion time? What factors determine the conversion time of an A/D converter?

9-4 | TRANSFER OF DIGITAL INFORMATION

Discussion

Digital information can be transferred from one system to another over wire connections or radio waves. This section introduces the transfer of digital signals over wires. Digital information can be transferred over telephone lines from one computer to another. Information can be processed in a computer and retransmitted for storage or user information.

Schmitt Trigger

Figure 9-14 shows a simple transmission line connection for transmitting digital data from one point to another. A detailed discussion of transmission lines will be presented in the next chapter. Because of the effects of a transmission line on a digital signal, the received signal will not appear as a "clean" pulse, as shown in Figure 9-14.

This difficulty is overcome with a device known as a *Schmitt trigger*. The action of a Schmitt trigger is shown in Figure 9-15. Two key values affect the output of the Schmitt trigger: the *upper threshold point* (UTP) and the *lower threshold point* (LTP). These values are never equal. When the input signal reaches the UTP, the output of the Schmitt trigger goes true. When the input signal reaches the LTP, the output of the Schmitt trigger goes false. The result is a clean square wave, free of noise, a replication of the original pulse.

The schematic of a Schmitt trigger along with some of the standard gates that have Schmitt trigger properties is shown in Figure 9-16. Note that the *hysteresis* symbol is used to signify the UTP and LTP action of the Schmitt device.

Digital Transmission Lines

A transmission line consists of two or more conductors that carry electrical signals from one point to another. The most basic kind of digital transmission system is shown in Figure 9-17. Note the line driver and line receiver. The *line driver* prepares the signal for trans-

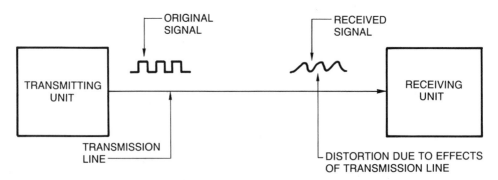

Figure 9-14 Simple digital transmission.

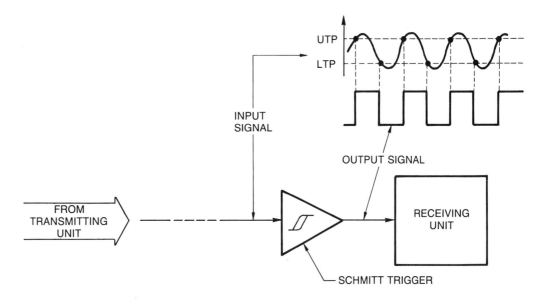

Figure 9-15 Action of a Schmitt trigger.

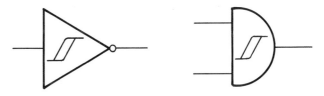

Figure 9-16 Symbols for Schmitt trigger inverter and gate.

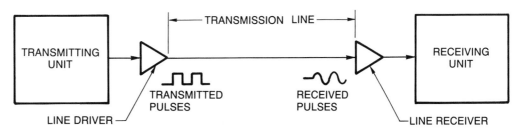

Figure 9-17 Transmission line connecting two digital information systems.

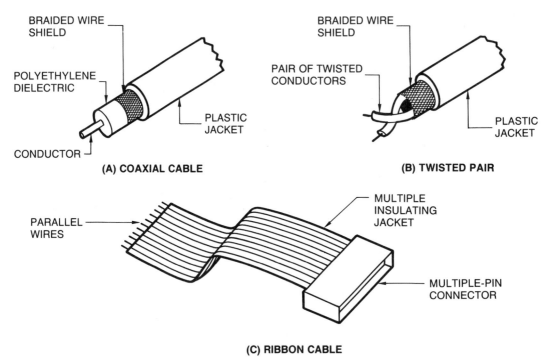

Figure 9-18 Most common forms of digital transmission lines.

mission; the *line receiver* prepares the signal for reception (such as the action of the Schmitt trigger).

The most commonly used transmission lines for digital systems are shown in Figure 9-18. Digital transmission lines consist of *coax cable, ribbon cable,* and the *twisted pair*. Ribbon cable is used for the simultaneous transmission of digital data (such as information from an A/D converter). The advantage of using a ribbon cable is that digital information can be transmitted very quickly, but this kind of transmission is usually limited to distances of 1 meter or less because of transmission line effects on the digital signals. Thus, long-distance digital communications do not use ribbon cables.

Transmission Types

The two different types of digital transmission are *single-ended* and *differential*. Figure 9-19 shows a single-ended digital transmission using a coaxial (coax) cable.

Single-ended transmission requires only a single line and is therefore the simplest form. However, it suffers from noise pickup. The differential form reduces the effects of noise pickup.

Differential transmission of digital information is shown in Figure 9-20. Differential transmission uses a differential line driver and a differential receiver. A *differential line driver* produces two signals, where one signal is the opposite of the other (when one output is true, the other is false). The operation of the *differential receiver* is similar to that of a differential amplifier. The differential receiver will not amplify changes common to both

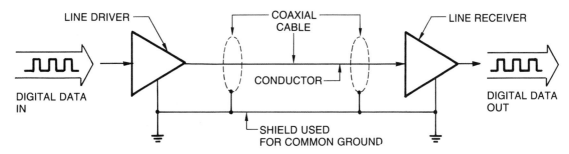

Figure 9-19 Single-ended transmission of digital signals.

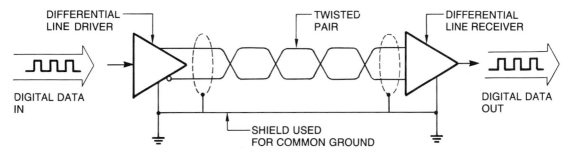

Figure 9-20 Differential transmission of digital signals.

lines (such as noise); this is called *common mode rejection*. Only signals with different polarities between the twisted pair leads, as produced by the differential line driver, will be amplified.

Operational Modes

In the transmission of digital information, there are three modes of operation:

1. Simplex mode
2. Half-duplex mode
3. Full-duplex mode

The *simplex mode* is shown in Figure 9-21. It has data flow in only one direction—from the transmitting device to the receiving device.

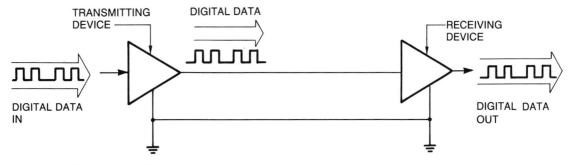

Figure 9-21 Simplex mode of operation.

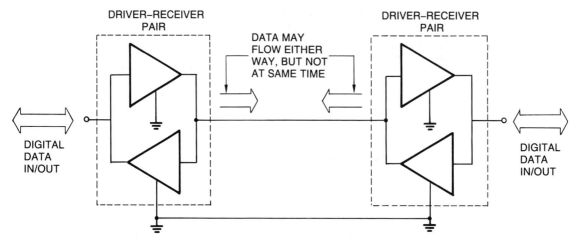

Figure 9-22 Half-duplex mode of operation.

The *half-duplex mode* is shown in Figure 9-22. It allows for two-way communication between two digital systems. However, the two-way communication cannot occur at the same time.

Figure 9-23 shows the *full-duplex mode*. It can be achieved by using FDM (frequency division multiplexing), where both systems may transmit and receive at the same time using different transmission frequencies. The details of FDM were presented in the last chapter.

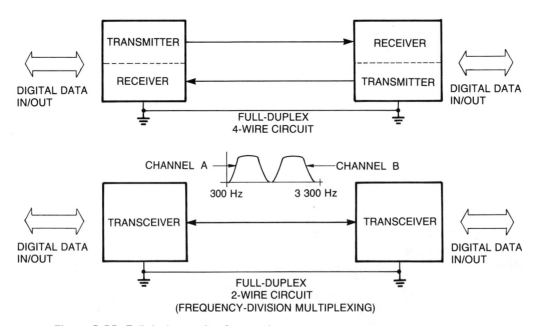

Figure 9-23 Full-duplex mode of operation.

Baud Rate

The *baud rate* is the maximum number of *signals* (more specifically, carrier states) that a device can transmit in 1 second. Thus a system that has a maximum baud rate of 1 kbaud means that it can handle 1 000 signals in 1 second.

Conclusion

This section presented various methods of transmitting digital information over wires. In the next section, you will be introduced to interfacing devices that process the transmitted and received digital signals.

9-4 Review Questions

1. Explain the operation of a Schmitt trigger. Give an application.
2. Define line driver and line receiver.
3. Describe the three most common types of transmission lines used in digital communications. Which one is not used for long-distance connections?
4. Explain the difference between single-ended and differential transmissions.
5. Describe the following modes of operation:
   (A) simplex,   (B) half-duplex,   (C) full-duplex.
6. Define baud rate.

9-5 | DIGITAL INTERFACING

Discussion

As presented in the last section, the most practical type of digital data transmission does not use a ribbon cable. Thus, digital information must be transmitted *serially* rather than in *parallel*. This section introduces the UART and the RS-232C standard for digital interfacing.

Serial and Parallel Data

Figure 9-24 shows the two different methods used to transmit digital information. *Serial transmission,* Figure 9-24(A), consists of transmitting and receiving one bit at a time. *Parallel transmission,* Figure 9-24(B), consists of transmitting and receiving groups of digital bits at the same time. Serial transmission is used for distances of 1 meter or more.

Asynchronous and Synchronous Waveforms

Two waveforms are *asynchronous* if there is no timing relationship between them. Two or more waveforms are *synchronous* if there is a timing relationship between them. Both waveforms are shown in Figure 9-25. Observe from Figure 9-25(A) that there is no timing relationship between the two waveforms; thus they are asynchronous. There is a timing relationship between the two waveforms in Figure 9-25(B); thus they are synchronous. In communications between computers, "asynchronous" means not synchronized with a clock signal.

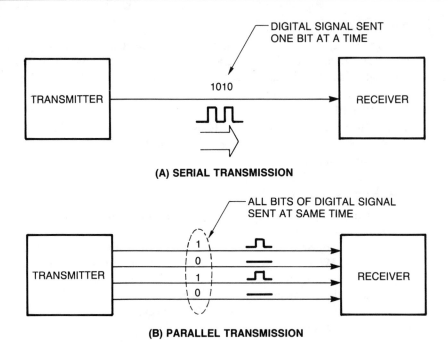

(A) SERIAL TRANSMISSION

(B) PARALLEL TRANSMISSION

Figure 9-24 Serial and parallel transmission.

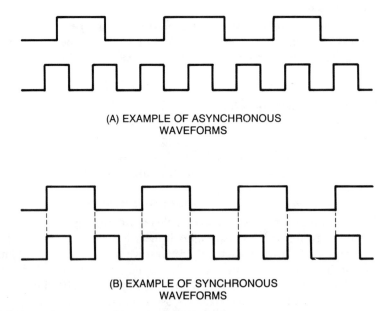

(A) EXAMPLE OF ASYNCHRONOUS
WAVEFORMS

(B) EXAMPLE OF SYNCHRONOUS
WAVEFORMS

Figure 9-25 Asynchronous and synchronous waveforms.

UART

Computers process digital information in parallel, where groups of bits represent words, numbers, or other useful information. Since the practical transmission of digital information over wires requires that this information be transmitted serially, a device at the computer is required to convert parallel information into serial form for transmission. This same device must be capable of converting a received serial transmission into parallel digital information. Such a device is called a universal asynchronous receiver transmitter (UART). The basic idea of a UART is shown in Figure 9-26.

The UART receives data in serial form, then converts it to parallel form. The parallel form of information is then put into the computer where the digital information is to be stored or processed. The UART transmits data in serial form. It takes the information from the computer in its parallel form and converts it into serial form for transmission. The process of receiving and transmitting digital information using the UART is shown in Figure 9-27.

A simplified block diagram of the UART is shown in Figure 9-28. The receiver shift register takes the serial data in and converts it into parallel information where it is stored in the receiving data register. This is in turn connected to internal registers inside the computer called *buffers*.

Parallel data from the computer goes from the buffers to the transmitting data register. This data is then converted to serial form by the transmitter shift register and sent out along the transmission line.

The RS-232C Interface

The most commonly used interface specification for the serial transfer of digital data is the *RS-232C* standard as defined by the Electronics Industries Association (EIA). The RS-232C is used to transfer digital information between computers and communications equipment. An example would be interfacing a computer to a printer or a display screen.

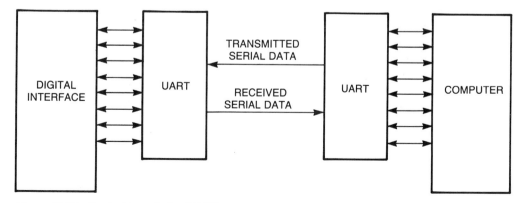

Figure 9-26 Basic idea of the UART.

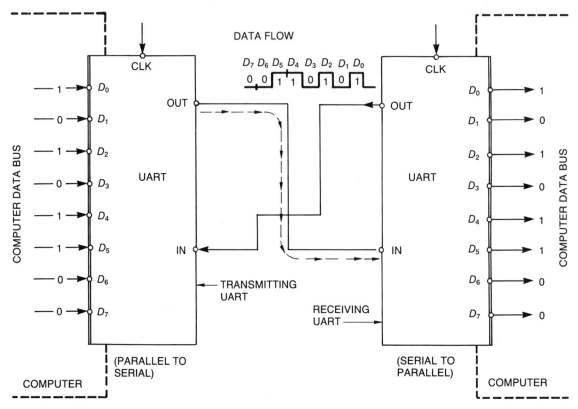

Figure 9-27 Basic operation of the UART.

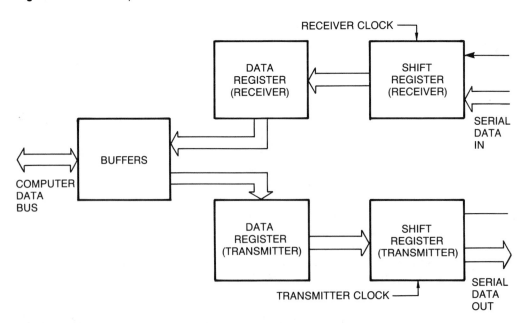

Figure 9-28 Simplified block diagram of the UART.

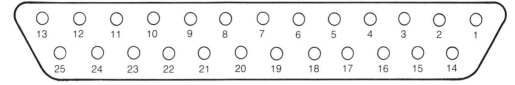

Figure 9-29 Pin diagram of 25-pin connector used for interfacing.

The RS-232C requires a 25-pin connector and specifies which pins of the connector shall carry serial information. See Figure 9-29. The RS-232C signal specifications are shown in Table 9-2.

Technically, the standard classifies all computer equipment and peripherals in one of two categories:

1. Data terminal equipment (DTE): printers and terminals
2. Data communications equipment (DCE): modems

Table 9-2	RS-232C SIGNAL SPECIFICATIONS	
Pin No.	**Type of Signal**	**Purpose of Signal**
1	Ground	Protective ground
2	Data	Transmitted data
3	Data	Received data
4	Control	Request to Send
5	Control	Clear to Send
6	Control	Data Set Ready
7	Ground	Signal ground
8	Control	Received signal detector
9		Reserved for testing
10		Reserved for testing
11		No assignment
12	Control	Secondary received signal Detector
13	Control	Secondary Clear to Send
14	Data	Secondary transmission data
15	Timing	Transmission signal timing
16	Data	Secondary received data
17	Timing	Receiver signal timing
18		No assignment
19	Control	Secondary Request to Send
20	Control	Data Terminal Ready
21	Control	Signal quality detector
22	Control	Ring indicator
23	Control	Data signal rate selector
24	Timing	Transmitting signal timing
25		No assignment

The term "microcomputer" is not in either category because the standard was set before microcomputers became popular. The microcomputer can be viewed as a DTE when interfaced to a modem or as a DCE when connected to a printer. It is important to understand that, from Table 9-2, pin 2 transmitted data is from the DTE to the DCE. With pin 3, received data is from the DCE to the DTE. Thus, the definitions are from the point of view of the DTE.

With the RS-232C interface specifications, there are four types of connecting lines:

1. Data signals
2. Control signals
3. Timing signals
4. Grounds

The voltage levels for control signals are −3 V to −25 V for an on condition and +3 V to +25 V for an off condition. For any piece of equipment using the RS-232C standard, all pins do not have to be used.

The maximum transmission rate permitted is 10 kbaud with a maximum recommended line length of about 15 meters. There is another standard called the *RS-422*, which allows a transmission rate of up to 10 Mbaud.

Conclusion

This section introduced some important concepts used for the interfacing of digital equipment. The electronic communications technician must understand these basic concepts because communication systems rely more and more upon digital processing and storage of information.

9-5 Review Questions

1. State the difference between serial and parallel transmission of digital data. Which is the most practical for long distance?
2. Explain the difference between asynchronous and synchronous waveforms.
3. Describe the basic operation of a UART. What do the initials UART mean?
4. What is meant by the RS-232C standard? How is this standard used?
5. How many types of data lines are used with the RS-232C standard? Name each type.

9-6 DATA MODULATION TECHNIQUES

Discussion

This section uses what you have learned about the three different forms of modulation (amplitude, frequency, and phase) as they apply to digital signals. In previous chapters, you were primarily concerned with continuously changing (analog) waveforms as the modulating signal. This section introduces you to the special requirements of the digital waveform as the modulating signal.

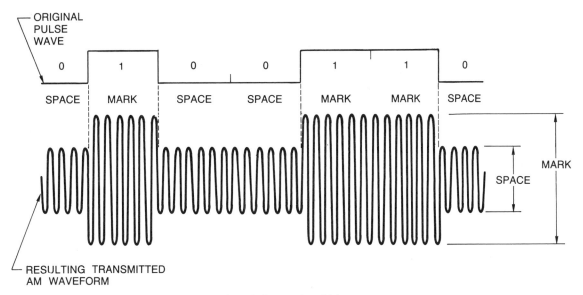

Figure 9-30 Transmitting digital data using AM.

Amplitude Modulation

For digital data, there are only two conditions that must be transmitted: on (called a *Mark*) and off (called a *Space*). Figure 9-30 shows an AM signal used to transmit digital data. A Space has an amplitude lower than that of a Mark. This kind of transmission suffers from amplitude noise.

Frequency Shift Keying

Another method of transmitting digital data uses FM. Since there are only two signals that must be transmitted (a Mark and a Space), the carrier need only change between two frequencies. See Figure 9-31. A frequency deviation method produces a frequency of 1 270 Hz for a Space and 1 070 Hz for a Mark. This method is called *frequency shift keying* (FSK).

Phase Modulation

Figure 9-32 shows an example of *phase modulation* for transmitting digital data. A Mark and a Space are 180° out of phase. This produces a phase shift of 180° whenever there is a transition in the digital information. The problem with this method of transmission is that there is no reference phase transmitted, making the received signal difficult to detect.

Differential Phase Shift Keying

Differential phase shift keying (DPSK) is illustrated in Figure 9-33. This modulation technique effectively transmits a synchronizing pulse along with the digital data. If there is a Mark, the phase of the signal is shifted +90°. If there is a Space, the phase of the signal is shifted −90°. These phase shifts are periodic and can be used to derive the RS-232C Receive Clock signal in synchronous modems.

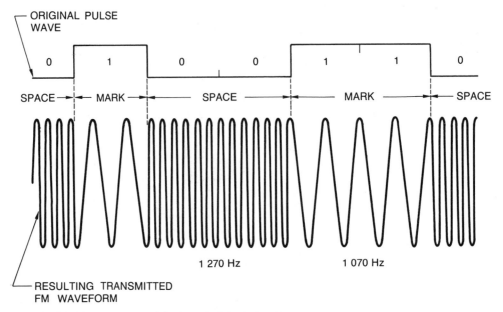

Figure 9-31 Frequency modulation of digital signal.

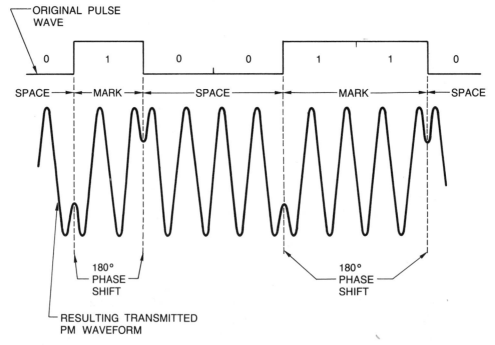

Figure 9-32 Phase modulation for digital signal.

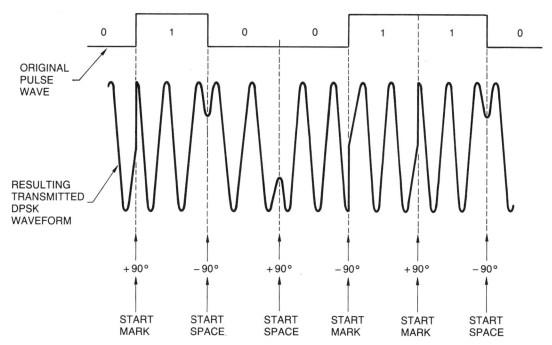

Figure 9-33 Differential phase shift keying.

This method differs from pure phase modulation in that the phase shift must occur at the end of each bit interval—even if the state of the information is not changing (i.e., two consecutive Marks).

Multilevel Modulation

The modulation methods presented in this section have been *digital*. That is, only two possible conditions were transmitted: two levels for AM, two frequencies for FM (FSK), and two phases for PM.

Multilevel modulation of a digital signal allows more than just two modulation levels. Figure 9-34 illustrates a multilevel modulated signal. The signal samples 2 bits at a time and produces four different amplitudes, depending on the binary value of the bits. The tremendous advantage in this is that the number of bits per second between systems can be increased without increasing the baud rate.

Here is how it's done. Recall that the baud rate was defined as the maximum *signal* frequency. You can think of a signal as a time interval during which the parameters of a carrier remain constant (appears as a pure sine wave). If modulated with *pairs* of bits, the minimum *signal* time is twice as great as it would be if modulated by 1 bit. Thus the number of bits per second (BPS) that can now be transmitted is 2 times the baud rate.

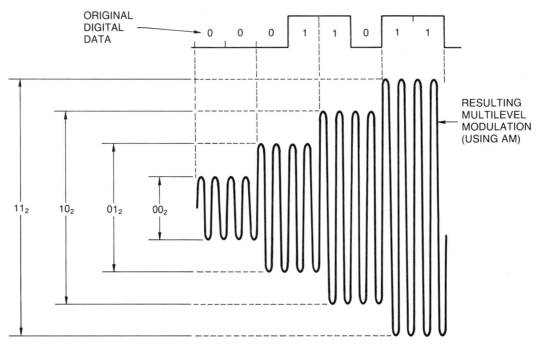

Figure 9-34 Multilevel modulation of digital data.

The general formula is

$$BPS = N(\text{baud rate}) \qquad \textbf{(Equation 9-1)}$$

where   BPS = Bits per second
        N = Bits per signal

Table 9-3 gives the values for the most common signal values used in telephone modems. A *modem* is any interface device that modulates a transmitted signal and demodulates a received signal. Hence, the origin of the word "modem"—*mo* for *mo*dulate and *dem* for *dem*odulate.

Table 9-3	BITS PER SIGNAL VERSUS SIGNAL VALUES
Number of Signal Values	**Bits per Signal**
2	1
4	2
8	3
16	4

Example 1

A differential phase shift keying (DPSK) system uses four different phase shift values of 0°, +90°, +180°, +270°. If this system uses an RS-232C system that is 2 500 bits per second, determine the resulting baud rate of the modulating carrier.

Solution

The relationship between BPS and bits per signal is

$$BPS = N(\text{baud rate}) \qquad \textbf{(Equation 9-1)}$$

Table 9-3 gives the number of bits per signal for a four-signal-valued system as 2. Hence

$$BPS = 2(\text{baud rate})$$
$$2\ 500 = 2(\text{baud rate})$$
$$\text{baud rate} = \frac{2\ 500}{2} = 1\ 250$$

Conclusion

This section introduced you to various modulation techniques for digital information. You saw how the baud rate was related to the modulation technique used for the transmission of digital data.

9-6 Review Questions

1. Define Mark and Space as they apply to the transmission of digital data.
2. Describe the process of FSK.
3. Describe the process of phase modulation for digital information. What is the difficulty with this method of modulation?
4. Explain the process of DPSK. What is the advantage of this kind of modulation?
5. Give an example of multilevel modulation. What is one advantage?
6. Explain the relationship between baud rate and number of bits per second with respect to multilevel modulation.

TROUBLESHOOTING AND INSTRUMENTATION

Digital Meters

Digital meters are quite popular. A digital multimeter is shown in Figure 9-35. A digital multimeter must take an analog quantity and convert it to an equivalent digital readout. This section introduces the essentials of digital meters and then compares them to the conventional analog meters.

Figure 9-35 Typical digital multimeter. *Courtesy* of Simpson Electric Co.

Basic Idea

Figure 9-36 illustrates the basic system used in digital meters. The quantity to be measured must first be converted to a form usable by the A/D converter. Since the quantity to be measured may be current, resistance, or some form of AC, the *signal processor* circuit must convert this input to a DC voltage that is within the measurement range of the A/D converter.

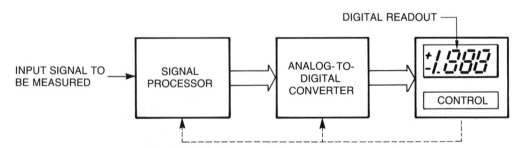

Figure 9-36 Basic digital meter system.

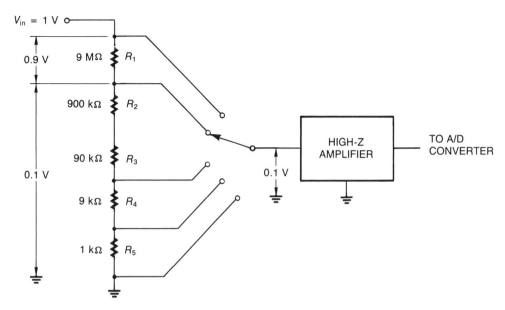

Figure 9-37 Voltage divider network for DVM input.

Signal Processor

The typical A/D converter used in a digital voltmeter (DVM) is capable of reading a DC voltage in the range of 0 V to 0.1 V. This limited voltage range must actually measure a much wider range of DC voltages. A circuit that will provide such a range is shown in Figure 9-37. An input voltage of 1 V is to be measured by the meter. Note that the total input impedance is greater than 10 MΩ. The input impedance to the amplifier is about 100 000 MΩ; thus the position of the *range* switch will not have any noticeable effect on the DVM input impedance.

This system lets the DVM user select the proper DVM range. Thus, if the input signal to be measured causes the input voltage to the amplifier to be greater than 0.1 V, the digital meter will indicate an *overflow,* which tells the user to change the range switch to a higher setting. This means turning the switch toward R_5.

Autoranging

Figure 9-38 illustrates the basic concepts of a system for *autoranging*. If the input signal causes the voltage to the A/D converter to exceed 0.1 V, the comparator will cause the binary counter to increment. This in turn will cause the bilateral switch to automatically change the resistance of the voltage divider until the correct input voltage to the A/D converter is produced.

If the input signal is lower than a given *threshold* level, then the second comparator is activated, causing the binary counter to count down until the correct input is achieved.

Autoranging can be used to measure quantities other than just DC voltages. The quantity need only be converted to a DC voltage.

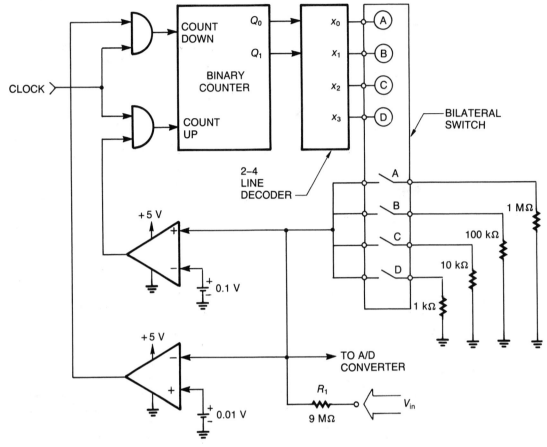

Figure 9-38 Basic autoranging circuit.

True rms Converter

An AC voltage can be converted to a true rms DC voltage by a *thermocouple*. Recall from your study of basic electricity that a thermocouple produces a DC voltage that is proportional to its temperature. See Figure 9-39. The voltage produced by the thermocouple is proportional to the heat produced by the input AC voltage. Thus, the larger the AC voltage, the greater will be the corresponding DC voltage.

Measuring Resistance

For a digital multimeter (DMM) to be a true multimeter, it must be capable of measuring resistance. Since any input quantity to be measured must be converted to a voltage, a system for developing a DC voltage that is proportional to the amount of resistance being

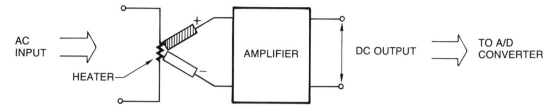

Figure 9-39 Thermocouple input to measure RMS.

measured must be used. Figure 9-40 shows such a basic system. Its basic operation is as follows:

1. The voltage drop across R_x is equal to the product of R_x and the value of the current I_s from the constant-current source (Ohm's law).
2. When the constant-current source is set at 1 mA, the input voltage will be 0 V for 0 Ω and 1 V for 1 kΩ. Any value over 1 kΩ will produce an *overrange* condition.
3. Larger values of resistors are measured by switching the constant-current source to its smaller output current state. This may be done by the operator or an autoranging circuit. This means that the maximum resistance that can be read by the digital ohmmeter is 10 kΩ with the 100-μA source and 1 MΩ with the 1-μA current source.

Comparison of Analog and Digital Meters

Figure 9-41 shows an analog multimeter. Table 9-4 gives a comparison of digital and analog meters.

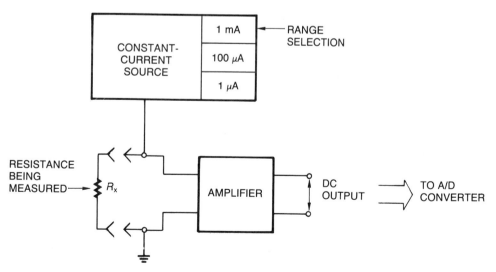

Figure 9-40 System for converting resistance reading to a DC voltage.

Figure 9-41 Typical analog meter. *Courtesy* of Simpson Electric Co.

Table 9-4	COMPARISON OF DIGITAL AND ANALOG METERS
Digital Meters	**Analog Meters**
Display Easy to read. Many displays give the decimal point and the polarity of the reading.	**Display** Must interpret reading on scales that are usually nonlinear. Accurate readings require a great deal of practice and are subject to parallax errors as well as operator reading errors.
Resolution A simple 3-digit display gives a resolution of one part in 1 000, a 4-digit display gives one part in 10 000. These are resolutions of 0.1% and 0.01%, respectively.	**Resolution** At best, an experienced operator can read one part in 100. That is a resolution of 1%.
Reading time An inexpensive DMM makes about 5 readings per second. In many applications, the reading rate exceeds 100 readings per second.	**Reading time** It usually takes about 1 second or more to settle after each reading.
Noise immunity Requires special circuits to reduce input noise interference of the digital reading.	**Noise immunity** Relatively immune to circuit noise. Thus readings are easier in a noisy environment.

Conclusion

Digital meters are playing an increasingly important role in electronic communications. This section introduced you to the basic operating systems of these meters. You also compared the digital meter with the more conventional analog meter movement.

9-7 Review Questions

1. Describe the basic operation of a digital multimeter.
2. What is the purpose of the signal processor circuit used in a DMM?
3. Explain autoranging.
4. Describe the operation of a true rms converter as it applies to a DMM.
5. Explain the process used by a DMM to read resistance.
6. What are some advantages and disadvantages of a DMM compared with a VOM?

MICROCOMPUTER SIMULATION

The ninth troubleshooting simulation introduces the skill of making bandwidth measurements. Here you will use a simulated RF generator and RF voltmeter to assist you in these measurements. By adjusting the frequency of the RF generator that is applied to the input of an amplifier, you will then be able to read the resulting output on the RF voltmeter. From these measurements you will be able to calculate the resulting bandwidth of the amplifier and determine if it is within the stated specifications.

Elements of a Digital Filter

The mathematics involved in the design of digital filters is beyond the scope of this book. The basic idea of a digital filter is to have a mathematical equation that keeps track of previous readings and biases future readings on the average of these past readings.

CHAPTER PROBLEMS

(Answers to odd-numbered problems appear at the end of the text.)

TRUE/FALSE
Answer the following true or false.
1. A D/A converter changes a digital code into an analog waveform.
2. A decoder with three binary inputs can decode a maximum of eight outputs.
3. When a D/A converter is controlled by a binary counter, the resulting output waveform is a square wave.
4. Waveform synthesis is the process of creating a specified waveform.
5. The circuit that actually determines the shape of a synthesized wave in a waveform synthesizer can be a ROM.

MULTIPLE CHOICE

Answer the following by selecting the most correct answer.

6. The output waveform for the circuit in Figure 9-42 is a
   (A) sawtooth.
   (B) staircase.
   (C) triangle.
   (D) pulse.

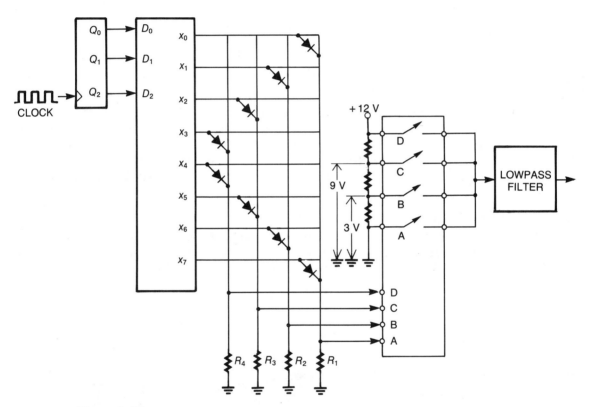

Figure 9-42

7. To change the frequency of the resulting waveform of the circuit in Figure 9-42, you would have to
   (A) increase the number of counts in the binary counter.
   (B) change the repetition rate of the clock pulse.
   (C) change the voltage applied to the resistive voltage divider.
   (D) do none of the above.

8. For the circuit shown in Figure 9-42, reducing the value of the voltage applied to the resistive voltage divider will cause the output waveform to have a
   (A) lower amplitude.
   (B) larger amplitude.
   (C) higher frequency.
   (D) lower frequency.
9. Which diode matrix in Figure 9-43 will produce an up-down staircase?
   (A) A   (B) B   (C) C   (D) D

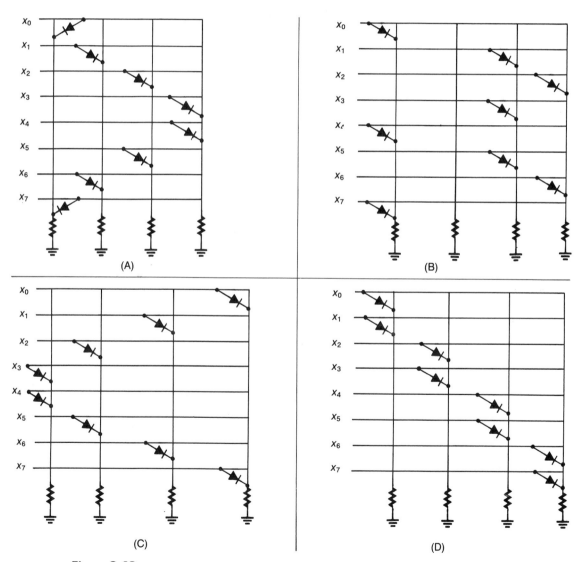

Figure 9-43

MATCHING

Match the most correct figures from Figure 9-44 to the following statements.

10. Half-duplex mode
11. Simplex mode
12. Full-duplex mode
13. Differential transmission
14. Single-ended transmission
15. Schmitt trigger

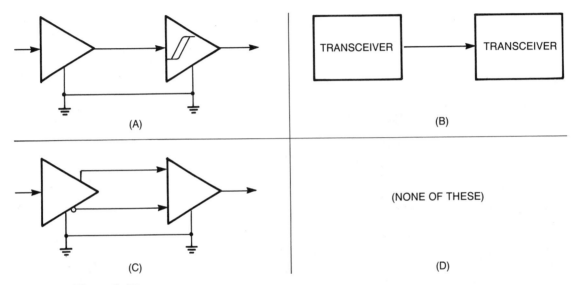

Figure 9-44

FILL IN

Fill in the blanks with the most correct answer(s).

16. _____ transmission consists of transmitting and receiving one bit at a time.
17. Parallel transmission consists of transmitting and receiving _____ bit(s) at a time.
18. Waveforms without any timing relationship between them are _____.
19. _____ waveforms have a timing relationship between them.
20. An interfacing device called the _____ converts parallel digital information into a serial stream for transmission.

OPEN ENDED

Answer the following as indicated.

21. What pins on the RS-232C standard are used for
    (A) data signals?
    (B) control signals?
    (C) timing signals?
    (D) grounds?

22. For the RS-232C standard, what are the voltage values for
    (A) data signals?
    (B) control signals?
    What is the maximum transmission rate?
23. Briefly describe the resulting carrier developed from FSK.
24. Sketch the waveforms resulting from
    (A) FSK
    (B) DPSK
    Describe the major difference.
25. Explain how multilevel modulation of digital data can increase the bit rate without affecting the baud rate.
26. For a DPSK system that uses $+45°$, $+90°$, $+135°$, $+180°$, $+225°$, $+270°$, $+315°$, and $+360°$, determine the baud rate for (A) 2 000 BPS and (B) 4 000 BPS.
27. Describe the basic difference in the appearances of a digital meter and an analog meter.
28. Sketch a basic digital meter measurement system.
29. Describe the operation of a basic autoranging system used for a DMM.
30. Compare the advantages of the DMM to those of a VOM. State some of the disadvantages.

CHAPTER 10

Transmission Lines and Antennas

OBJECTIVES

In this chapter, you will study:

- [ ] Some of the basic concepts of transmission lines.
- [ ] The impedance considerations of transmission lines and how to work with them.
- [ ] The fundamental properties of standing waves as they apply to transmission lines.
- [ ] The basic theory of antennas and some applications.
- [ ] Advantages and disadvantages of various antennas.
- [ ] How radio waves are transmitted through space and the effect of space on their propagation.
- [ ] Impedance-matching considerations when using the RF signal generator.

INTRODUCTION

The transmission of radio energy through wires, antennas, and space is a major part of any communication system. This chapter presents a foundation of the many considerations and phenomena that occur in this part of the system.

10-1 | INTRODUCTION TO TRANSMISSION LINES

Discussion

Electromagnetic waves travel in free space at a rate of 3×10^8 meters per second (this is 186 000 miles per second). When these waves travel in any other medium, they move slower. No medium has yet been found in which they will move faster. If such a medium were discovered, then the basic foundations of Einstein's theories of relativity would be violated.

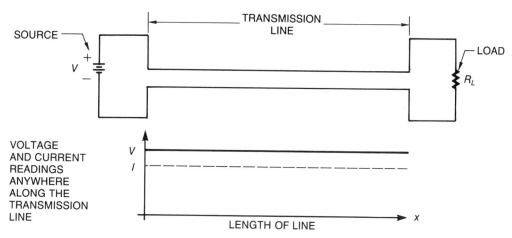

Figure 10-1 Reading DC voltages in a DC transmission line.

When radio frequencies travel in wires, they travel slower. However, as a first approximation, you can assume that they travel at the speed of light. As you will discover in this section, many apparently strange and unusual phenomenon occur when radio frequencies are sent along wires.

Transmission of DC Signals

When DC was used, the time it took for the DC current to reach the load was not a consideration, because in DC, you assume that the voltage and current have been there "forever." The DC voltage anywhere between the two wires making up the circuit will be everywhere the same. See Figure 10-1.

As Figure 10-1 shows, the *transmission line* is simply the wires that connect the source (in this case a battery) to the load (in this case a resistor). You can think of a transmission line as any material structure that forms a continuous path from one place to another that is used for directing the transmission of electromagnetic energy along this path.

From Figure 10-1, note the graph of the voltage and current readings along the length of the line. For an ideal (no resistance) line, the readings will be the same everywhere. Understand what this graph means. It will be used to explain the action of high-frequency signals along the same kind of transmission line. Note that the *x*-axis of the graph represents the *distance, not* the time along the line.

Transmission Time

The time it takes for a signal to travel the length of a transmission line is

$$t_s = \frac{l}{\text{Vel}_s}$$

(Equation 10-1)

where t_s = Time for the signal to travel the given length
l = Length of the line
Vel_s = Velocity of the signal in the transmission line

Example 1

How long will it take for an electromagnetic signal to travel the length of a 100-meter transmission line?

Solution

$$t_s = \frac{l}{\text{Vel}_s}$$ **(Equation 10-1)**

$$t_s = \frac{100 \text{ m}}{3 \times 10^8 \text{ m/s}}$$

$$33.3 \times 10^{-8} \text{ s} = 333 \text{ ns}$$

Pulse Transmission

Consider the case when the source is a pulse generator as shown in Figure 10-2. The length of this transmission line is 100 meters. As shown in Example 1, it takes 333 ns for a signal to travel from the source to the load. If the pulse is less than 333 ns, then the source will have returned to zero before the leading edge of the pulse has reached the load! This means that the transmission line itself is capable of storing as well as transmitting energy. Therefore, a transmission line can be thought of as having the characteristics of an actual circuit. These characteristics are shown in Figure 10-3.

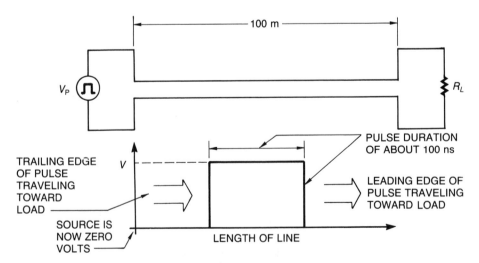

Figure 10-2 Using a pulse generator as the signal source.

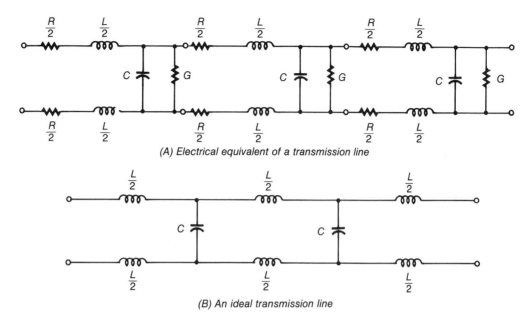

(A) Electrical equivalent of a transmission line

(B) An ideal transmission line

Figure 10-3 Electrical characteristics of a transmission line.

Capacitance exists between the transmission lines because the signals in the wires are 180° out of phase, thus creating an *electrostatic field* that can be represented by the parallel capacitors. The transmission line can be represented as inductors in series because whenever current flows in a conductor, a magnetic field is produced around the wire. Series resistance exists because of the resistance of the wire. The shunt conductance is due to the leakage between the wires.

At very high frequencies, the inductive reactance becomes much larger than the series resistance and the capacitive susceptance becomes much larger than the shunt conductance. Thus the transmission line can be represented as "ideal," as shown in Figure 10-3(B).

A phenomenon called the *skin effect* is illustrated in Figure 10-4. Because of the skin effect, most current flow takes place along the surface of the wire at very high frequencies. Little, if any, current flows inside the wire. This is not true for low frequencies or DC because the magnetic fields created by the currents are not changing rapidly enough to offer much opposition near the center of the wire.

Characteristic Impedance

There are two types of transmission lines: balanced and unbalanced. In a *balanced* transmission line both wires carry RF energy and the current in one wire is 180° out of phase with the current in the other wire. In an *unbalanced* transmission line one wire is at "ground" potential while the other wire carries the RF energy.

One of the most familiar types of transmission lines is the TV twin-lead type, Figure 10-5. The TV twin lead is a balanced transmission line. The *characteristic impedance* of any transmission line is the impedance that a theoretically infinite length of this line would

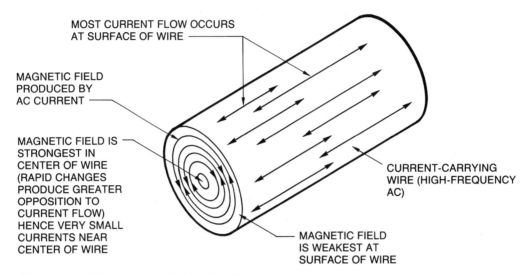

MOST CURRENT FLOW OCCURS AT SURFACE OF WIRE

MAGNETIC FIELD PRODUCED BY AC CURRENT

MAGNETIC FIELD IS STRONGEST IN CENTER OF WIRE (RAPID CHANGES PRODUCE GREATER OPPOSITION TO CURRENT FLOW) HENCE VERY SMALL CURRENTS NEAR CENTER OF WIRE

CURRENT-CARRYING WIRE (HIGH-FREQUENCY AC)

MAGNETIC FIELD IS WEAKEST AT SURFACE OF WIRE

Figure 10-4 Phenomenon of the skin effect.

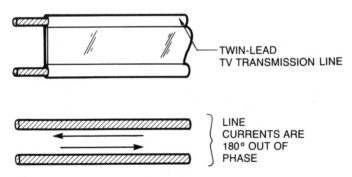

TWIN-LEAD TV TRANSMISSION LINE

LINE CURRENTS ARE 180° OUT OF PHASE

Figure 10-5 TV twin-lead transmission line.

present at its input (source) end. The characteristic impedance of a balanced transmission line is

$$Z_0 = 276 \log_{10}\left(\frac{d}{r}\right)$$

(Equation 10-2)

where Z_0 = Characteristic impedance of the line

$\log_{10}$ = Logarithm to the base 10

d = Distance of the conductor from center to center

r = Radius of the conductors

Example 2

What is the characteristic impedance of a standard TV lead-in for the dimensions shown in Figure 10-6?

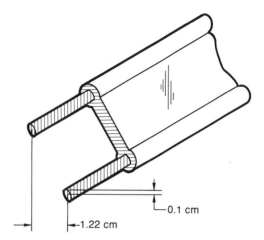

0.1 cm

1.22 cm

Figure 10-6 Dimensions of balanced TV transmission line.

Solution

$$Z_0 = 276 \log_{10}\left(\frac{d}{r}\right)$$

(Equation 10-2)

$$Z_0 = 276 \log_{10}\left(\frac{1.22}{0.1}\right)$$

$$Z_0 = 276 \log_{10}(12.2)$$

$$Z_0 = 276 \, (1.086) = 300 \, \Omega$$

The characteristic impedances of some common transmission lines are shown in Table 10-1.

Table 10-1	CHARACTERISTIC IMPEDANCE OF TYPICAL TRANSMISSION LINES	
Type of Line Number	**Description of Transmission Line**	**Characteristic Impedance (Ω)**
RG 8/U	Coaxial cable	52
RG11 A/U	Coaxial cable	75
214-056	Twin lead (common TV lead-in)	300

Deriving Characteristic Impedance

The relationship between a unit-length capacitance and unit-length inductance to the characteristic impedance of a transmission line is

$$Z_0 = \sqrt{\frac{L}{C}}$$

(Equation 10-3)

where Z_0 = Characteristic impedance of the transmission line in ohms
L = Inductance per unit length
C = Capacitance per unit length

Example 3

Determine the characteristic impedance of a transmission line with an inductance of 0.2 μH per foot and a capacitance of 40 pF per foot.

Solution

$$Z_0 = \sqrt{\frac{L}{C}}$$

(Equation 10-3)

$$Z_0 = \sqrt{\frac{0.2 \ \mu H/ft}{40 \ pF/ft}}$$

$$Z_0 = \sqrt{\frac{200 \times 10^{-9}}{40 \times 10^{-12}}}$$

$$Z_0 = \sqrt{5 \times 10^3} = 70.7 \ \Omega$$

Conclusion

This section introduced the basic ideas of transmission line theory. The next section introduces some important transmission line considerations that are essential to make communication systems function properly.

10-1 Review Questions

1. How fast do electromagnetic waves travel in free space? Can they travel slower or faster? Explain.
2. Describe a transmission line. What are the current and voltage characteristics along the length of a transmission line with a single DC source?
3. Explain what happens when a pulse is sent along a transmission line. Do the leading and trailing edges of the pulse arrive at the load at the same time?
4. Describe the equivalent circuit of a transmission line.

5. Explain what is meant by the characteristic impedance of a transmission line. How long does this transmission line have to be?
6. State the difference between an unbalanced transmission line and a balanced transmission line. Give an example of each.

| 10-2 | STANDING WAVES

Discussion

When a transmission line has infinite length, all of the power transmitted by the source is equal to all of the power consumed by the load. Obviously it isn't practical to have transmission lines of infinite length. When transmission lines are of some practical length, special considerations must be made if all of the source power is to be consumed by the load. This section concerns itself with these important transmission line characteristics.

Line Length

Figure 10-7 shows an infinite transmission line and a finite transmission line terminated in a specific resistive load. Figure 10-7(A) shows a 300-Ω transmission line of infinite length. All of the source energy is delivered to the load. In Figure 10-7(B), the transmission line has been cut, but its end has been *terminated* with a 300-Ω resistor. Again all of the source energy will be transmitted to the load (which is now the 300-Ω resistor) because the transmission line still "sees" the same 300-Ω resistance as if the rest of it were still infinitely long.

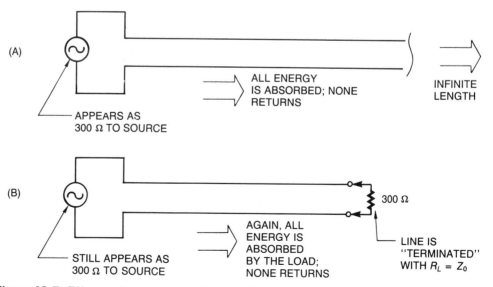

Figure 10-7 Effects of transmission line termination.

Standing Waves

Whenever a finite transmission line is terminated in a load that is different from its characteristic impedance (its impedance when it is infinitely long), then not all of the power delivered by the source is consumed by the load. Some of this power is *reflected* back from the load to the source.

Thus there are *two* waves created: the original transmitted wave, sometimes called the *incident* wave, and the *reflected wave*. The phase relationships of these two waves produce *standing waves*. Standing waves are the distribution of voltage and current on a transmission line formed by two sets of waves traveling in opposite directions and characterized by the points of successive maxima and minima along the line.

Open Transmission Line

Consider a finite transmission line that is open, Figure 10-8. The voltage across the open end of the transmission line must be a maximum while the current must be zero. This is what you would expect from any open circuit. However, what is not expected is the wave pattern seen along the transmission line. At every half-wavelength, the voltage is a maximum while the current is zero. At every quarter-wavelength, the values of the current and voltage alternate from maximum to zero. This occurs because the load is not equal to the characteristic impedance of the line (Z_0). What is observed are *standing waves*. These two sets of waves, called *incident* and *reflected* waves, are shown in Figure 10-9.

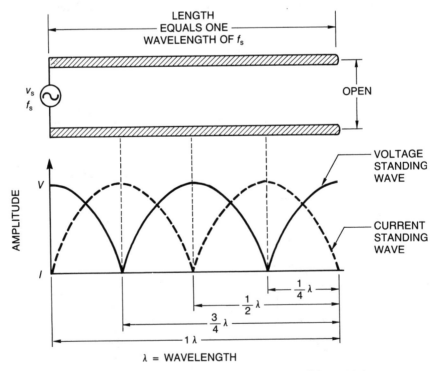

Figure 10-8 Standing-wave pattern of an open transmission line.

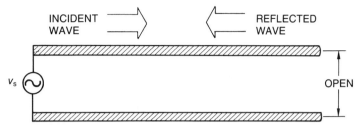

Figure 10-9 Incident and reflected waves.

Shorted Transmission Line

The standing-wave pattern of a *shorted transmission line* is shown in Figure 10-10. As you would expect, the current is a maximum at the short while the voltage is zero. Note again that the voltage and current values repeat themselves every half-wavelength and alternate every quarter-wavelength.

Since the standing-wave patterns for a shorted transmission line are different from those of an open transmission line, you can correctly conclude that the value of the load does determine the standing-wave pattern. Since the standing-wave pattern is the result of

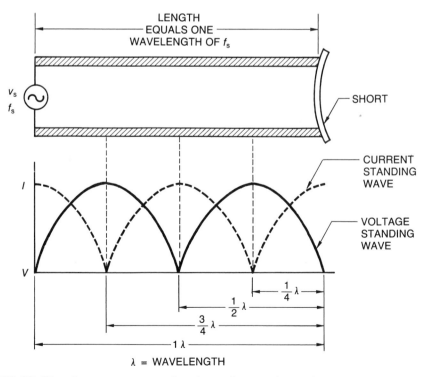

Figure 10-10 Standing-wave pattern for shorted transmission line.

the sum of two waves (the incident and the reflected), and since there is no reflected wave when the load equals the characteristic impedance of the transmission line, there are no standing waves when $Z_0 = Z_L$.

Standing-Wave Ratio

The optimal condition for transmitting power to a load over a transmission line occurs when there is no standing wave—when $Z_0 = Z_L$. The maximum voltage and minimum voltage along the line are equal, and the maximum current and minimum current along the line are also equal. This ideal condition is shown in Figure 10-11.

When Z_0 does not equal Z_L, standing waves that can be measured are produced. This measurement is called the *standing-wave ratio* (SWR). The standing-wave ratio for current is called the ISWR. The standing-wave ratio for voltage is called the VSWR. In symbols,

$$\text{VSWR} = V_{\text{rms max}} : V_{\text{rms min}}$$ **(Equation 10-4)**

where   VSWR = Voltage standing-wave ratio
$V_{\text{rms max}}$ = Maximum value of voltage standing wave
$V_{\text{rms min}}$ = Minimum value of voltage standing wave

and

$$\text{ISWR} = I_{\text{rms max}} : I_{\text{rms min}}$$ **(Equation 10-5)**

where   ISWR = Current standing-wave ratio
$I_{\text{rms max}}$ = Maximum value of current standing wave
$I_{\text{rms min}}$ = Minimum value of current standing wave

It can be shown that the standing-wave ratios, ISWR and VSWR, are equal for the same transmission line.

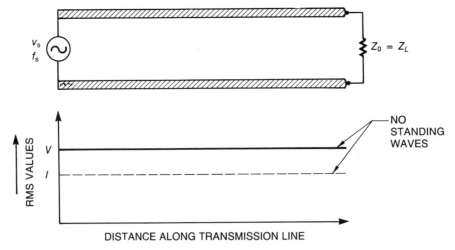

Figure 10-11 Voltage and current relations when $Z_0 = Z_L$.

A very useful relation for transmission lines is

$$SWR = Z_L : Z_0 \qquad \textbf{(Equation 10-6A)}$$

or

$$SWR = Z_0 : Z_L \qquad \textbf{(Equation 10-6B)}$$

where   SWR = Standing-wave ratio
Z_L = Impedance of the load
Z_0 = Characteristic impedance of the line

Use the ratio that makes the SWR greater than 1.

Example 1

A transmission line with a characteristic impedance of 50 Ω is delivering power to a 150-Ω load. Calculate the SWR and determine the minimum voltage reading if the maximum voltage is 30 V.

Solution

Since SWR, VSWR, and ISWR are all equal, finding the value of one gives the value of the others.

$$SWR = Z_L : Z_0 \qquad \textbf{(Equation 10-6A)}$$

$$SWR = \frac{150\ \Omega}{50\ \Omega}$$

$$SWR = 3 : 1$$

$$VSWR = SWR = \frac{V_{rms\ max}}{V_{rms\ min}}$$

$$3 = \frac{30}{V_{rms\ min}}$$

$$V_{rms\ min} = \frac{30}{3} = 10\ V$$

Conclusion

This section introduced the concept of standing waves and how they can be measured. You also saw the relationship of the load to the standing-wave pattern and how fractional parts of the wavelength determined the values of the standing wave. The next section introduces transmission line power considerations and how a transmission line can be made to act exactly like a resonant circuit.

10-2 Review Questions

1. Describe how a finite transmission line can deliver all of the source power to the load.
2. Explain what causes standing waves along a transmission line. What do standing waves indicate?

3. Describe the standing-wave pattern of a transmission line terminated in an open. Do the same for a transmission line terminated by a short.
4. Define incident wave and reflected wave.
5. Explain the optimal conditions for transmitting power from source to load along a transmission line.

10-3 | UNMATCHED TRANSMISSION LINES

Discussion

When the characteristic impedance of a transmission line is not matched to the load, two important phenomena occur: (1) Not all of the transmitted power is used by the load. (2) The transmission line behaves as a resonant circuit. These two characteristics of an *unmatched* transmission line are introduced in this section.

Reflection Coefficient

When the load is not matched to the characteristic impedance of the transmission line, a *mismatch* is said to occur. When a mismatched condition does occur, a quantity called the *reflection coefficient* will be other than zero. The reflection coefficient is

$$K_{ref} = \frac{V_{ref}}{V_{inc}}$$ **(Equation 10-7)**

where K_{ref} = Reflective coefficient
V_{ref} = Voltage of the reflective wave
V_{inc} = Voltage of the incident wave

Both V_{ref} and V_{inc} must be measured at the same point on the transmission line.
The relationship between the SWR and K_{ref} is

$$SWR = \frac{K_{ref} + 1}{1 - K_{ref}}$$ **(Equation 10-8)**

where SWR = Standing-wave ratio
K_{ref} = Reflective coefficient

Sometimes it is convenient to express the reflection coefficient in terms of the characteristic impedance of the transmission line and the load:

$$K_{ref} = \left| \frac{Z_L - Z_0}{Z_L + Z_0} \right|$$ **(Equation 10-9)**

Example 1

As shown in Figure 10-12, a 50-Ω load is being fed from a 72-Ω transmission line. From this mismatch, determine the reflection coefficient.

$Z_0 = 72\ \Omega$

Figure 10-12

Solution

$$K_{\text{ref}} = \left| \frac{Z_L - Z_0}{Z_L + Z_0} \right| \qquad \text{(Equation 10-9)}$$

$$K_{\text{ref}} = \left| \frac{72\ \Omega - 50\ \Omega}{72\ \Omega + 50\ \Omega} \right|$$

$$K_{\text{ref}} = \left| \frac{22}{122} \right| = 0.180$$

Reflected Power

The main reason for using a transmission line is to transfer RF power from one point to another. There are times when you will want to know how much power is being delivered to the load for an unmatched transmission line. This can be derived as follows:

$$\text{General power formula } P = \frac{V^2}{R}$$

$$\text{Reflected power } P_{\text{ref}} = \frac{V_{ref}^2}{R_L}$$

$$\text{Incident power } P_{\text{inc}} = \frac{V_{inc}^2}{R_L}$$

Ratio of reflected and incident power:

$$\frac{P_{\text{ref}}}{P_{\text{inc}}} = \frac{V_{\text{ref}}^2/R_L}{V_{\text{inc}}^2/R_L} = \frac{V_{\text{ref}}^2}{V_{\text{inc}}^2}$$

Since

$$K_{ref} = \frac{V_{ref}}{V_{inc}}$$

Then

$$K_{ref}^2 = \frac{P_{ref}}{P_{inc}} \qquad \textbf{(Equation 10-10)}$$

It is sometimes convenient to express the amount of reflected power as a percentage:

$$\%P_{ref} = K_{ref}^2 \times 100\% \qquad \textbf{(Equation 10-11)}$$

Thus, the power absorbed by the load as a percentage is

$$\%P_{abs} = 100 - \%P_{ref} \qquad \textbf{(Equation 10-12)}$$

Example 2

For the transmission line in Figure 10-12, determine:

(A) The percentage of incident power reflected from the load.

(B) The percentage of incident power absorbed by the load.

(C) If 10 W were being transmitted, what would be the power absorbed by the load and the power reflected by the load?

Solution

(A) From Example 1, $K_{ref} = 0.180$

$\%P_{ref} = K_{ref}^2 \times 100\%$ **(Equation 10-11)**

$\%P_{ref} = 0.180^2 \times 100\%$

$\%P_{ref} = 0.032\,4 \times 100\% = 3.24\%$

(B) For the power absorbed,

$\%P_{abs} = 100 - \%P_{ref}$ **(Equation 10-12)**

$\%P_{abs} = 100 - 3.24 = 96.8\%$

(C) Total power absorbed by the load is

$P_L = 96.8\%$ of 10 W $= 9.68$ W

Total power reflected by the load is

$P_{ref} = 3.24\%$ of 10 W $= 0.324$ W

Transmission Line Resonant Circuits

A transmission line terminated in its characteristic impedance has an SWR of 1 and is called a *nonresonant* or *flat* transmission line. This means that it will present the same impedance to any frequency. A transmission line that is not terminated in its characteristic impedance will have an SWR greater than 1. This means that the impedance presented to the source will vary with the source frequency. This is therefore called a *resonant* transmission line.

The equivalent circuits of a quarter-wave ($\frac{\lambda}{4}$) short-circuit transmission line are shown in Figure 10-13. The line appears as a parallel resonant circuit to the source because *at*

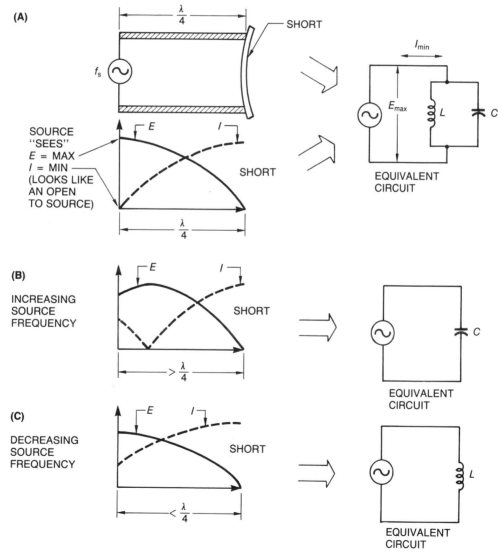

Figure 10-13 Equivalent circuits of $\frac{\lambda}{4}$ shorted transmission lines.

the source the voltage is maximum and the current is minimum; this makes the source "see" an open. This is equivalent to a parallel resonant circuit at resonance. As the frequency is increased, the line becomes capacitive. This is because as the frequency is increased above resonance for a parallel resonant circuit, it begins to appear capacitive. When the frequency is decreased below resonance for a parallel resonant circuit, it begins to look inductive.

Figure 10-14 shows the equivalent circuit of a quarter-wave open-circuit transmission line. The source sees a maximum current and a minimum voltage. This appears as a *series*

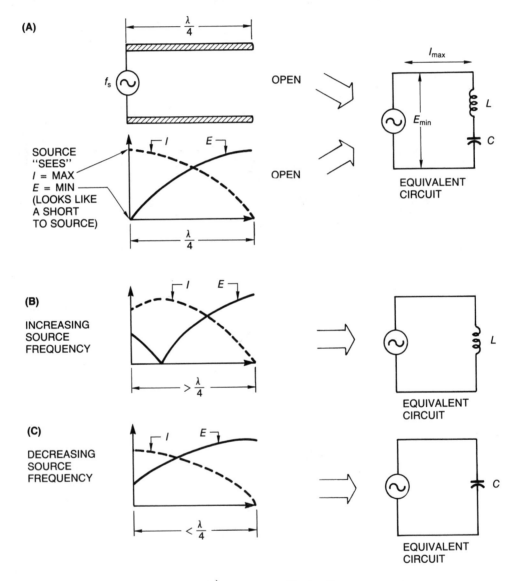

Figure 10-14 Equivalent circuits of $\frac{\lambda}{4}$ open transmission lines.

resonant circuit to the source. Increasing the source frequency causes the equivalent circuit of the transmission line to appear as an inductor. This is because as the source frequency of a series resonant circuit is increased above resonance, it begins to appear inductive. As the source frequency is decreased, the transmission line now appears capacitive. This is because a series resonant circuit below resonance will appear capacitive.

At only one frequency will the length of the transmission line be exactly $\frac{1}{4}$ of a wavelength. The wavelength of any wave is

$$\lambda = \frac{Vel}{f}$$ **(Equation 10-13)**

where λ = Length of the wave in meters
         Vel = Velocity of the wave in meters per second
           f = Frequency of the wave

Thus the frequency for a given wavelength is

$$f = \frac{Vel}{\lambda}$$ **(Equation 10-14)**

Example 3

For what frequency is the shorted transmission line in Figure 10-15 (A) one-quarter wavelength? (B) one-half wavelength?

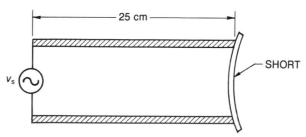

Figure 10-15

Solution

First find the frequency where four times the length of the given transmission line will equal the wavelength.

(A) λ = 4 × 25 cm = 100 cm = 1 m

Thus

$$f = \frac{Vel}{\lambda}$$ **(Equation 10-14)**

$$f = \frac{3 \times 10^8 \text{ m/s}}{1 \text{ m}}$$

$$f = 3 \times 10^8 \text{ Hz} = 300 \text{ MHz}$$

This is the frequency where the given transmission line is a quarter-wavelength.

(B) Find the frequency where two times the length of the given transmission line will equal the wavelength.

$$\lambda = 2 \times 25 \text{ cm} = 50 \text{ cm} = 0.5 \text{ m}$$

Thus

$$f = \frac{\text{Vel}}{\lambda}$$ **(Equation 10-14)**

$$f = \frac{3 \times 10^8 \text{ m/s}}{0.5 \text{ m}}$$

$$f = 6 \times 10^8 \text{ Hz} = 600 \text{ MHz}$$

Thus at 600 MHz, the transmission line in Figure 10-15 is a half-wavelength.

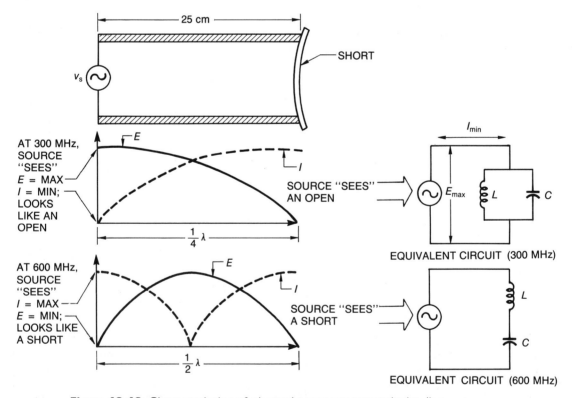

Figure 10-16 Characteristics of shorted resonant transmission line for two different frequencies.

Table 10-2	CHARACTERISTICS OF RESONANT TRANSMISSION LINES		
Transmission Line Wavelength	**Type of Termination**	**Characteristic Circuit**	**Z (ohms)**
$\frac{1}{4}$	Open	Series resonant	0
$\frac{1}{2}$		Parallel resonant	∞
$\frac{1}{4}$	Short	Parallel resonant	∞
$\frac{1}{2}$		Series resonant	0

The significance of Example 3 is shown in Figure 10-16. When the transmission line appears as a quarter-wave, the source sees maximum voltage and minimum current. Thus, for this frequency, the transmission line looks like an *open* to the source (parallel resonant circuit). As the frequency is increased to 600 MHz, the source sees zero volts and maximum current. Now the transmission line appears as a *short* to the source (series resonant circuit).

The characteristics of open or shorted transmission lines are summarized in Table 10-2.

Conclusion

This section presented transmission lines that were not terminated in their characteristic impedance. You saw how absorbed and reflected power could be calculated. This section also demonstrated that the impedance of an open or a shorted transmission line depended on the frequency of the source. Remember that a transmission line terminated by a load that is equal to the characteristic impedance of the line always presents the same impedance to the source regardless of the line frequency.

10-3 Review Questions

1. What is a transmission line mismatch?
2. Describe the relationship between the amount of transmission line mismatch and the amount of transmitted power absorbed by the load.
3. Define (A) reflected power and (B) transmitted power.
4. Explain the difference between a flat transmission line and a resonant transmission line.
5. When does a transmission line have an SWR equal to 1?
6. When does an open transmission line look like (A) a short to the source? (B) an open to the source?
7. When does a transmission line terminated in a short look like (A) a short to the source? (B) an open to the source?
8. Explain when a transmission line looks like a series resonant or a parallel resonant circuit to the source.

10-4 | PRINCIPLES OF ANTENNAS

Discussion

An *antenna* is that portion of a communications system that is used for radiating waves into space or receiving them from space. In Chapter 3, you learned that an antenna was any conductor that received radio waves and converted them into movements of electricity. This section offers a broader definition of an antenna. You will see that an antenna can be used to transmit or receive electromagnetic waves.

Transmission Line

Antenna analysis can begin with what you already know about transmission lines. Observe the quarter-wave open-ended transmission line in Figure 10-17. Note how some of the electrical energy of the transmission line "shoots" into the space at the open end. In practice, not all of the energy is reflected back to the source in an open-ended quarter-wave transmission line. The reason is that the surrounding space is, in effect, a load for the transmission line. Thus, some power is dissipated in this load (the space) in the form of *radiated energy*. The amount of wave energy radiated is very small for two reasons:

1. There is a mismatch between the transmission line and the load (surrounding space).
2. Since the two wires of the transmission line are close together and 180° out of phase, the radiation from one tends to cancel the radiation from the other.

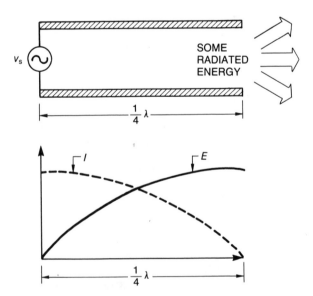

Figure 10-17 Radiation of quarter-wave transmission line.

Half-Wave Dipole

The most basic antenna is a quarter-wave transmission line that has been "spread apart." When this happens, the antenna is fully coupled to the load of free space and there is no canceling effect of parallel leads. Such an antenna, called a *half-wave dipole* or *Hertz* antenna, is shown in Figure 10-18. Both ends of the antenna appear as an open; hence the voltage at the ends is a maximum while the current is a minimum. This means that there is maximum current at the source, so a half-wave antenna appears as a very low impedance to the source. Actually, the impedance of a half-wave dipole is 73 Ω.

The *radiation resistance* of an antenna is defined as the ratio of total power dissipated to the square of the effective value of antenna current:

$$R_{\text{rad}} = \frac{P}{I^2}$$

(Equation 10-15)

where R_{rad} = Radiation resistance of antenna
        P = Power converted into electromagnetic radiation
        I = rms value of antenna current

Note that the radiation resistance of an antenna is a *fictitious* resistance that would dissipate as much power as the antenna is radiating if the resistance were connected to the same transmission line in place of the antenna. Keep in mind, radiation resistance is not a true resistance. A true resistance causes heat losses. In any antenna, there are some heat losses, but they are not accounted for in radiation resistance.

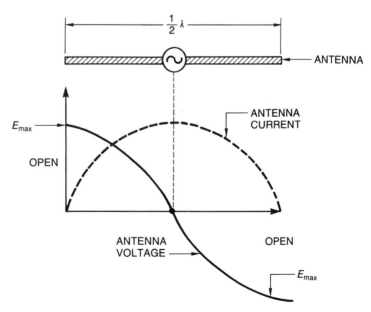

Figure 10-18 Half-wave dipole or hertz antenna.

Example 1

How much power will a half-wave dipole radiate when it has an antenna rms current of 10 A?

Solution
Using the relationship for antenna radiation resistance:

$$R_{\text{rad}} = \frac{P}{I^2}$$

(Equation 10-15)

Solving for P gives

$P = R_{\text{rad}}I^2$
$P = 73 \ \Omega \times 10 \ \text{A}^2$ (half-wave dipole $R_{\text{rad}} = 73 \ \Omega$)
$P = 7\ 300 \ \text{W}$

Radiation Pattern

The energy radiated from an antenna is not the same in all directions. The reason is shown in Figure 10-19. Since a dipole has high voltage potentials at both ends, an *electrostatic* field exists between these points. Unlike a capacitor, the electrostatic field is not confined to the space between the plates. For an antenna, this field is radiated into space. Since there is current flow in the antenna, a *magnetic* field also exists. Note from Figure 10-19 that the magnetic field is perpendicular to the electrostatic field.

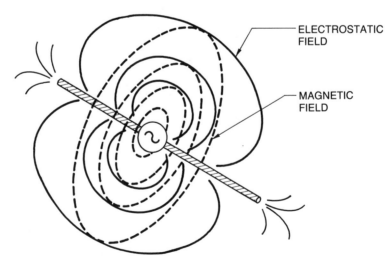

ELECTROSTATIC FIELD

MAGNETIC FIELD

Figure 10-19 Half-wave dipole electrostatic and magnetic fields.

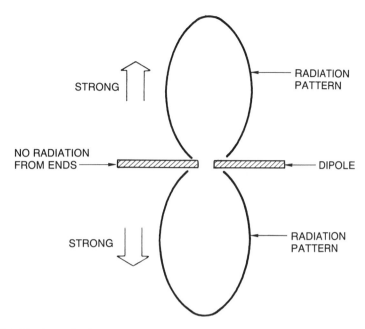

Figure 10-20 Dipole radiation pattern.

The resulting *radiation pattern* of the half-wave dipole is shown in Figure 10-20. You can see that the radiation pattern for a half-wave dipole is strongest perpendicular to the antenna and almost zero at its ends.

Antenna Polarization

The combination of the *electrostatic* and *electromagnetic* fields creates the transmitted *electromagnetic* wave. See Figure 10-21. When the electric field is *horizontal,* the antenna is said to be *horizontally polarized.* When the electric field is *vertical,* the antenna is said to be *vertically polarized.* In practice, a vertically polarized receiving antenna will only receive signals from a vertically polarized transmitting antenna. The same is true of a horizontally polarized transmitting and receiving antenna. A constructed dipole antenna is shown in Figure 10-22. Note the location of the insulators and the "feed-in" (transmission line).

Electrical and Physical Size

Because the wires used in an antenna have a physical thickness and the velocity of the wave traveling along the antenna is actually less than the speed of light in free space, the physical length of the antenna is about 95% of its calculated electrical length.

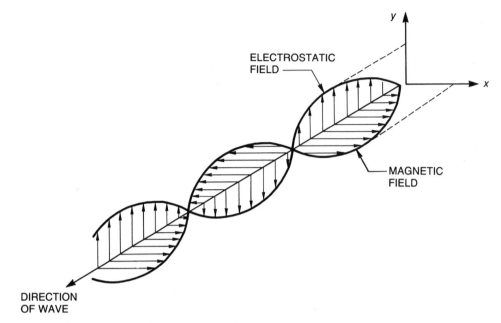

Figure 10-21 Transmitted electromagnetic wave.

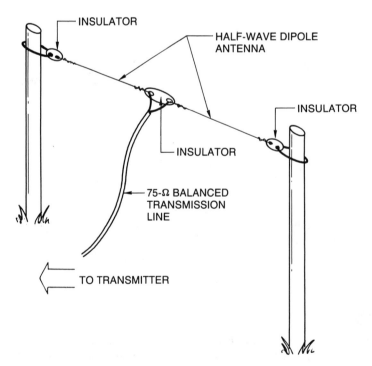

Figure 10-22 Constructed half-wave dipole antenna.

Example 2

Calculate the physical length of a half-wave dipole antenna for a 100-MHz transmitter.

Solution

First calculate the wavelength of 100 MHz.

$$\lambda = \frac{\text{Vel}}{f}$$ **(Equation 10-13)**

$$\lambda = \frac{3 \times 10^8}{100 \text{ MHz}}$$

$$\lambda = 3 \text{ m}$$

The electrical length is 3 m. By the 95% rule, the physical length is

$$\lambda = 0.95 \times 3 \text{ m} = 2.85 \text{ m}$$

Antenna Q

Antennas have a Q just as a resonant circuit does. The Q of an antenna is defined as

$$Q = \frac{f_0}{\text{BW}}$$ **(Equation 10-16)**

Example 3

Determine the Q of an antenna with a bandwidth of 0.6 MHz that is cut for a frequency of 30 MHz.

Solution

$$Q = \frac{f_0}{\text{BW}}$$ **(Equation 10-16)**

$$Q = \frac{30 \text{ MHz}}{0.6 \text{ MHz}}$$

$$Q = 50$$

Conclusion

This section introduced you to the principles of antennas. The antenna presented was the half-wave dipole, Hertz antenna. There are many other kinds of antennas. These will be presented in the next section.

10-4 Review Questions

1. What is the purpose of an antenna?
2. Describe the basic properties of a half-wave dipole. How is it related to a quarter-wave transmission line?
3. Discuss the voltage and current characteristics of a half-wave dipole.
4. Explain radiation resistance. Does such a resistance actually exist?
5. Describe the radiation pattern of a half-wave dipole.
6. Describe the properties of an electromagnetic wave. What do horizontal and vertical polarization mean?
7. Explain how you would actually construct a dipole antenna. What materials would you need?
8. What is antenna Q?

10-5 TYPES OF ANTENNAS

Discussion

In this section, you will see the differences between various types of antennas. In the last section, you were introduced to antenna principles through the half-wave dipole or Hertz antenna. There are other kinds of antennas, whose differences are seen in

- Physical appearance
- Impedance
- Propagation pattern

Full-Wave Dipole

A half-wave dipole can operate on harmonics of its fundamental frequency. This means that a dipole can operate as a full-wave dipole. Figure 10-23 shows the radiation pattern and current distribution of a full-wave dipole antenna. There are four lobes produced by the radiation pattern of the full-wave dipole. This is the result of the current distribution on the dipole. Note from the current distribution that the current in the center is low, meaning that a full-wave dipole has a high input impedance.

$1\frac{1}{2}$-Wave Dipole

The third harmonic of a half-wave dipole antenna produces the current distribution and radiation pattern of Figure 10-24. For the $1\frac{1}{2}$-wave dipole, the radiation pattern consists of four major lobes and two minor lobes. The lower portion of Figure 10-24 indicates the antenna current.

The harmonic operation of a dipole antenna is used for designing an antenna to operate on two harmonically related frequencies. For example, the 7-MHz and 21-MHz amateur radio bands allow the same dipole to be operated as a half-wavelength dipole at 7 MHz and a $1\frac{1}{2}$-wave dipole at 21 MHz.

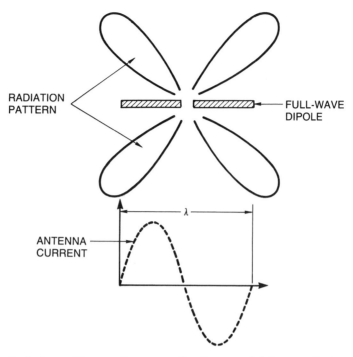

Figure 10-23 Radiation pattern and current distribution of full-wave dipole antenna.

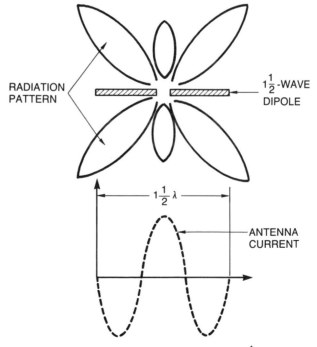

Figure 10-24 Radiation pattern and current distribution of $1\frac{1}{2}$ wave dipole.

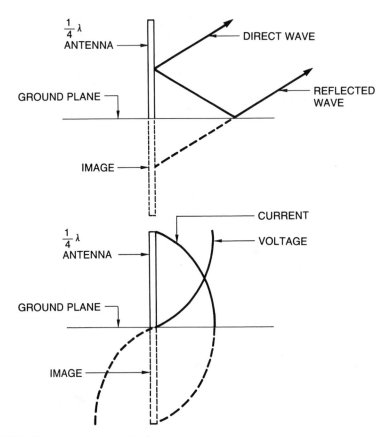

Figure 10-25 Quarter-wave vertical antenna.

Vertical Antenna

A quarter-wave vertical dipole antenna constructed above a perfect ground has the same characteristics as a vertical half-wave dipole. The reason for this is because of the reflection of the wave from the ground. This phenomenon is illustrated in Figure 10-25.

An antenna of the type in Figure 10-25 is sometimes called a Marconi antenna. Note from the figure that a perfect ground forms an *image antenna*. The radiation pattern for the quarter-wave vertical antenna is shown in Figure 10-26.

Antenna Arrays

Sometimes it is desirable to change the radiation pattern to favor a certain direction. An example is a commercial radio station that is to transmit most of its broadcasting power toward a large metropolitan area. A pattern that accomplishes this is shown in Figure 10-27.

The direction of a radiated wave can be influenced by an *antenna array*. An array can be created by adding more elements to the antenna. These extra elements, sometimes referred to as *exciters*, come in two categories: (1) *parasitic array*, (2) *driven array*.

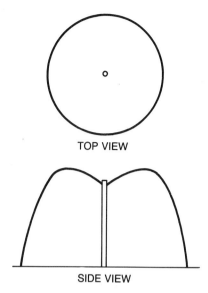

Figure 10-26 Radiation pattern for the quarter-wave vertical antenna.

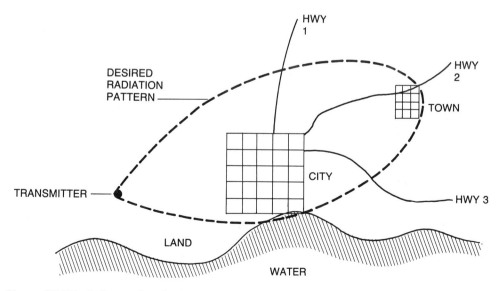

Figure 10-27 A directed radiation pattern.

Parasitic Array

An antenna element not connected to the transmission line develops a voltage through induction and is called a *parasitic element* or a *reflector*. The most elementary antenna array is shown in Figure 10-28. The parasitic element is located $\frac{1}{4}$ wavelength from the driven element. The unidirectional pattern presented by such an array is shown in Figure 10-29. The measurement of a unidirectional antenna's directional properties is its *beamwidth*. This is the measurement of the angle θ from the radiation pattern between two points on either side of maximum radiation where the field strength drops 3 dB.

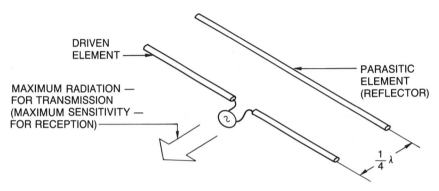

Figure 10-28 Most elementary antenna array.

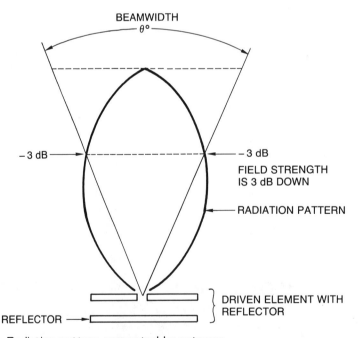

Figure 10-29 Radiation pattern presented by antenna.

Antenna Gain

Antenna gain is a comparison of the output strength of the antenna radiation in a particular direction and a *reference antenna*. The reference antenna is an antenna, such as the Hertz, that radiates an omnidirectional wave. Thus the gain of an antenna is the increase in power because of the unidirectional signal compared to a omnidirectional signal. If an antenna has a 3-dB gain in a certain direction, it means that the antenna has improved upon the reference antenna by 3 dB. Power is radiated in a particular direction by stealing it from other directions, as with the half-wave-dipole-driven element with a parasitic reflector, for example. This antenna gain refers not to any increase in *total* radiation power, just in a specified direction. This relationship is expressed mathematically as

$$A(dB) = 10 \log_{10}\left(\frac{P_2}{P_1}\right)$$
   (Equation 10-17)

where $A(dB)$ = Antenna gain in dB
   P_1 = Power of actual antenna
   P_2 = Power of reference antenna

Example 1

An antenna is radiating 500 W and has a 6-dB gain over a reference antenna. How much power must the reference antenna radiate in order to be equally effective in the most preferred direction?

Solution

$$A(dB) = 10 \log_{10}\left(\frac{P_2}{P_1}\right)$$
   (Equation 10-17)

$$6 = 10 \log_{10}\left(\frac{P_2}{500}\right)$$

$$0.6 = \log_{10}\left(\frac{P_2}{500}\right)$$

$$\text{antilog}\ (0.6) = \frac{P_2}{500}$$

$$4 = \frac{P_2}{500}$$

$$P_2 = 4 \times 500 = 2\ 000\ \text{W}$$

Since any transmitting antenna can also be used as a receiving antenna, receiving antennas also have gain. The gain is measured as before with respect to a *reference* antenna. Usually it is convenient to measure the signal voltage of a receiving antenna. When this is done, the antenna gain in dB is

$$A(\text{dB}) = 20 \log_{10}\left(\frac{V_2}{V_1}\right) \qquad \textbf{(Equation 10-18)}$$

where $A(\text{dB})$ = Power in dB
V_2 = Voltage of reference antenna
V_1 = Voltage of actual antenna

Example 2

Determine the dB gain of a receiving antenna that delivers a 60-μV signal to a transmission line over that of an antenna that delivers a 20-μV signal under identical conditions.

Solution

$$A(\text{dB}) = 20 \log_{10}\left(\frac{V_2}{V_1}\right) \qquad \textbf{(Equation 10-18)}$$

$$A(\text{dB}) = 20 \log_{10}\left(\frac{60 \times 10^{-6}}{20 \times 10^{-6}}\right)$$

$$A(\text{dB}) = 20 \log_{10}(3) = 20(0.477)$$

$$A(\text{dB}) = 9.54 \text{ dB}$$

Yagi-Uda Antenna

The *Yagi-Uda* antenna is named after the Japanese scientists who developed it. It has two parasitic elements, a *reflector* and a *director*. Such an antenna is usually referred to as a Yagi and is illustrated in Figure 10-30. A Yagi antenna is highly directional. As Figure 10-31 shows, the antenna gain in a particular direction increases with the number of elements added.

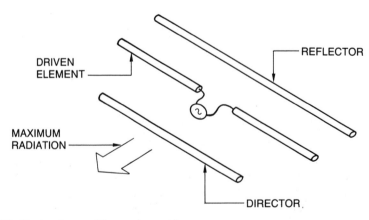

Figure 10-30 Yagi antenna (three-element).

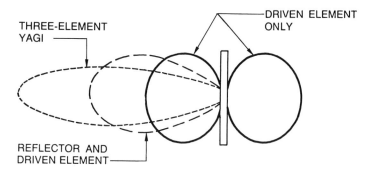

Figure 10-31 Yagi gain vs. number of elements.

Driven Collinear Array

A *driven collinear array* is shown in Figure 10-32. In this antenna, all the elements are placed in a line. This gives omnidirectional coverage and directs the energy down toward the horizon. Doing this increases the coverage at VHF and UHF.

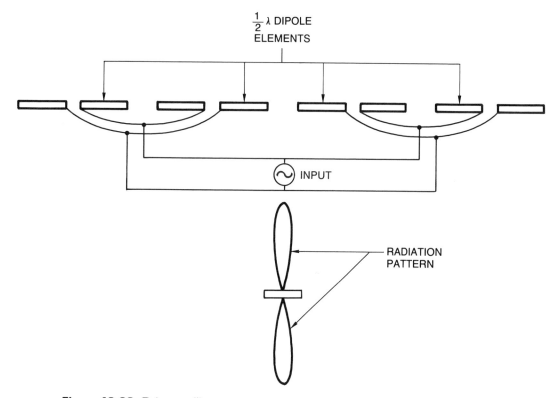

Figure 10-32 Driven collinear array.

Conclusion

This section introduced and compared various kinds of antennas. You also saw how to measure antenna gain and learned what it means. The next section explores some of the properties of space and its effect on the transmitted wave.

10-5 Review Questions

1. Describe the radiation pattern of a full-wave dipole.
2. Discuss the difference in radiation patterns and construction of a full-wave dipole and a one-and-one-half-wave dipole antenna.
3. Explain how a full-wave and a one-and-one-half-wave dipole antenna can be used by the same transmitter.
4. Describe the radiation pattern of a vertical quarter-wave antenna.
5. Explain how beam width is measured.
6. Give an example of how a directional signal can be used.
7. How is antenna gain measured?
8. Give some examples of antenna arrays.

10-6 WAVE PROPAGATION

Discussion

After the radio wave leaves the antenna, it has three natural ways of traveling from transmitter to receiver:

1. Along the ground (a *ground wave*).
2. Along a straight line (a *line-of-sight wave* or *space wave*).
3. Up to the ionosphere and back to earth (a *sky wave*).

A fourth method is satellite communications. They will be presented in the next chapter. This section introduces the concepts involved in the three natural ways of wave propagation in space.

General Observations

Figure 10-33 illustrates the relationship between the frequency of the transmitted radio wave and its predominant method of traveling. Each method of wave propagation will be discussed separately in this section. What is important for now is to note from Figure 10-33 that the way in which an electromagnetic wave is propagated depends on its frequency.

Ground Wave

A *ground wave* is an electromagnetic wave that travels along the earth's surface. It is sometimes called a *surface wave*. The ground wave follows the curvature of the earth and can be transmitted beyond the horizon, as shown in Figure 10-34.

Ground-wave propagation is confined to low radio frequencies (usually 300 kHz or less) and travels much further when the surface is a good conductor. Thus a ground wave will travel very far over water (especially salt water). Surfaces such as sand tend to absorb the wave energy, so transmission over desert terrain is not as efficient as over water.

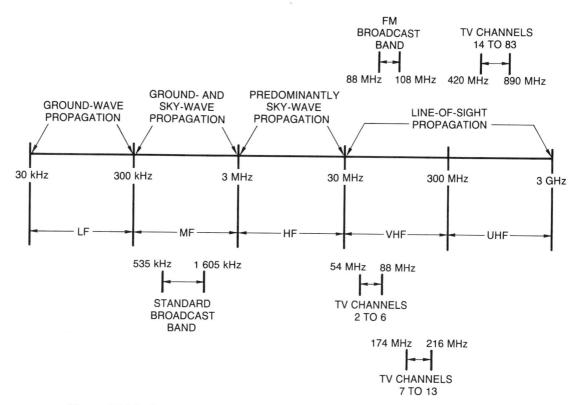

Figure 10-33 Frequency spectrum and predominant method of propagation.

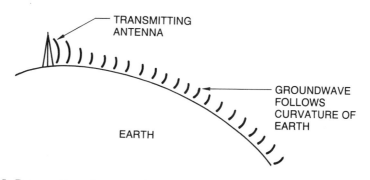

Figure 10-34 Propagation of a ground wave.

For low frequencies, such as 30 kHz, where the wavelength is long (10 000 meters or 6.2 miles for this frequency), terrain features such as mountains have little effect on the wave transmission. However, at higher frequencies, where the wavelength approaches the size of mountain ranges and hills, these terrain features do affect the propagation distance.

Ground-wave propagation is achieved through vertical polarization. This does present a problem at lower frequencies because of the required height of the antenna. If a horizontally polarized antenna were used, the ground wave would be absorbed by the conductivity of the earth.

Line-of-Sight-Communications

At frequencies above 3 MHz, the ground wave is reduced much quicker than the space wave. Line-of-sight communications is used primarily in the VHF, UHF, and higher-frequency bands. Line-of-sight wave propagation is shown in Figure 10-35.

An equation for finding the maximum transmitting distance of a line-of-sight wave between two antennas of known height is

$$d = \sqrt{2h_{\text{trans}}} + \sqrt{2h_{\text{rec}}} \qquad \textbf{(Equation 10-19)}$$

> *where* d = Radio horizon (miles)
> h_{trans} = Transmitter antenna height in feet
> h_{rec} = Receiver antenna height in feet

The *radio horizon* is not the physically measured straight-line distance between antennas, but is about $\frac{4}{3}$ greater than this value.

Example 1

How far from a 2 000-ft transmitting antenna can a 25-ft receiving antenna be in order to ensure line-of-sight communications between them?

Solution

$$d = \sqrt{2h_{\text{trans}}} + \sqrt{2h_{\text{rec}}} \qquad \textbf{(Equation 10-19)}$$
$$d = \sqrt{2(2\,000)} + \sqrt{2(125)}$$
$$d = \sqrt{4\,000} + \sqrt{50}$$
$$d = 63.24 + 7.07 = 70.3 \text{ miles}$$

The solution assumes there are no obstructions between the antennas.

Sky Waves

Sky-wave propagation depends on the characteristics of the ionosphere. These waves occur between frequencies of 300 kHz and 30 MHz. The ionosphere is composed of layers of ionized particles many miles above the earth. These ionized layers can reflect higher-frequency radio waves and direct them back to earth. The ionosphere is not consistent. Its height, thickness, and reflectivity can change from hour to hour as well as from day

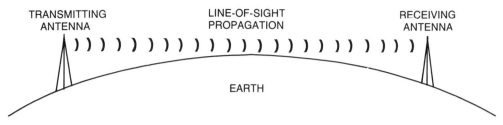

Figure 10-35 Line-of-sight propagation.

to day, month to month, year to year, and decade to decade. This change comes from the heating and radiation effects of the sun, which causes hourly changes. Sunspot activity affects decade changes. Figure 10-36 illustrates the ionospheric layers and their changes for a typical 24-hour period.

As shown in the figure, the lowest layer of the ionosphere is called the *D layer*. This layer is furthest from the sun and is 25 to 55 miles above sea level. It is present only during the day, is fairly weak, and usually doesn't cause reflection. However, during the day, this layer does absorb some radio energy. This is a problem for medium-frequency (MF) signals, which can be entirely absorbed by the D layer. Hence, at night, when there is no D layer, transmission by MF is much more reliable. The *E layer* is about 55 to 90 miles above the earth's surface. It has a maximum density at noon and is weakly ionized

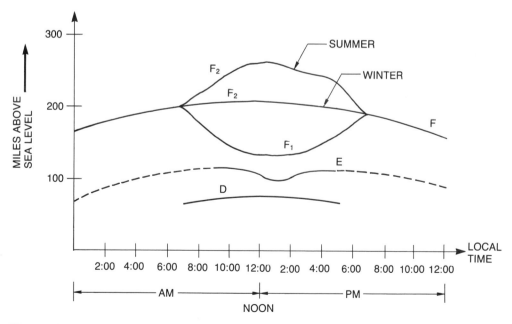

Figure 10-36 Changes in ionospheric layers.

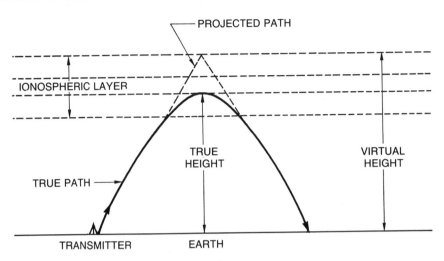

Figure 10-37 Effects of sky-wave propagation.

at night. It can refract signals up to about 20 MHz. The *F layer* is from 90 to 155 miles high. It splits into two layers during the day, called the F_1 and F_2 layers.

The effect of sky-wave propagation is shown in Figure 10-37. It shows how a radio wave can be reflected back to the earth. Note the difference between *true* and *virtual* heights. The radio wave actually follows a curved path. The height of this curved path above earth is the true height. The virtual height is the distance above earth if the radio wave actually did reflect at an angle.

Skip Distance

The *skip distance* is the distance that a radio wave travels over the earth. Skip distance is illustrated in Figure 10-38. Skip distance is influenced by the transmission angle and the transmission frequency as well as the state of the ionization layer.

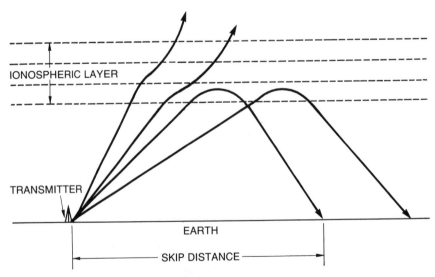

Figure 10-38 Illustration of skip distance.

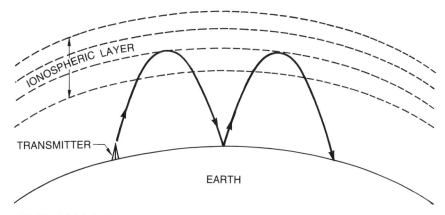

Figure 10-39 Multiple hop of a sky wave.

Multiple Hop

A phenomenon called *multiple-hop* transmission occurs when the sky wave is reflected back to the earth to be bent again by the ionosphere. See Figure 10-39.

For optimum conditions, a single hop can travel about 2 000 miles. Thus multiple hops can greatly increase the distance traveled by sky waves. However, each hop increases the attenuation of the signal, thus limiting the maximum distance that can be traveled.

Fading

When sky-wave propagation is used, the received signal often will periodically become stronger, then weaker. This can occur because of the changes in the ionosphere or because of *multiple-path* transmission. This phenomenon of multiple path transmission is shown in Figure 10-40.

Multiple-path transmissions can cause *fading* because the receiver is actually receiving two versions of the same carrier. Since one signal path is longer than the other, these signals can become out of phase and thus begin to cancel each other at the receiving end. For this reason, a receiver closer or *further* away from the transmitter may receive a stronger signal than one located at the point where multiple-path waves meet.

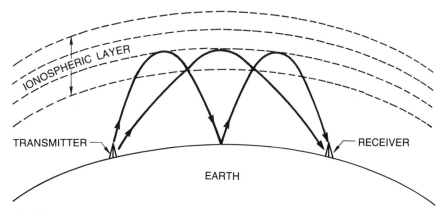

Figure 10-40 Multiple-path transmission.

Conclusion

This section introduced the properties of wave propagation. You saw that at certain frequencies, wave propagation was very reliable. The effects of the ionosphere were presented here. The next section on instrumentation and troubleshooting presents some important concepts concerning the proper use of RF test equipment and impedance matching.

10-6 Review Questions

1. State the three natural ways a radio wave travels from transmitter to receiver.
2. Explain the properties of a ground wave. Describe the effect of a horizontally polarized antenna on ground-wave transmission.
3. Describe the difference between the radio horizon and the actual horizon.
4. Discuss what conditions effect the transmission of sky waves. Explain how often these transmission conditions change.
5. Explain the phenomenon of skip distance. Explain multiple hops.
6. Describe the process of fading and its cause.

TROUBLESHOOTING AND INSTRUMENTATION

Radio Frequency Generator Impedance Matching

The *radio frequency* (RF) generator is a necessary instrument for designing, troubleshooting, and aligning communications equipment. As with any piece of laboratory equipment, the RF generator must be connected properly to the circuit under test. In this section, you will be introduced to the importance and methods of *impedance matching* the RF generator and the circuit under test. An RF generator used in the communications laboratory is shown in Figure 10-41.

Figure 10-41 Laboratory RF generator. *Courtesy* of Hewlette-Packard Company

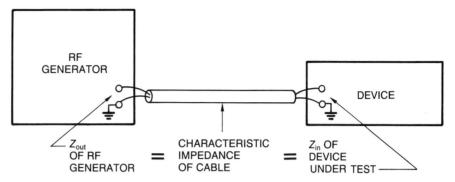

Figure 10-42 Impedance-matching requirements of RF generator.

Basic Idea

For reliable testing to occur, the load impedance must equal the generator impedance. This concept is shown in Figure 10-42. The RF generator is usually connected to the circuit under test through a connecting coaxial cable. This coax cable must have a characteristic impedance (Z_0) equal to the output impedance of the generator. Equally important, the cable must be terminated in its characteristic impedance. Any impedance mismatch will cause signal reflections and result in standing waves on the connecting cable.

Resistive Pad

When the RF generator is to be connected to a circuit whose input impedance is different from the impedance of the generator, an impedance-matching device, such as an impedance-matching transformer, should be used. If such a device is not available, then a *resistive pad* can be constructed and used.

The design equations for a resistive pad are

$$R_1 = Z_1 \sqrt{1 - \frac{Z_2}{Z_1}} \qquad \text{(Equation 10-20)}$$

$$R_2 = \frac{Z_2}{\sqrt{1 - \frac{Z_2}{Z_1}}} \qquad \text{(Equation 10-21)}$$

where R_1 = Resistor in series with the load
R_2 = Resistor in parallel with the source
Z_1 = Larger of the load or source impedance
Z_2 = Smaller of the load or source impedance

A typical resistive pad is shown in Figure 10-43.

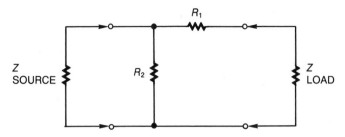

Figure 10-43 Typical resistive pad.

Example 1

Design a resistive pad for connecting an RF generator with a 50-Ω output impedance to a circuit with a 75-Ω input impedance.

Solution

$$R_1 = Z_1 \sqrt{1 - \frac{Z_2}{Z_1}} \qquad \textbf{(Equation 10-20)}$$

$$R_2 = \frac{Z_2}{\sqrt{1 - \frac{Z_2}{Z_1}}} \qquad \textbf{(Equation 10-21)}$$

Since Z_1 must represent the larger of the two impedances, solving for R_1 gives

$$R_1 = 75 \ \Omega \sqrt{1 - \frac{50 \ \Omega}{75 \ \Omega}}$$

$$R_1 = 75\sqrt{1 - 0.666}$$

$$R_1 = 75\sqrt{0.333} = 75(0.577)$$

$$R_1 = 43.3 \ \Omega$$

Solving for R_2 gives

$$R_2 = \frac{50 \ \Omega}{\sqrt{1 - \frac{50 \ \Omega}{75 \ \Omega}}}$$

$$R_2 = \frac{50}{\sqrt{1 - 0.666}}$$

$$R_2 = \frac{50}{\sqrt{0.333}} = \frac{50}{0.577}$$

$$R_2 = 86.6 \ \Omega$$

The completed and connected pad is shown in Figure 10-44, using practical values of R_1 and R_2.

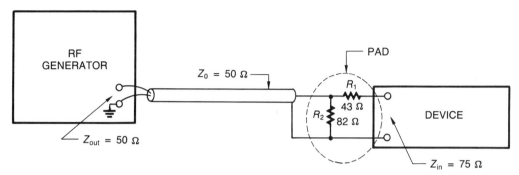

Figure 10-44

Pad Analysis

To demonstrate that the source sees 50 Ω looking toward the load and that the load sees 75 Ω looking back toward the source, refer to the steps in Figure 10-45. Note that when the equivalent circuit for the source is solved, the impedance is almost 50 Ω. When the equivalent circuit is solved for the load, the impedance is almost 75 Ω. The pad will cause a 4-dB loss, which you must consider when making measurements.

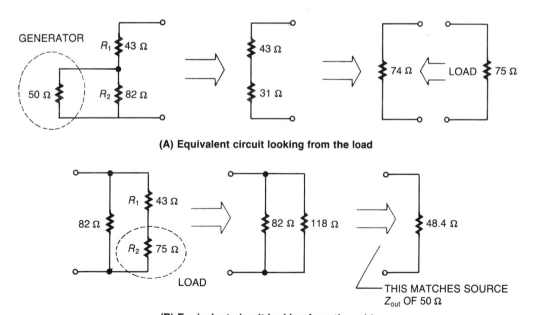

Figure 10-45 Analysis of resistive pad.

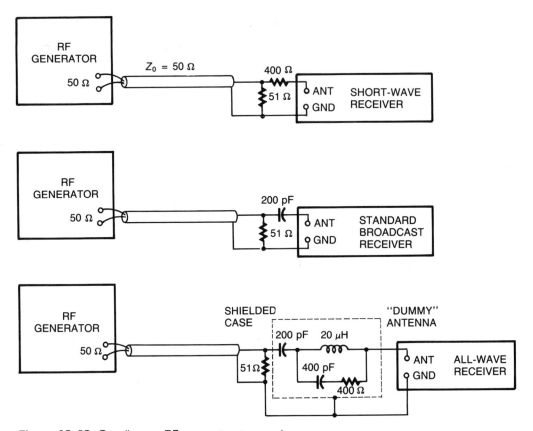

Figure 10-46 Coupling an RF generator to receivers.

RF Generator Coupling Circuits

Some standard methods of coupling an RF generator to communications equipment are shown in Figure 10-46. For an RF generator with a 50-Ω output, the coax cable is always terminated in a 51-Ω resistor (closest standard value to 50 Ω). For the all-wave receiver, a standard *dummy antenna* recommended by the Institute of Radio Engineers (IRE) is used.

Conclusion

This section showed how to properly terminate an RF generator to ensure accurate measurements and calibrations. Make sure that all RF test equipment is properly terminated in order to obtain accurate readings. The best rule is to consult the manual that comes with the test equipment—especially under the section on impedance matching. The next section presents a structured BASIC program that simulates a portion of the *Smith chart*.

10-7 Review Questions

1. State the relationship between the RF generator output impedance, characteristic impedance of the connecting cable, and the input impedance of the circuit under test in order for reliable measurements to occur.

2. Explain the use of a resistive pad when an RF generator is used.
3. Discuss what the input impedance of a properly designed resistive pad looks like to the source.
4. Discuss what the output impedance of a properly designed resistive pad looks like to the load.

Quarter-Wave Matching Transformers

When the characteristic impedance of a transmission line is not equal to the impedance of the load, a *quarter-wave matching transformer* can be used. Such an arrangement is shown in Figure 10-47. Note that there is no actual transformer involved, only a quarter-wave section. The word "transformer" is used because the results are the same as if an impedance-matching transformer were used.

The relationship for the matching transformer is

$$Z_T = \frac{Z_0^2}{Z_L}$$

(Equation 10-22)

where Z_T = The impedance seen looking into the quarter-wave section when it is attached to the load
Z_0 = Characteristic impedance of the quarter-wave matching section
Z_L = Impedance of the load

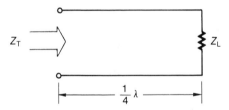

Figure 10-47 Quarter-wave matching transformer.

Smith Chart Simulation

Occasionally, the impedance from the input end of a transmission line can have a reactive component. This happens if the transmission line is not matched to the load or if the load itself has a reactive component. To eliminate standing waves on this type of line, it is necessary to eliminate the reactive component. This can be accomplished by using a short-circuit or open-circuit transmission line *stub*. These circuits were introduced in Section 10-3. A review of Figures 10-13 and 10-14 may be helpful.

To use these stubs to eliminate standing waves, place them in parallel with the transmission line close to the load end of the line. As explained in Section 10-3, these quarter-wave stubs can be made to appear capacitive or inductive, depending on their length and whether they are open ended or terminated by a short. This procedure is illustrated in Figure 10-49.

Example 1

As shown in Figure 10-48, a 50-Ω transmission line is used to feed a signal to a 150-Ω load. What must be the characteristic impedance of a quarter-wave matching transformer used to correct this mismatch?

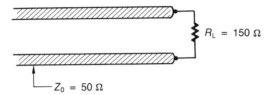

$R_L = 150\ \Omega$

$Z_0 = 50\ \Omega$

Figure 10-48

Solution

$$Z_T = \frac{Z_0^2}{Z_L}$$

(Equation 10-22)

Solving for Z_0 gives

$$Z_0 = \sqrt{Z_T Z_L}$$
$$Z_0 = \sqrt{(50\ \Omega)(150\ \Omega)}$$
$$Z_0 = \sqrt{7\ 500} = 86.6\ \Omega$$

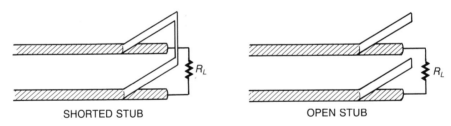

SHORTED STUB   OPEN STUB

Figure 10-49 Using a quarter-wave stub to prevent standing waves on unmatched transmission line.

MICROCOMPUTER SIMULATION

Troubleshooting simulation number ten gives you the opportunity of making bandwidth measurements on multi-stage amplifiers. Here you will again use the simulated RF generator and RF voltmeter. However, this time you will first make an overall system evaluation. From this, you will determine if the system meets specifications. If the system does not meet specifications, then you will determine the stage that is causing the problem.

CHAPTER PROBLEMS

(Answers to odd-numbered problems appear at the end of the text.)

1. Find the required pulse duration of a pulse that will have its leading edge reach the load at the same time that its trailing edge occurs at the generator. Refer to Figure 10-50.

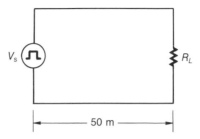

Figure 10-50

2. What is the pulse duration for the circuit in Figure 10-50 if the leading edge of the pulse is to arrive at the load when the trailing edge is at the halfway point of the transmission line?
3. Determine the wavelength of a 300-MHz sine wave in meters.
4. Determine the wavelength of a 100-MHz sine wave in feet and in miles.
5. As shown in Figure 10-51(A), the on time for a given pulse train is 15 ns, and the off time is 15 ns. For these conditions, what is the maximum number of pulses that can be on the line at a time if the transmission line is 100 meters long?
6. For the pulses in Figure 10-51(B), if the on time is 15 ns and the off time is 25 ns, how long must the transmission line be to allow 25 on pulses to be on the line at a time?

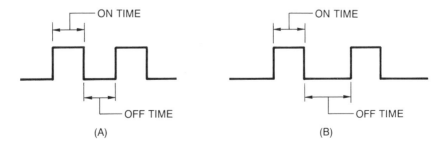

Figure 10-51

7. How many cycles of a 1-MHz wave can a 6-mile transmission line accommodate?
8. How long must a power transmission line be to accommodate one complete cycle of a 60-Hz sine wave?
9. A transmission line has a capacitance of 25 pF/ft and an inductance of 0.15 μH/ft. Determine the characteristic impedance of the line.
10. Determine the characteristic impedance of a transmission line that has a capacitance of 15 pF/m and an inductance of 0.20 μH/m.
11. For a transmission line that has an inductance of 0.4 μH/m with a characteristic impedance of 75 Ω, what is the capacitance per meter?
12. A transmission line has a capacitance of 50 pF/ft. If the line has a characteristic impedance of 50 Ω, what is the inductance per foot?
13. For a transmission line giving a maximum rms voltage reading of 50 V at one point and a minimum rms voltage reading of 25 V at another point, determine:
    (A) the VSWR of the line.
    (B) the ISWR of the line.
    (C) the SWR of the line.
14. The maximum rms voltage reading of a transmission line is 12 V and the minimum is 4 V. Determine:
    (A) the ISWR of the line.
    (B) the VSWR of the line.
    (C) the SWR of the line.
    (D) If the minimum line current is 12 mA, what would the largest line current reading be?
15. What would be the SWR of a transmission line if power is being delivered to a 100-Ω load by a line with a characteristic impedance of 50 Ω?
16. Determine the SWR of a transmission line with a characteristic impedance of 300-Ω delivering power to a 75-Ω load. If the maximum rms line voltage is 100 V what is the minimum line voltage?
17. A 75-Ω transmission line feeds a 25-Ω load. Determine:
    (A) SWR.
    (B) reflection coefficient.
    (C) percent of incident power reflected from the load.
    (D) percent of incident power absorbed by the load.
18. A 50-Ω transmission line feeds a 100-Ω load. Find:
    (A) SWR.
    (B) reflection coefficient.
    (C) percent of incident power reflected from the load.
    (D) percent of incident power absorbed by the load.
19. At a frequency of 120 MHz, a 250-Ω resistive load is fed by a 50-Ω transmission line. Determine the length of a quarter-wave matching section that will eliminate the standing waves.
20. Determine the length of a quarter-wave matching section that will eliminate the standing waves for a transmission line with a characteristic impedance of 72 Ω feeding a load of 100 Ω at a frequency of 300 MHz.

21. (Use the Smith chart simulation program to solve this problem.) Using a short-circuit stub, determine the stub length and location on the line for a SWR of 5 and a frequency of 150 MHz.

22. (Use the Smith chart simulation program to solve this problem.) For an open-circuit stub, determine the stub length and location on the line for an SWR of 3 and a frequency of 200 MHz.

23. For an antenna with a radiation resistance of 50 Ω, how much power will be radiated if the antenna current is 50 A?

24. How much power will be radiated if the antenna current is 20 A and the antenna radiation resistance is 75 Ω?

25. For an antenna that radiates 500 W with an antenna current of 5 A, what is the antenna radiation resistance?

26. Determine the radiation resistance of an antenna that radiates 150 W when the antenna current is 15 A.

27. An antenna has a radiation resistance of 50 Ω and radiates 250 W. What must be the antenna current?

28. Find the antenna current for an antenna that radiates 1 kW with a radiation resistance of 73 Ω.

29. Determine the Q of an antenna with a bandwidth of 0.5 MHz that is cut to a frequency of 20 MHz.

30. What is the Q of an antenna with a bandwidth of 0.75 MHz that is cut to a frequency of 50 MHz?

31. Calculate the length of a half-wave dipole to transmit a 10-MHz signal.

32. Calculate the length of a half-wave dipole used to receive a 15-MHz signal.

33. What should be the height of a Marconi antenna to transmit a 200-MHz signal? (Use the 95% rule.)

34. To receive a vertically polarized 500-MHz signal, what kind of antenna must be used? Assuming the 95% rule, find its length.

35. What frequency is being transmitted by a half-wave dipole if its length is 3.25 meters?

36. What frequency could be transmitted by the 3.25-meter dipole in problem 35 if it were treated as a $1\frac{1}{2}$-wave dipole?

37. What is the dB gain of an antenna that delivers a 100-μV signal over that of an antenna that delivers 75 μV?

38. Determine the dB gain of an antenna that delivers 150 μV over that of an antenna that delivers 75 μV.

39. An antenna has a gain of 12 dB over a reference antenna that radiates 100 W. Calculate how much power must be radiated by the reference antenna in order to be equally effective in the most preferred direction.

40. Solve problem 39 if the antenna gain is 6 dB.

41. To install an antenna with a line-of-sight transmission at a distance of 75 miles, determine the height of the receiving antenna if the transmitting antenna is 250 ft.

42. What is the maximum line-of-sight distance between two 300-ft antennas?

43. Which radio wave is primarily transmitted along the surface of the earth?

44. What frequency range is not affected by atmospheric disturbances?

45. The major mode of propagation for antennas that are line of sight starts at about what frequency?
46. Which ionospheric layers are present during the day?
47. Which ionospheric layers are present at night?
48. Describe which frequencies travel better over water than sand.
49. Describe the primary cause of fading.
50. Design a resistive pad that will match an RF generator using a 75-Ω transmission line into a 150-Ω load.

CHAPTER 11

Waveguides and Satellite Communications

OBJECTIVES

In this chapter, you will study:

- [ ] Why waveguides are used and what they do.
- [ ] Characteristics of rectangular and circular waveguides.
- [ ] Various waveguide modes of operation.
- [ ] What waveguide couplers are.
- [ ] Cavity resonators.
- [ ] Microwave antennas.
- [ ] Radar principles.
- [ ] Fundamentals of satellite communications.
- [ ] Antenna instrumentation.

INTRODUCTION

Waveguides are used in high-frequency communications in place of wires. Technically, any conductor (for example, a transmission line) is acting as a waveguide, but it is common practice to call a specific kind of transmission line a waveguide. Waveguides look like rectangular metal boxes or round metal tubing. Both kinds are hollow on the inside. These strange looking funnels of radio frequency energy are used in radar, satellite communications, and cross-country microwave communications. Sometimes technicians will refer to waveguides as "plumbing."

11-1 INTRODUCTION TO WAVEGUIDES

Discussion

As the frequency of the carrier is increased to 1 GHz and above, transmission by wires for short distances becomes very inefficient. A *waveguide* is the most efficient form of connecting these high radio frequencies from one place to another for short distances. Figure 11-1 shows what some typical waveguides look like. In this section you will see how energy is conducted along waveguides and how the physical dimensions of these guides determine the frequency they guide.

Waveguide Operation

Recall the *skin effect* discussed in Chapter 10. This showed that very little current flowed in the interior of a solid conductor at high radio frequencies. Thus, you could correctly conclude that a hollow conductor could also conduct high radio frequencies.

Also recall from the last chapter, that a quarter-wave shorted stub appeared as a parallel resonant circuit to the source. This fact can be used in the initial analysis of a waveguide. A transmission line can be 'transformed' into a waveguide using multiple quarter-wave shorted stubs, as shown in Figure 11-2.

Recall from the last chapter that a quarter-wave shorted stub appeared as a parallel resonant circuit to the source and thus as a *high-impedance* point. Thus transmission down the center of the waveguide offers minimum reduction of the signal. This relationship is true for any waveguide width that is a whole multiple of half-wavelengths.

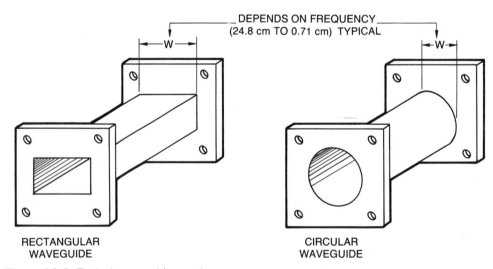

Figure 11-1 Typical waveguide sections.

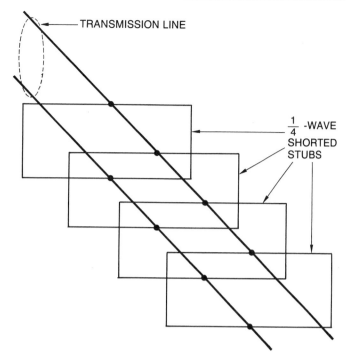

Figure 11-2 Using quarter-wave shorted stubs to create a waveguide.

The cross section of a standard *rectangular waveguide* is shown in Figure 11-3. The internal *width* of the waveguide is determined by the frequency at which it will be used. Standard waveguides have a 2:1 aspect ratio so that the b dimension is half the a dimension.

As shown in Figure 11-3, the width is the a dimension, and the height is the b dimension. Because of the skin effect, the internal walls of the waveguide are coated with silver or gold, which are better conductors than copper.

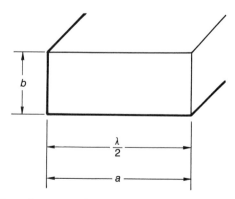

Figure 11-3 Cross section of rectangular waveguide.

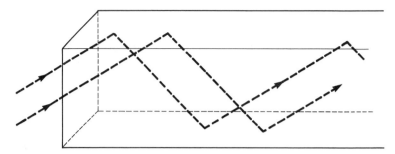

Figure 11-4 Method of wave propagation in a waveguide.

Waveguide Modes

When a signal is transmitted down the waveguide, it must be transmitted at an angle. If it were transmitted parallel to the waveguide walls, the walls would absorb all of the signal energy. The method of propagating a signal in a waveguide is shown in Figure 11-4.

Two types of waves can be generated in a waveguide:

1. *Transverse electric:* called the *TE mode*.
2. *Transverse magnetic:* called the *TM mode*.

Table 11-1 shows the difference between the two modes of waveguide operation.

Table 11-1	MODES OF WAVEGUIDE OPERATION
Mode Type	**Propagation properties**
TE	Electric field is perpendicular to the direction of wave propagation
TM	Magnetic field is perpendicular to the direction of wave propagation

Recall from the last chapter that an electromagnetic wave consists of two perpendicular components: a magnetic field and an electrostatic field. The TE and TM modes are illustrated in Figure 11-5.

Dominant Mode of Operation

The most natural mode of operation for a waveguide is called the *dominant mode of operation*. This mode is the lowest possible frequency that can be propagated in a given

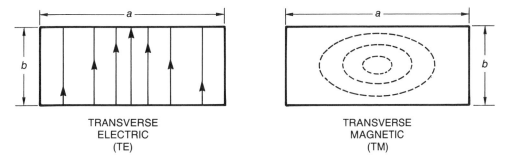

TRANSVERSE
ELECTRIC
(TE)

TRANSVERSE
MAGNETIC
(TM)

Figure 11-5 Illustration of TE and TM modes in a rectangular waveguide.

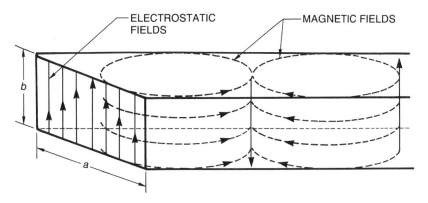

ELECTROSTATIC
FIELDS

MAGNETIC FIELDS

Figure 11-6 Dominant mode of waveguide operation.

waveguide. As shown in Figure 11-6, $\frac{1}{2}$ wavelength is the lowest frequency where the waveguide will still present the properties just discussed.

The mode of operation of a waveguide is further divided into two submodes:

1. TE_{mn} for the transverse electric mode
2. TM_{mn} for the transverse magnetic mode

> *where* m = Number of half-wavelengths across waveguide
>   width (the a dimension)
> n = Number of half-wavelengths along the waveguide
>   height (the b dimension)

Figure 11-7 shows examples of the different modes of waveguide operation.

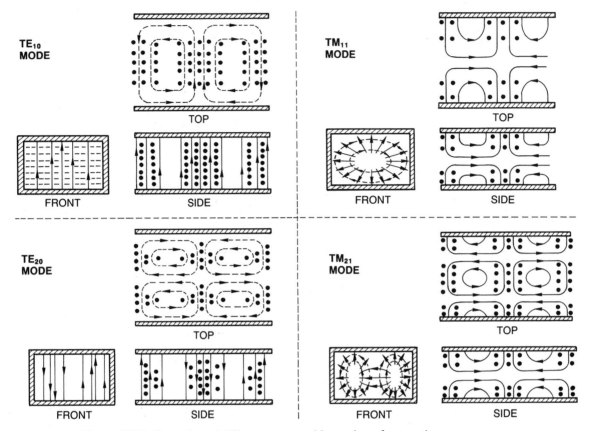

Figure 11-7 Examples of different waveguide modes of operation.

Conclusion

This section introduced the fundamental concepts of waveguides. The next section presents some important mathematical relationships in the analysis of waveguides.

11-1 Review Questions

1. Explain why waveguides are used.
2. Describe the skin effect.
3. Discuss the relationship between the width of a waveguide and the frequency of transmission.
4. State the two major modes of waveguide operation.
5. Explain the dominant mode of waveguide operation.
6. Discuss the submodes of waveguide operation.

| 11-2 | WAVEGUIDE CHARACTERISTICS |

Discussion

This section presents some of the important relationships between the physical characteristics of the waveguide and measurements of the electromagnetic wave within the guide.

Cutoff Wavelength

The *cutoff wavelength* is the wavelength of the lowest frequency that can be accommodated in a given waveguide. It is expressed mathematically as

$$\lambda_0 = \frac{2a}{m}$$ **(Equation 11-1)**

where λ_0 = Cutoff wavelength of waveguide
   a = Waveguide width
   m = Number of half-wavelength fields

Example 1

Determine the cutoff wavelength of a 5-cm × 2.5-cm rectangular waveguide. Assume the dominant mode of operation.

Solution

$$\lambda_0 = \frac{2a}{m}$$ **(Equation 11-1)**

In the dominant mode of operation, $m = 1$. Thus,

$$\lambda_0 = 2 \times 5 \text{ cm} = 10 \text{ cm}$$

The cutoff wavelength must be larger than the free-space wavelength of the signal in order for the signal to propagate down the waveguide. The reason is that the wave must travel at an angle to the waveguide surface. When the wavelength is measured, it is measured along the waveguide surface. This useful relationship is demonstrated in Example 2.

Example 2

A 10-GHz signal is to be propagated along a waveguide with a width of 5 cm. What is the largest value of m that can be accommodated by the waveguide?

Solution

First calculate the free-space wavelength of the signal.

$$\lambda = \frac{3 \times 10^{10} \text{ m/s}}{10 \times 10^9 \text{ Hz}}$$

$$\lambda = 3 \text{ cm}$$

Using the relationship

$$\lambda_0 = \frac{2a}{m} \qquad \text{(Equation 11-1)}$$

gives for $m = 1$

$$\lambda_0 = 2 \times 5 \text{ cm} = 10 \text{ cm (propagation takes place)}$$

for $m = 2$

$$\lambda_0 = \frac{2 \times 5 \text{ cm}}{2} = 5 \text{ cm (propagation takes place)}$$

for $m = 3$

$$\lambda_0 = \frac{2 \times 5 \text{ cm}}{3} = 3.33 \text{ cm (propagation takes place)}$$

and for $m = 4$

$$\lambda_0 = \frac{2 \times 5 \text{ cm}}{4} = 2.5 \text{ cm (propagation does not take place because } \lambda_0 < \lambda)$$

Guide Wavelength

Reflecting the electromagnetic wave inside the waveguide at an angle produces a measured wavelength along the length of the waveguide that is longer than the free-space wavelength of the wave. This wavelength is called the *guide wavelength* and is expressed as

$$\lambda_G = \frac{\lambda}{\sqrt{1 - \left(\dfrac{\lambda}{\lambda_0}\right)^2}} \qquad \text{(Equation 11-2)}$$

where λ_G = Guide wavelength—wavelength of
          signal measured along the length
          of the waveguide
     λ = Free-space wavelength
     λ_0 = Cutoff wavelength

The reason for this phenomenon is illustrated in Figure 11-8.

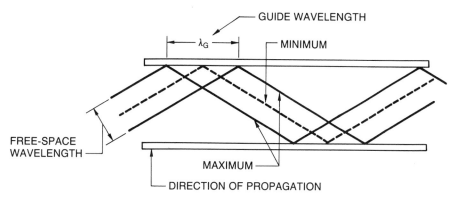

Figure 11-8 Illustration of guide wavelength.

Example 3

A rectangular waveguide with internal dimensions of 4.0 cm × 2.0 cm has a frequency of 10 GHz. Determine the guide wavelength if $m = 2$.

Solution

First solve for the cutoff frequency.

$$\lambda_0 = \frac{2a}{m}$$ **(Equation 11-1)**

For $m = 2$

$$\lambda_0 = \frac{(2 \times 4.0 \text{ cm})}{2}$$

$$\lambda_0 = 4.0 \text{ cm}$$

Solving for the guide wavelength gives

$$\lambda_G = \frac{\lambda}{\sqrt{1 - \left(\dfrac{\lambda}{\lambda_0}\right)^2}}$$ **(Equation 11-2)**

$$\lambda_G = \frac{3}{\sqrt{1 - \left(\dfrac{3}{4}\right)^2}} = \frac{3}{\sqrt{1 - 0.562}}$$

$$\lambda_G = \frac{3}{\sqrt{0.437\,5}} = \frac{3}{0.66}$$

$$\lambda_G = 4.53 \text{ cm}$$

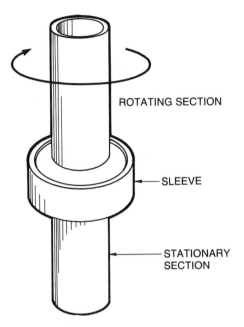

Figure 11-9 Use of circular waveguide.

Circular Waveguides

Circular waveguides are used whenever a rotating element, such as a radar antenna, must be attached to the communications system. Such a coupling system is illustrated in Figure 11-9.

For circular waveguides, the TE_{mn} and TM_{mn} modes are used, except that m indicates the number of *full periods* of the field in the angular direction about the waveguide perimeter. The n subscript shows the number of half-period variations of the field along the radial direction. Some of the different field patterns for circular waveguides are illustrated in Figure 11-10.

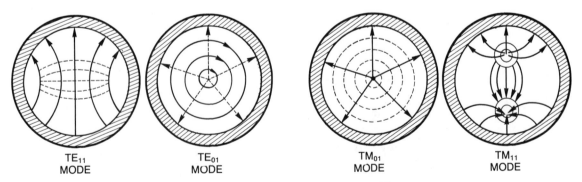

Figure 11-10 Typical field patterns for circular waveguides.

Other Types of Waveguides

A waveguide with circular to rectangular taper is shown in Figure 11-11. This waveguide is used for making the transition from a rectangular to a circular waveguide.

A *ridged* waveguide is shown in Figure 11-12. It is used to operate at lower frequencies than dictated by its outside dimensions. This has practical uses where space is limited, such as in a space satellite.

A *flexible* waveguide is illustrated in Figure 11-13. It is constructed from spiral-wound ribbons of brass or copper. The outside of the waveguide is covered with a soft dielectric material such as rubber. These waveguides are used in the laboratory where continuous flexing may be necessary.

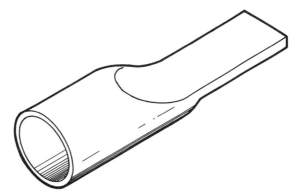

Figure 11-11 Waveguide with circular to rectangular taper.

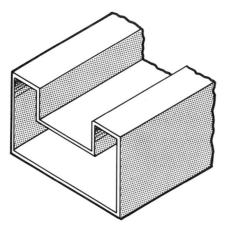

Figure 11-12 A ridged waveguide.

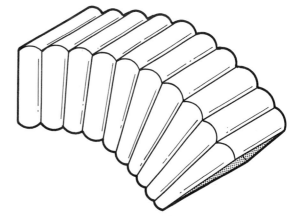

Figure 11-13 A flexible waveguide.

Conclusion

You have seen some of the relationships between the physical properties of a waveguide and the transmitted frequency. You also saw types of waveguides available and their possible uses. The next section shows how energy is coupled to and from a waveguide.

11-2 Review Questions

1. Explain what is meant by the cutoff wavelength of a waveguide.
2. What factors determine the cutoff wavelength of a waveguide?
3. Discuss the relationship between the free-space wavelength and the cutoff wavelength of a waveguide signal.
4. Describe the relationship between the guide wavelength and the free-space wavelength.
5. Name a use for a circular waveguide.
6. Explain the difference between the m and n designation (i.e., TE_{mn}) for circular and rectangular waveguides.
7. What is the advantage of a ridged waveguide?
8. Describe the construction of a flexible waveguide and state its use.

11-3 WAVEGUIDE COUPLING AND ATTENUATION

Discussion

This section presents ways to determine modes of waveguide operations. Also included are methods of coupling or connecting waveguides, impedance matching, and attenuation.

Waveguide Coupling

To inject a signal into a waveguide, you can use several methods. One way is to insert a short piece of wire in the waveguide to act as an antenna. The method by which this is done will determine the mode of operation of the waveguide. See Figure 11-14.

Waveguide connectors can be used to change the direction of the signal. A direction change may be needed to accommodate the requirements of the communications equipment building or some other structure. The most common types are shown in Figure 11-15.

Waveguide Tuning

A waveguide can be *tuned* much like a transmission line can be tuned. When a piece of metal is inserted into a waveguide slot, the waveguide can be tuned. The Q of this tuned circuit depends on the diameter of the piece of metal: the smaller the diameter, the higher the Q. This method of waveguide tuning is shown in Figure 11-16.

Waveguide Termination

A waveguide is another form of a transmission line, and the impedance of its load is an important consideration. There is no place on a waveguide to connect a physical resistor for proper termination. However, other methods of terminating waveguides are used, as

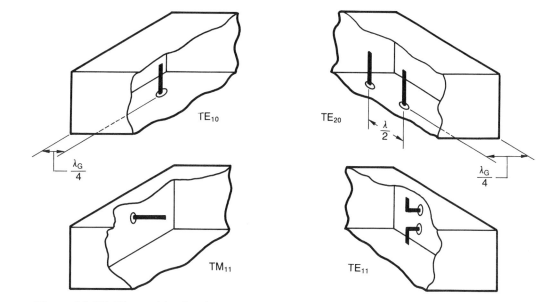

Figure 11-14 Waveguide signal injection.

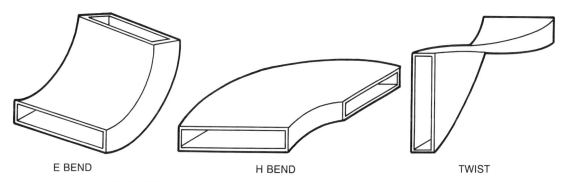

E BEND   H BEND   TWIST

Figure 11-15 Waveguide bends and twists.

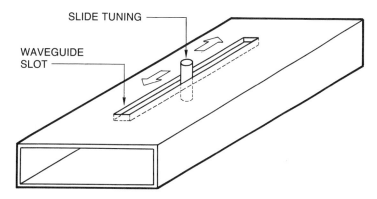

Figure 11-16 Waveguide tuning.

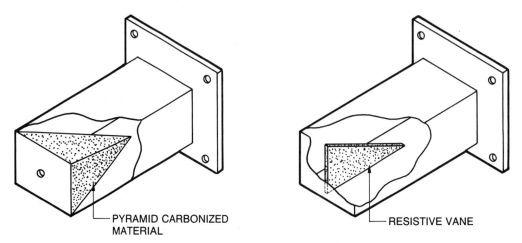

Figure 11-17 Waveguide terminations to minimize reflections.

shown in Figure 11-17. These terminations reduce standing waves by absorbing most of the energy and dissipating it in the form of heat.

Other methods of waveguide termination are shown in Figure 11-18. These methods are used when you want to have all of the energy reflected from the end of the waveguide.

Variable Attenuators

Variable attenuators are available for waveguides. These are useful for calibration and other measurement techniques. See Figure 11-19. A *vane attenuator* consists of two metal vanes that are moved from side to side. The attenuation is minimum when the vanes are alongside the waveguide and maximum when they are near the center. The *flap attenuator* produces maximum attenuation when the flap is pushed into the slot, and minimum attenuation when the flap is out of the slot.

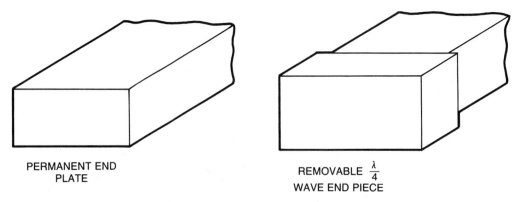

PERMANENT END
PLATE

REMOVABLE $\frac{\lambda}{4}$
WAVE END PIECE

Figure 11-18 Other methods of waveguide termination.

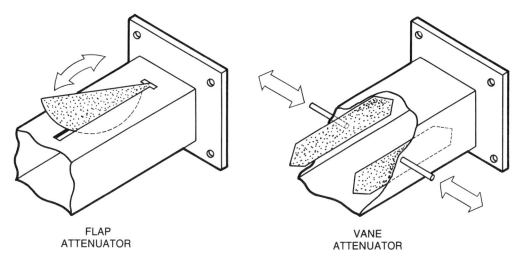

FLAP
ATTENUATOR

VANE
ATTENUATOR

Figure 11-19 Variable waveguide attenuators.

Conclusion

This section illustrated different methods of coupling electromagnetic energy within waveguides and terminating waveguides. The next section introduces more devices, especially for microwave energies.

11-3 Review Questions

1. Describe the methods used to couple a signal into a waveguide.
2. State some of the methods used to change the direction of waveguide signals.
3. Explain a method of tuning a waveguide. State what physical properties of the tuner determine the Q.
4. Discuss some of the methods used for terminating waveguides.

11-4 MICROWAVE MIXERS AND RESONATORS

Discussion

This section presents some of the fundamental microwave devices used for frequency mixing and oscillation. The use of waveguides can be extended to include waveguide techniques for mixers as well as oscillators.

The word "microwave" literally means "small (micro) wave." The word is used to describe a range of frequencies, generally from 1 GHz and above. Because these frequencies are so high, their wavelength is very small, hence the term *microwave*. In the next section, you will see that these frequencies exhibit some distinct properties.

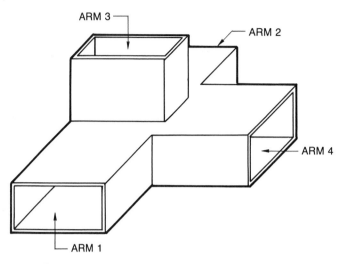

Figure 11-20 Hybrid-T junction.

Microwave Mixers

A microwave mixer consists of a *hybrid-T junction* sometimes called a *magic-T*. Such a construction is shown in Figure 11-20. One of the peculiar properties of the hybrid-T is that arms 3 and 4 are both electrically connected to arms 1 and 2 but not to each other. Thus when a signal is applied to arm 3, it will be divided at the junction leaving arms 1 and 2, but none will leave arm 4. A signal applied to arm 4 produces a signal out of arms 1 and 2 but not arm 3.

The reason for this phenomenon is shown in Figure 11-21. The electric field for the dominant mode is evenly symmetrical about the AB plane in arm 4. This is not so in arm 3. Thus the electric field in arm 4 is a mirror image of the electric field on either side. This is not so with arm 3. Therefore no signal can propagate down arm 3.

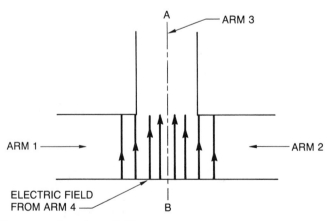

Figure 11-21 Cross section of magic-T.

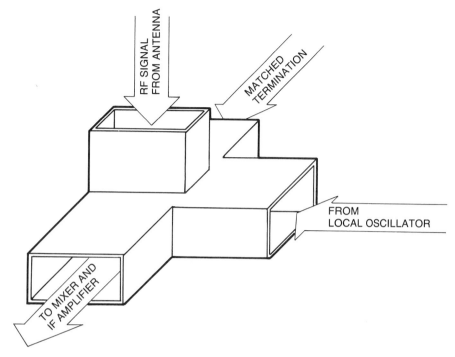

Figure 11-22 Magic-T used in microwave receiver.

One use of a magic-T is for the front end of a microwave receiver. As shown in Figure 11-22, the received signal and local oscillator are fed into arms 3 and 4. This produces two signals: one from arm 1 and the other from arm 2. Arm 1 is terminated in a matched impedance, but arm 2 is attached to the microwave mixer producing an IF output frequency.

Resonant Cavities

A resonant circuit can be produced at microwave frequencies by using what is called a *quarter-wave hairpin*. Such a device can be constructed from a wire. The inductance consists of one turn and the capacitance consists of the capacitance across the ends of the wire, as shown in Figure 11-23.

Figure 11-23 Quarter-wave hairpin.

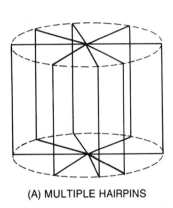

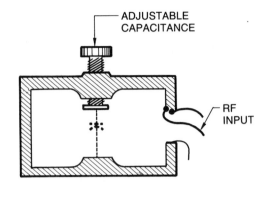

(A) MULTIPLE HAIRPINS

(B) RESONANT CAVITY

Figure 11-24 Resonant cavity.

This configuration appears as a parallel resonant circuit; if replicated as shown in Figure 11-24(A), the Q of the circuit will increase. This concept can be increased until the device appears as a solid cylinder called a *resonant cavity*. The frequency of the resonant cavity can be adjusted by a variable capacitor in the center of the resonant cavity.

Cavity Wavemeter

Figure 11-25 illustrates a *cavity wavemeter*. The movable plunger can be moved up and down the *tunable cavity*. If the size of the cavity becomes resonant to the frequency to be measured, some energy will be absorbed. This absorbed energy can be measured by a wattmeter. If no energy is being absorbed, then the cavity is not resonant.

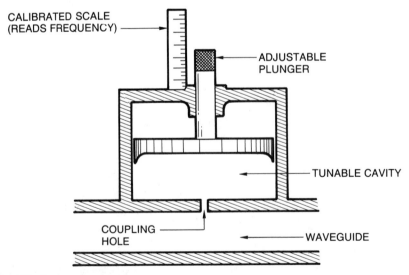

Figure 11-25 Cavity wavemeter.

Conclusion

This section introduced several useful microwave devices. Waveguides are used in many different ways to work with microwave energies. The next section introduces microwave antennas.

11-4 Review Questions

1. Define microwave. What frequency ranges are considered to be in the microwave range?
2. Describe the construction of a microwave mixer. What unique property does this device exhibit?
3. Describe the construction of a quarter-wave hairpin.
4. What is the analogy between a quarter-wave hairpin and a resonant cavity?
5. Describe the construction of a cavity wavemeter.

11-5 │ MICROWAVE COMMUNICATIONS

Discussion

Microwaves are radio frequencies with such short wavelengths that they exhibit some of the properties of light. As stated in the last section, their frequency range is from 1 GHz and up. Generally, microwaves are excellent for line-of-sight communications because they can be focused into a beam by using the proper antennas. This section presents some of the fundamentals of microwave communications.

Microwave Antennas

Because of the high frequencies used in microwave communications, the resulting short wavelength allows highly directional antennas to be used. Some of the antennas can be produced by an open-ended section of a waveguide. Such antennas are categorized as *horn antennas* and are illustrated in Figure 11–26.

Recall that transmission over long distances at microwave frequencies is more economical with antennas. Waveguides are more economical for short distances. The basic idea behind the horn antenna is to provide a gradual flare at the end of a waveguide, thus producing maximum radiation into space with minimum reflection back to the source.

As with waveguides, the relationship of the physical dimensions of the horn antenna to the microwave frequency is important. The actual design of these kinds of antennas is beyond the scope of this book. Suffice it to say that they do allow for line-of-sight transmission of electromagnetic energies at microwave frequencies.

Another form of microwave antenna is the *parabolic* or *dish* antenna. Its basic construction is illustrated in Figure 11-27.

The parabolic antenna is popular at microwave frequencies because of its ability to *focus* electromagnetic waves. This focusing helps concentrate the electromagnetic energies into a beam, thus allowing maximum reception with minimum power. These antennas are used equally well in transmitting or receiving.

A series of microwave transmitters and receivers using line-of-sight transmission forms a *microwave link*. See Figure 11-28. The advantages of a microwave link are (1) minimum

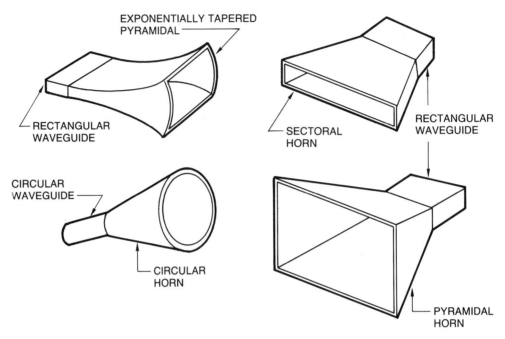

Figure 11-26 Typical horn antennas.

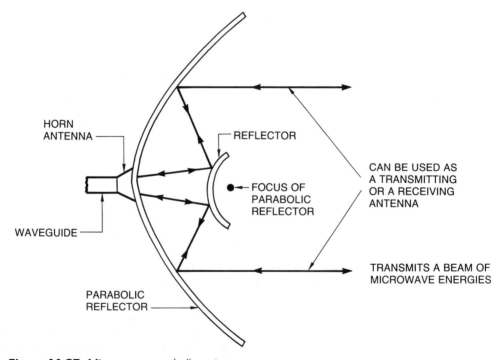

Figure 11-27 Microwave parabolic antenna.

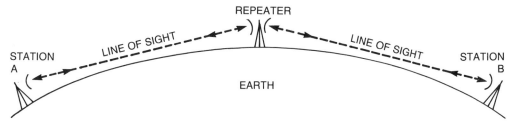

Figure 11-28 A microwave link.

atmospheric disturbance, (2) some degree of privacy, (3) no transmission line maintenance. Microwave links are used at 30- to 50-mile intervals and carry telephone, TV, and other signals.

Conclusion

This section introduced the basics of a microwave system. In the circuits section of this text, you will see how specific devices work to produce microwave frequencies.

11-5 Review Questions

1. Discuss the unique property of microwave transmission.
2. Explain how microwave antennas differ from antennas used at lower frequencies.
3. Describe the construction of a horn antenna.
4. Discuss the unique characteristics of a microwave parabolic antenna.
5. What are the characteristics of a microwave link?
6. What are the advantages of a microwave link?

11-6   RADAR

Discussion

Radar comes from the word "radio" and can be read backward as well as forward. Basically, this describes how radar works. A radio wave is transmitted, and its reflected signal is received back at the transmitter. The amount of time it takes for this to happen is directly proportional to the distance the reflecting object is from the transmitter.

Basic Concept

A radar station consists of a transmitter, a directional antenna, and a receiver, Figure 11-29. A *pulse* of electromagnetic energy is sent from the antenna. After a short time delay, another pulse is sent. If an object capable of reflecting the electromagnetic energy is in the path of the radar beam, then some of this energy is reflected back to the radar antenna, which receives and amplifies the received signal. Since the velocity of electromagnetic radiation is known, it is easy to have the radar system calculate for the observer the distance to the object.

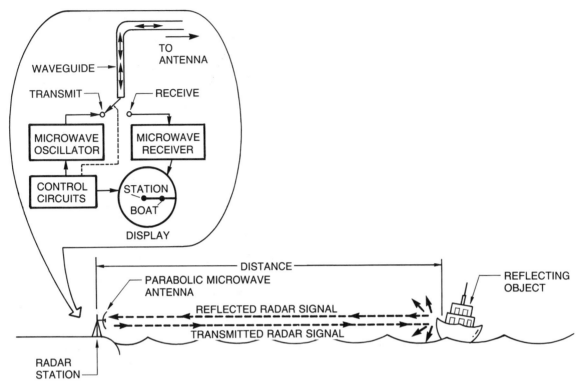

Figure 11-29 Basic radar system.

Radar Signal

Figure 11-30 shows a typical radar signal. It consists of an initial *pulse* followed by a period of *receive time*. During the receive time, the radar system waits for a return of its signal. This process is repeated again, and one complete cycle is called the *pulse repetition*

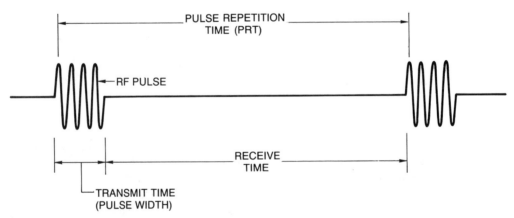

Figure 11-30 Typical radar signal.

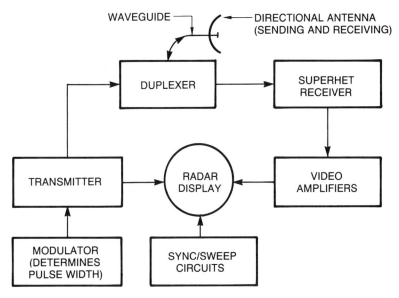

Figure 11-31 Block diagram of a radar signal.

time (PRT) of the radar system. The PRT of the radar system must be long enough to allow for the longest range of the radar system. The minimum range of the radar system is dictated by the width of the radar pulse.

Radar System

The block diagram of a radar system is shown in Figure 11-31. The unique feature of this radar system is the duplexer. A *duplexer* is an electronic switch that allows the same antenna to be used for transmitting and receiving, and it routes the signal accordingly. The *indicator* is usually a cathode ray tube with a display that gives the operator an easy visual indication of distances. Usually the directional antenna is rotated, with the display synchronized to the rotation. A typical resulting display is shown in Figure 11-32.

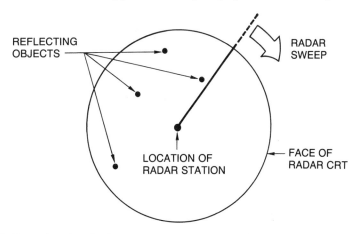

Figure 11-32 Typical radar display.

Conclusion

This section presented the basic concepts of a radar system. The next section discusses some of the important concepts of satellite communication systems.

11-6 Review Questions

1. Explain why the word "radar" was chosen to describe a system that is used to electronically determine distance.
2. Describe the basic operation of a radar system.
3. State the main sections of a radar system.
4. Describe a typical radar signal. What components of the signal determine the range?
5. Describe the operation of a duplexer.

11-7 | SATELLITE COMMUNICATIONS

Discussion

Satellite communications allow line-of-sight communication signals to be transmitted anywhere in the world. This is done with satellites that orbit the earth in such a way that they appear stationary over one point of the earth. Satellite communications are also used to explore extraterrestrial geographies, such as the moons of Saturn and beyond. Both methods of satellite communications are illustrated in Figure 11-33.

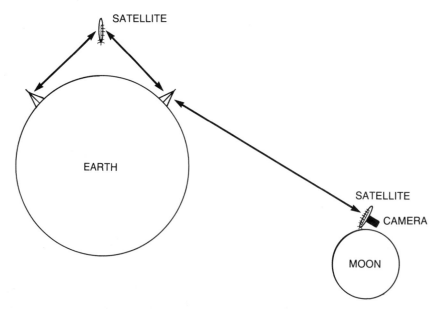

Figure 11-33 Methods of satellite communications.

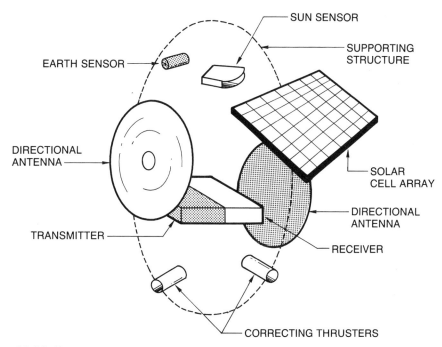

Figure 11-34 Basic communications satellite.

Basic Satellite System

Figure 11-34 is a basic satellite communication system. It consists of

1. A solar cell array for a source of continuous power
2. Directional antennas for receiving and transmitting signals
3. Correcting thrusters
4. Environmental sensors
5. Communications receiver and transmitter

Communications Satellite

It is the function of a communications satellite to orbit high enough above the earth so that it may serve as a microwave link that connects widely separated points on the earth. Such an orbit is called a *geosynchronous* or *geostationary* orbit. Many of these satellites are in orbit above the earth's equator at intervals of about 180 miles. These satellites have microwave receiving and transmitting systems.

The operation of this system consists of receiving a signal from the earth (such as a TV program), amplifying it, and retransmitting it to earth, usually at a lower microwave frequency.

Satellites are equipped with multiple repeater units called *transponders*. The latest satellite systems consist of 46 transponders. The capacity of a single transponder can be as high as 3 000 channels with a total bandwidth of 3 460 MHz.

Multiuser Satellite Communication Systems

The frequency allocation for satellite systems is given in Table 11-2.

Table 11-2	SATELLITE SYSTEM FREQUENCY ALLOCATIONS	
Uplink Frequencies (GHz)		**Downlink Frequencies (GHz)**
5.925–6.425		3.700–4.200
7.900–8.400		7.250–7.750
14.00–14.5		11.70–12.20
27.50–30.00		17.70–20.20

For example, a country, say Mexico, may use 132 voice-grade channels for transmitting and receiving voice and other low-frequency data to and from other countries. Assume the required bandwidth is 10 MHz to meet all these requirements. With a bandwidth of 500 MHz for one transponder, 50 different users each requiring 132 channels could be accommodated. Assume that the Mexico user is assigned the uplink frequency band of 6 000 MHz to 5 990 MHz (the corresponding downlink frequency band would be 3 775 MHz to 3 765 MHz). This leaves room for other users in the same satellite transponder. For example, a Japanese user may be assigned the 5 940–5 930-MHz uplink with the corresponding 3 715–3 705-MHz downlink. An Australian user may also be included, as shown in Figure 11-35. Such a system allows two-way communications between any two of three different locations, and only a part of one transponder is used.

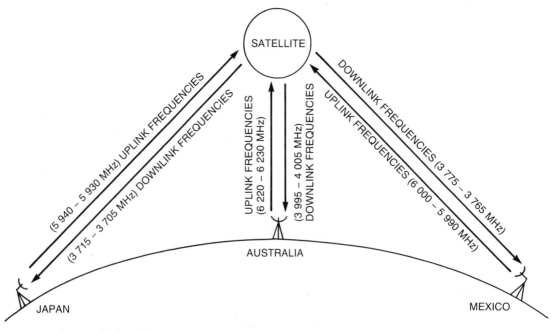

Figure 11-35 Multiuser satellite system.

Conclusion

Communications satellites have made isolated regions of the earth accessible to TV, telephone, and radio signals. Televised events can be globally viewed, and international telephone communication is now commonplace.

11-7 Review Questions

1. Explain what kind of transmission is used by a satellite communication system.
2. Describe the two major satellite communication systems.
3. State the major parts in a communications satellite.
4. Describe the kind of planetary orbit used by a communications satellite.
5. Explain why many users can use one satellite at the same time.
6. Explain the difference between a downlink and an uplink.

TROUBLESHOOTING AND INSTRUMENTATION

Antenna Measurements

There are many ways to measure the resonant frequency of an antenna. To begin these measurements, you should know the approximate resonant frequency of the antenna. You can find it from the physical dimensions of the antenna. The method presented in this section uses the VSWR bridge.

VSWR Bridge

A *VSWR bridge* is a device that separates the forward wave from the reflected wave. See Figure 11-36. With the VSWR bridge, the frequency at which the reflected wave drops to a minimum is considered to be the resonant frequency.

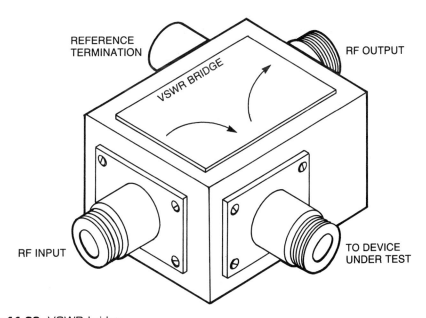

Figure 11-36 VSWR bridge.

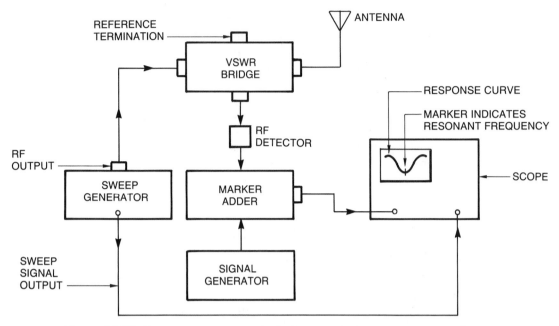

Figure 11-37 Sweep generator method of measuring antenna resonant frequency.

Sweep Generator Method

One method of using the VSWR bridge to determine the resonant frequency of an antenna is illustrated in Figure 11-37. This method produces a scope pattern of the reflected signal amplitude with respect to the applied sweep frequency. The point of minimum response represents the resonant frequency of the antenna. A signal generator can be used to place a marker on the scope pattern to indicate the resonant frequency.

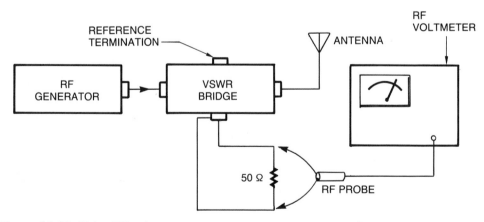

Figure 11-38 Using RF voltmeter to measure antenna resonant frequency.

RF Voltmeter Method

Figure 11-38 illustrates a method of measuring the resonant frequency of an antenna with the VSWR bridge and an RF voltmeter. Before performing this test, make sure that the response of the RF generator and RF voltmeter are reasonably flat over the range of frequencies to be measured. Then vary the output frequency of the RF generator around the expected resonant frequency of the antenna under test. The frequency at which the RF voltmeter shows a minimum reading indicates the resonant frequency of the antenna.

Conclusion

This section presented two methods for measuring the resonant frequency of an antenna. There are other methods, depending upon the type of equipment available and the type of antenna under test. The next section presents a structured BASIC program that simulates standard waveguides.

11-8 Review Questions

1. To use the measurements described in this section, state what you should already know about the antenna to be measured.
2. Describe a VSWR bridge.
3. Describe the equipment needed to measure the resonant frequency of an antenna by the sweep generator method.
4. Describe the equipment needed to measure the resonant frequency of an antenna by the RF generator method.

MICROCOMPUTER SIMULATION

The eleventh troubleshooting simulation requires you to use all of the troubleshooting skills you have acquired up to this point. Here you will be working with a multi-stage amplifier which may have a GO/NO-GO problem, or reduced gain or bandwidth. This simulation is approaching a real world troubleshooting situation where you, as the technician, must be prepared for a variety of potential problems.

CHAPTER PROBLEMS

(Answers to odd-numbered problems appear at the end of the text)

TRUE/FALSE
Answer the following problems true or false.
1. "Microwave" means small wave and is used to designate frequencies of 1 GHz and above.
2. Satellite communication uses line-of-sight transmission and reception.
3. A magic-T is a device used to select radar signals.

4. The width of a waveguide must always be equal to or larger than the wavelength of the lowest frequency to be transmitted in the guide.
5. The dominant mode in a waveguide represents the lowest frequency that can be transmitted by the waveguide.

MULTIPLE CHOICE
Answer the following questions by selecting the most correct answer.

6. In a waveguide, the TE mode means that the:
   (A) magnetic field is perpendicular to the direction of wave propagation.
   (B) electric field is perpendicular to the direction of wave propagation.
   (C) magnetic and electric fields are perpendicular to each other.
   (D) None of the above are correct.
7. In a waveguide, the TM mode means that the:
   (A) magnetic field is perpendicular to the direction of wave propagation.
   (B) electric field is perpendicular to the direction of wave propagation.
   (C) magnetic and electric fields are perpendicular to each other.
   (D) None of the above are correct.
8. Ridged waveguides are used:
   (A) to offset antenna losses.
   (B) so that higher frequencies can be used.
   (C) for impedance matching in waveguide couplers.
   (D) None of the above are correct.
9. The mode at which a waveguide operates is determined by the:
   (A) method used to couple the signal into the waveguide.
   (B) shape of the waveguide.
   (C) impedance of the waveguide.
   (D) length of the waveguide.
10. A waveguide can be terminated by:
    (A) placing a resistor across the end of the guide.
    (B) ending it with a capacitor or an inductor.
    (C) terminating it with an end plate.
    (D) Both (A) and (C) are correct.

MATCHING
Match the figures in Figure 11-39 to the correct modes on the left.
11. TE_{10}.
12. TE_{20}.
13. TM_{12}.
14. TM_{21}.

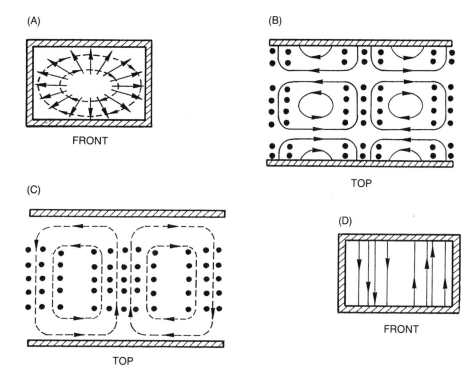

Figure 11-39

FILL IN

Fill in the blanks with the most correct answer(s).

15. A _____ antenna is essentially an extension of the end of a waveguide.

16. At microwave frequencies, communication for short distances is best achieved by using

    _____.

17. The antenna used at microwave frequencies to focus its waves is a(an) _____ antenna.

18. A radar transmitter sends a _____ of electromagnetic energy that can be reflected by certain objects.

19. The device that allows the same antenna to transmit and receive and routes the signal accordingly is a(an) _____.

20. A satellite that maintains an orbit above the same place on the earth is a(an) _____ orbit.

OPEN ENDED

Answer the following questions as indicated.

21. Determine the cutoff wavelength of a 6-cm × 3-cm waveguide. Assume the dominant mode of operation.

22. For the waveguide in problem 21, what would be the cutoff frequency for the TE_{20} mode of operation?

23. A 5-GHz signal is to be propagated along a waveguide with a width of 3 cm. What is the largest value of m that can be accommodated by the waveguide?

24. For the waveguide in problem 23, what would be the largest value of m if the signal frequency were 6 GHz?

25. Determine the guide wavelength if $m = 1$ for a 6.0-cm $\times$ 3.0-cm waveguide with a frequency of 12 GHz.

26. What would be the guide wavelength for the guide in problem 25 if $m = 2$?

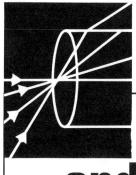

CHAPTER 12

Fiber Optics and Lasers

OBJECTIVES

In this chapter, you will study:

- ☐ The advantages and disadvantages of communication systems that use light.
- ☐ Fundamental properties of light propagation.
- ☐ Construction and measurement of fiber optics.
- ☐ Applications and limitations of fiber-optic communications.
- ☐ Theory of laser operation.
- ☐ Laser measurements and parameters.
- ☐ How laser light is created.
- ☐ The laser in communications.
- ☐ Laser safety.

INTRODUCTION

One of the newest and fastest-growing areas of electronic communications is optical transmission. The unique property of this type of communication is that it can be seen. The antennas that you and I call *eyes* are tuned to this part of the frequency spectrum, which exhibits some very unique properties. These unique properties will be used to overcome many of the limitations of microwave and lower-frequency transmissions.

12-1 LIGHT AS A CARRIER

Discussion

This chapter discusses the transmission of data using light. Recall the simple flashlight system introduced in Chapter 2. With that system, the voice changed the intensity of the flashlight beam. The system was crude, but it did illustrate the concept of a carrier.

You will be introduced to two new concepts in electronic communications: fiber optics and lasers. *Fiber optics* is the study of light transmission in a controlled path. The transmission is through light conductors. This is similar to the transmission of electrical currents through wires or microwave frequencies through waveguides. *Laser* stands for *l*ight *am*plification by *s*imulated *e*mission of *r*adiation. Essentially a laser is capable of producing light of a single frequency (much like an RF carrier wave) that travels in intense, almost perfect, parallel rays with very little scattering. This form of light is called *coherent*.

Background

The first recorded attempt to transmit voice with light as a carrier was by Alexander Graham Bell. In 1880, he patented a device called the *photophone*, Figure 12-1. With the photophone, a voice signal could be transmitted about 200 feet under ideal conditions. Voice waves caused the "speaking trumpet" to vibrate a mirror. The sunlight reflected from the mirror into a photovoltaic cell whose voltage changed in step with the changes in light intensity. Since the changes in light intensity were produced by the voice, the voice was reproduced by the vibrations of the headset, which were caused by the voltage changes from the photovoltaic cell. The weakness of this system was its poor and unreliable light source and transmission medium.

The generation of laser light was predicted by Albert Einstein in 1920. In May 1960, Dr. Theodore Mainman, a scientist at Hughes Research Laboratories, produced the first laser light. Thus the needed source for communications by light was demonstrated.

The application of practical glass fibers for light transmission was first proposed in 1966, and the first practical fiber optics were demonstrated in 1970. By 1979, glass fibers capable of carrying 400 000 000 bits per second were demonstrated in the laboratory. Hence the needed light carrier for communications by light was also made available.

Advantages

Using light as a carrier in a controlled environment (such as an optical fiber) has several advantages over using an RF carrier through space, wires, or waveguides:

- Wider bandwidth
- Immunity from electromagnetic interference
- Immunity from interception by external means (privacy)

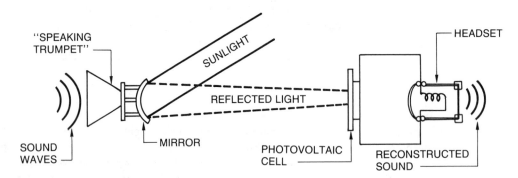

Figure 12-1 The photophone.

- Inexpensive and abundant materials (silicon is the most abundant element on the earth)
- Resistance to corrosion and oxidation
- Immune to atmospheric changes

The development of the laser has led to:

- Practical communications using light
- Accurate surveying instruments
- Medical applications where selected tissue can be treated by extremely small intense beams of light
- Welding, drilling, and machining of metals and other materials with microscopic accuracy
- A source of extremely high, fast-moving, accurate energy
- Optical scanners for reading bar codes (currently in use at grocery checkout counters), laser disks (audio and video), and credit card identification
- Holographs, where a two-dimensional surface is capable of reproducing three-dimensional information

Conclusion

This section introduced fiber optics and lasers. You were given a brief historical development of these systems. The advantages of the optical fiber and the laser were given. The next section defines some very important optical terms. Once you are familiar with these terms, you will understand the application of light to fiber optics and lasers.

12-1 Review Questions

1. Define fiber optics.
2. Define the word laser.
3. Describe the first device used to transmit voice using light as a carrier. Name some problems with this system.
4. State some advantages of using light as a carrier in a controlled environment.
5. Describe some uses of lasers.

12-2 | INTRODUCTION TO LIGHT

Discussion

This section presents the fundamental methods and units of measuring some properties of light. You may think of this section as the "basic electronics" part of light communications. When you have completed it, you will be prepared to further analyze the phenomenon of light.

Wave and Particle

Light can be thought of as *electromagnetic waves*. This *wave theory* of light adequately predicts many of its properties. Light can also be thought of as composed of tiny particles of energy called *photons*. This *particle* aspect of light is necessary to explain the inter-

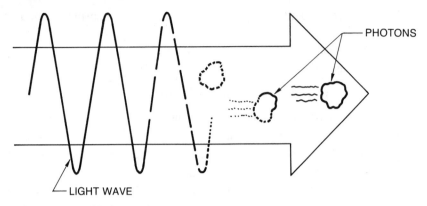

Figure 12-2 Duality of light.

action of semiconductor material and light, such as in a phototransistor. Thus, light can be thought of as propagating through space like a radio wave, but as composed of particles when it interacts with matter. This *duality* of light is illustrated in Figure 12-2.

Optical Spectrum

The *optical spectrum* is electromagnetic radiation in the wavelength range of infrared, visible, and ultraviolet rays. This radiation has a velocity of $2.997\ 925 \times 10^8$ meters/second in a vacuum (300 000 000 meters/second is generally used as a first approximation). The optical spectrum and the various units of measurements are illustrated in Figure 12-3. Note that *visible light* is a small part of the optical spectrum. Thus you may correctly conclude that fiber optics can conduct, and lasers can produce, energy that you and I cannot see, yet this energy has the same properties as visible light since it is within the optical spectrum.

Because the frequencies involved within the optical spectrum are so large

$$300\ 000\ 000\ 000 \text{ to } 30\ 000\ 000\ 000\ 000\ 000 \text{ Hz}$$

it is more convenient to use the wavelength. As shown in Figure 12-3, the most common units of measurement for wavelength are the *micron* (10^{-6} m), the nanometer (10^{-9} m) and the angstrom (10^{-10} m). As shown in Figure 12-4, the visible spectrum is between 390 nanometers and 770 nanometers.

In this chapter, when reference is made to *light,* it means the range of frequencies within the total *optical spectrum,* not just the *visible spectrum.*

Light Ray

For many applications, it is convenient to think of a light beam as a single straight line called a *light ray.* The difference between a light beam and a light ray is shown in Figure 12-5. The light ray concept easily shows the direction of the light beam.

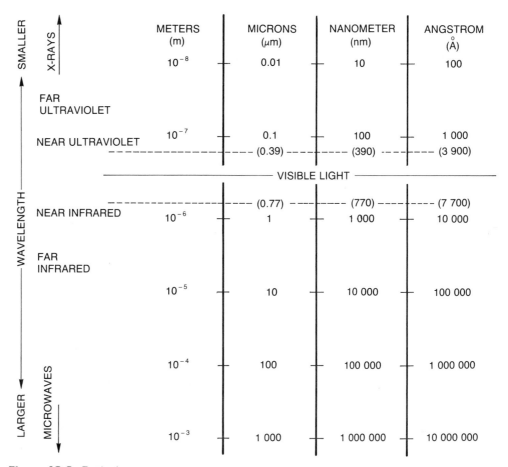

Figure 12-3 Optical spectrum.

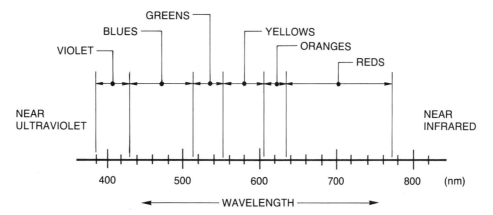

Figure 12-4 The visible spectrum.

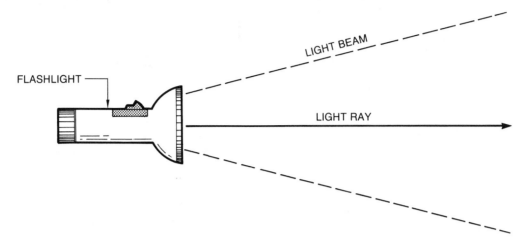

Figure 12-5 Light ray and light beam.

Reflection

A *reflected* wave is a light wave that meets a polished surface and returns back to space (or its originating medium). A reflected light wave can be thought of as having two distinct *rays:* the *incident ray* and the *reflected ray*. As Figure 12-6 shows, the angle of incidence

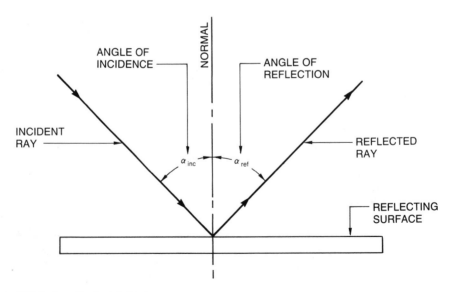

Figure 12-6 A reflected light ray.

is equal to the angle of reflection. Note how the angle is measured with respect to an imaginary perpendicular line to the reflecting surface called the *normal*.

$$\alpha_{inc} = \alpha_{ref}$$ **(Equation 12-1)**

where α_{inc} = Angle of incidence
            α_{ref} = Angle of reflection

Example 1

A light ray strikes a polished metal surface at an angle of 30° with respect to the normal. Construct a *ray diagram* that shows the angle of incidence and the angle of reflection.

Solution
Refer to Figure 12-7 for the solution.

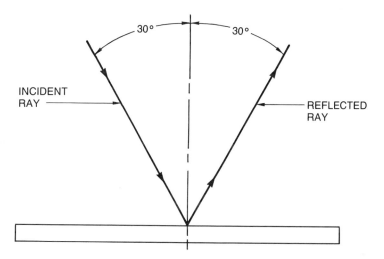

Figure 12-7

Refraction

Refraction is the bending of light as it passes from one medium to another. This is not the same as reflection. Refraction is illustrated in Figure 12-8.

Refraction occurs because light travels slower in water than in air. Thus the light waves bend as they cross the boundary between water and air. Note from Figure 12-8 the *true* and *apparent* positions of an object under water. This illusion is caused by refraction because the velocity of light in water is smaller than the velocity of light in air.

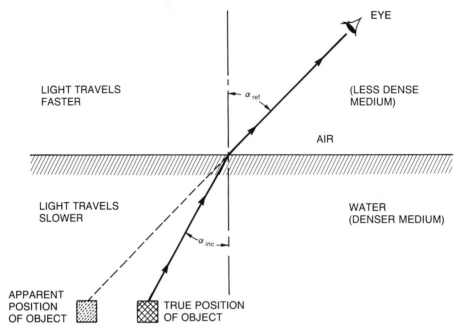

Figure 12-8 Illustration of refraction.

Snell's Law

The refraction of light in different mediums can be stated in the mathematical relationship called *Snell's law:*

$$\frac{\sin \alpha_{inc}}{\sin \alpha_{refr}} = \frac{Vel_1}{Vel_2}$$

(Equation 12-2)

where α_{inc} = Angle of incident wave
α_{refr} = Angle of refracted wave
Vel_1 = Wave velocity in medium 1
Vel_2 = Wave velocity in medium 2

Example 2

Referring to Figure 12-9, assume that the velocity of the incident ray is 3×10^8 m/s. Using Snell's law, determine the velocity of light in the liquid.

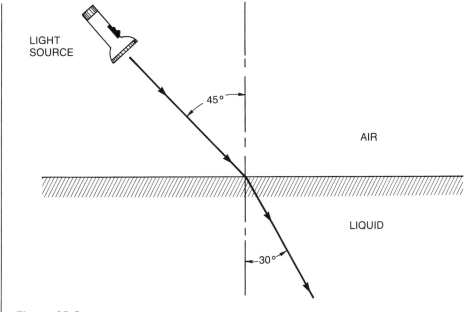

Figure 12-9

Solution

Using Snell's law

$$\frac{\sin \alpha_{inc}}{\sin \alpha_{ref}} = \frac{Vel_1}{Vel_2}$$

(Equation 12-2)

$$\frac{\sin 45°}{\sin 30°} = \frac{3 \times 10^8}{Vel_2}$$

$$\frac{0.707}{0.5} = \frac{3 \times 10^8}{Vel_2}$$

$$1.414 = \frac{3 \times 10^8}{Vel_2}$$

Solving for Vel_2 gives

$$Vel_2 = \frac{3 \times 10^8}{1.414}$$

$$Vel_2 = 2.12 \times 10^8 \text{ m/s}$$

Index of Refraction

The *index of refraction* of a material is an indication of how much refraction a light ray will experience in the material as compared to a vacuum. It is expressed as

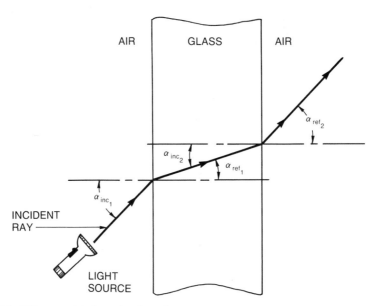

Figure 12-10 Effects of index of refraction.

$$n = \frac{\sin \alpha_{inc}}{\sin \alpha_{refr}}$$

(Equation 12-3)

where n = Index of refraction
  α_{inc} = Angle of incidence
  α_{refr} = Angle of refraction

In Figure 12-10, note that when the light wave goes from a material with a lower index of refraction to a material with a higher index of refraction, the ray bends inward. The opposite is true when the indexes are reversed.

Table 12-1 lists some typical indexes of refraction.

Table 12-1	TYPICAL INDEXES OF REFRACTION (APPROXIMATE VALUES)
Material	**Approximate Index of Refraction**
Air	1.000 3
Diamond	2.4
Glass	1.6
Vacuum	1.000 0
Water	1.3

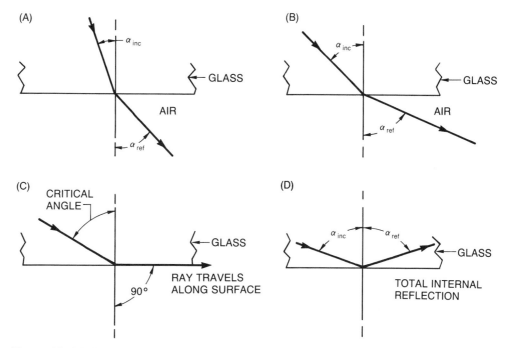

Figure 12-11 Concept of critical angle.

Critical Angle

An important concept in fiber optics is the *critical angle*. In Figure 12-11(A) when the ray passes from glass to air, the angle of refraction is greater than the angle of incidence. In Figure 12-11(B), the angle of incidence is increased further and, as expected, so is the resulting angle of refraction. However, a point is reached, Figure 12-11(C), where a further increase in the angle of incidence causes the angle of refraction to be exactly 90° to the normal. At this point, the light ray travels along the surface of the glass. The angle of incidence that causes the angle of refraction to be exactly 90° is called the *critical angle*. Figure 12-11(D) shows that increasing the angle of incidence to be greater than the critical angle causes the ray to be reflected back into the glass rather than being transmitted out of it.

Conclusion

This section introduced some fundamental properties and measurements of light. You are now ready to see the unique qualities of fiber optics.

12-2 Review Questions

1. What is the duality of light?
2. What are the limits of the optical spectrum?
3. What is the difference between the optical spectrum and the visible spectrum?
4. Discuss what the word "light" means in the context of this chapter.
5. Describe the difference between a light ray and a light beam.

6. Discuss the difference between a reflected ray and a refracted ray.
7. Explain the meaning of Snell's law.
8. Discuss the importance of the critical angle.

12-3 LIGHT TRANSMISSION IN GLASS

Discussion

The basic concept behind fiber optics is illustrated in Figure 12-12. The main idea is to have light from a light source injected into a glass fiber. Once the light is inside the fiber, its rays reflect off the inside walls of the glass toward the other end of the fiber. This process continues until the light emerges at the receiving end of the glass fiber.

Though this process appears simple, it wasn't until recently that the practical limitations of such a system were overcome. This section presents the principles of light transmission in glass. An understanding of these principles makes the application of fiber optics possible.

Development

The first recorded transmission of light in a transparent medium other than air was demonstrated in 1870 by John Tyndal, a British physicist, who demonstrated that light would follow a stream of water pouring from an illuminated container. The first use of glass fibers to carry an image was in the 1930s. Communication systems using fiber optics became practical in the late 1960s.

Glass-Air Boundary

Recall from the last section the discussion concerning the importance of *critical angle*. The critical angle depends on the refractive index between two materials. It may be expressed as

$$\sin \alpha_c = \frac{n_2}{n_1}$$

(Equation 12-4)

where α_c = Critical angle
n_1 = Index of refraction of the more dense material
n_2 = Index of refraction of the less dense material

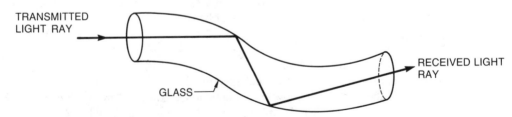

TRANSMITTED LIGHT RAY

RECEIVED LIGHT RAY

GLASS

Figure 12-12 Basic concept of fiber optic.

Example 1

Determine the critical angle for a glass-air interface. Explain what this means.

Solution

Using the relationship for critical angle

$$\sin \alpha_c = \frac{n_2}{n_1}$$

(Equation 12-4)

Table 12-1 gives n_{air} = 1.000 3 and n_{glass} = 1.6. Therefore n_1 = 1.6 and n_2 = 1.000 3. Hence,

$$\sin \alpha_c = \frac{1.000\ 3}{1.6} = 0.625$$
$$\alpha_c = \sin^{-1}(0.625) = 38.7°$$

This answer means if the angle of the incident light ray is less than 38.7° with respect to the normal, it will *refract* and leave the glass material. If the angle of the incident light ray is greater than 38.7° with respect to the normal, it will *reflect* and stay within the glass. See Figure 12-13.

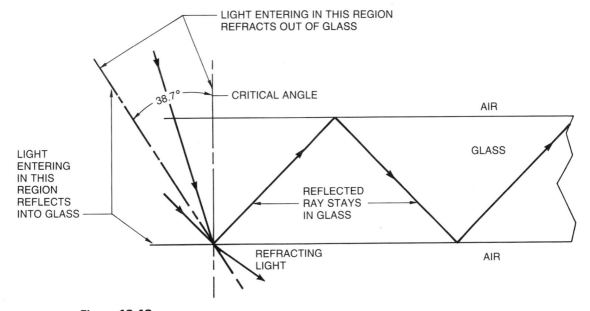

Figure 12-13

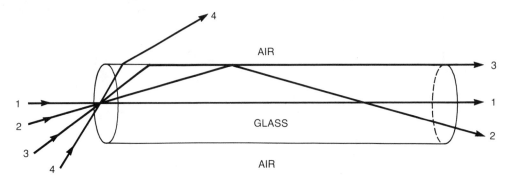

Figure 12-14 Light propagation in a glass rod.

Light in a Glass Rod

The critical angle between a glass-air interface depends on the actual index of refraction of the glass and the surrounding medium. This can have many different values, depending on the composition of the glass material. In general, the greater the density of the glass, the greater will be the index of refraction. For the purposes of discussion in this section, the index of refraction for glass will be 1.6.

Figure 12-14 illustrates the propagation of light within a glass rod. Recall from Example 1 that the critical angle for a glass-air interface is 38.7° for an index of refraction of 1.6 for glass with the surrounding medium.

Figure 12-14 shows several different light rays. Ray 1 transmits along the axis of the rod. Therefore ray 1 is called an *axial ray*. Table 12-2 summarizes the action of the rays in Figure 12-14.

Propagation in Glass Fiber

Optical fibers that propagate light by total internal reflection are commonly found. Observe the transmission of the axial ray in the glass fiber of Figure 12-15. Light will propagate within the fiber as long as the angle of incidence is greater than the critical angle: otherwise, the light ray will escape from the glass fiber.

Table 12-2	RAY PROPAGATION IN GLASS ROD
Ray Number	**Properties**
1	Axial ray: transmits along the axis of the glass rod
2	Reflected ray: strikes the surface greater than the critical angle of 38.7°
3	Surface ray: transmits along the surface of the glass-air interface and is equal to the critical angle
4	Refracted ray: less than the critical angle, so ray escapes from the glass rod

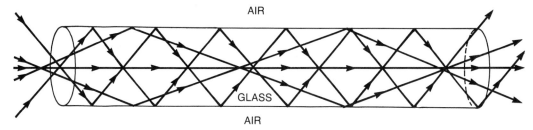

Figure 12-15 Light propagation in glass fiber.

Two major methods of transmitting light along a glass fiber are using a step-index fiber, and using a graded-index fiber. A *step-index* fiber is a glass fiber with a uniform index of refraction. A *graded-index* fiber has an index of refraction that is largest in the center and smaller toward the surface. See Figure 12-16.

To see how a graded-indexed fiber operates, consider a glass fiber constructed from several layers of glass with different indexes of refraction, as shown in Figure 12-17.

To understand how the multilayered glass fiber affects a light ray, recall that a light ray will bend toward the surface when going from a more dense material to a less dense material. It will do the opposite when going from a less dense material to a more dense material. This results in a light ray transmission as shown in Figure 12-18.

In actual practice, a graded-indexed fiber is made from a glass fiber that has a continuous change in its index of refraction. Instead of appearing as a multilayered glass fiber with distinct changes in its index of refraction, a graded-index fiber can be thought of as having an infinite number of layers.

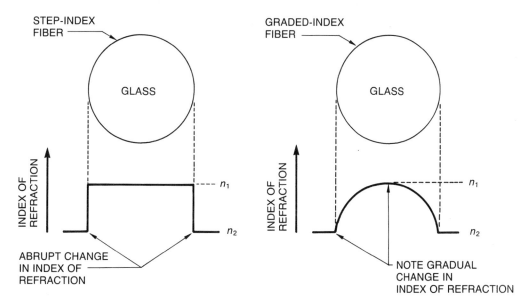

Figure 12-16 Step-index and graded-index fibers.

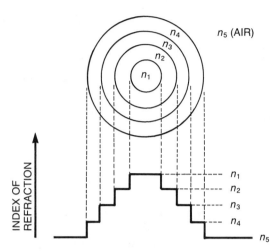

Figure 12-17 Example of a multilayered glass fiber.

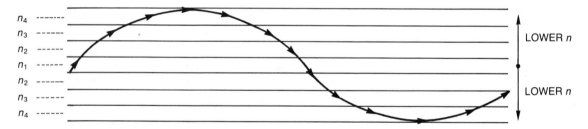

Figure 12-18 Ray travel in a multilayered glass fiber.

Conclusion

This section presented some important properties of the light transmission in glass. The discussion ranged from the glass-air interface to transmission in a glass rod and then a glass fiber. In the next section, you will see how fibers for light transmission are actually constructed.

12-3 Review Questions

1. Explain the significance of the critical angle in the glass-air interface of a glass rod.
2. Describe what will happen to light that enters a glass rod at (A) less than the critical angle, (B) more than the critical angle, (C) at the critical angle.
3. What are the differences between (A) an axial ray, (B) a reflected ray, (C) a surface ray, and (D) a reflected ray?
4. Describe the difference between a step-index optical fiber and a graded-index optical fiber.
5. Describe the propagation of a light ray in a multilayered glass fiber.

12-4 | CONSTRUCTION OF OPTICAL FIBERS

Discussion

The construction of an actual optical fiber includes more than just a simple strand of glass. For example, if a chip or scratch occurred on the surface of the glass fiber, transmission losses would occur. See Figure 12-19. This section presents the problems encountered with the application of fiber optics and the methods for minimizing them.

Kinds of Optical Fibers

An optical fiber is usually covered with a protective jacket to protect it from surface damage. The optical fiber can be made from a single strand of glass or from many strands of glass. The many-stranded form is less likely to break. Both forms are shown in Figure 12-20.

Glass fibers have superior optical properties but are fragile. Certain kinds of optical plastics are more durable but have poorer transmission properties. A good compromise is a combination glass-plastic optical fiber.

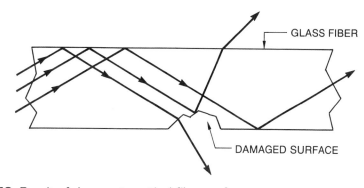

Figure 12-19 Result of damage to optical fiber surface.

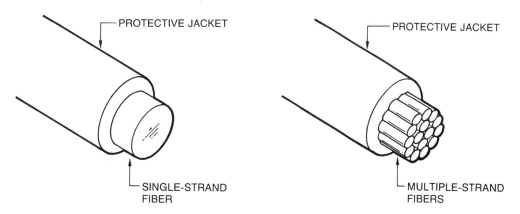

Figure 12-20 Single-strand and multiple-strand fibers.

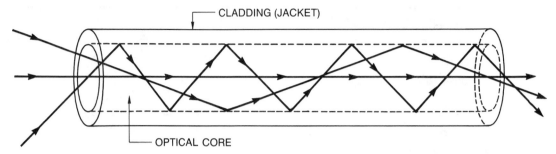

Figure 12-21 Light transmission in a step-indexed fiber.

Optical Dispersion

Consider a step-index fiber transmitting a beam of light as illustrated in Figure 12-21. The paths traveled by each light ray are not the same length. Thus each light ray will arrive at the end of the optical fiber at a different time. These various paths of light are called *modes*. The light transmitted in the optical fiber in Figure 12-21 is classified as a *multimode* step-indexed optical fiber.

The disadvantage in a multimode optical fiber is that the original input pulse will be "washed-out" or flattened. This pulse spreading is called *modal dispersion*. See Figure 12-22.

Modal dispersion can be reduced by surrounding the optical fiber with an optical cladding. The refractive index of the cladding must be smaller than that of the fiber. As the difference between the refractive indexes becomes smaller, the modal dispersion becomes smaller.

Another way of reducing modal dispersion is to use an optical fiber with a very small diameter. In practice, this is about 2.5 microns, but such fibers are hard to work with. Hence, designers use an optical cladding, a small difference in refraction index, and a core of about 15 microns. This construction gives single-mode transmission and eliminates optical dispersion. See Figure 12-23.

Refractive Index Profiles

A *refractive index profile* illustrates the relative indexes of refraction of each material in an optical fiber along with the thickness of each material. The refractive index profile of the three most common types of optical fibers is shown in Figure 12-24.

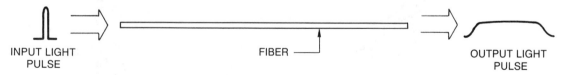

INPUT LIGHT
PULSE

FIBER

OUTPUT LIGHT
PULSE

Figure 12-22 Effects of modal dispersion.

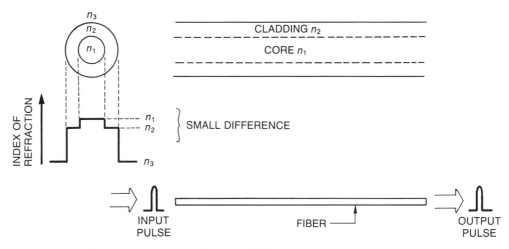

Figure 12-23 Single-mode step-index optical fiber.

The step-index multimode fiber, Figure 12-24(A), is the most economical of the three but suffers from modal dispersion, thus limiting the rate at which information can be transmitted. The step-index single-mode fiber, Figure 12-24(B), is more expensive, but it doesn't suffer from modal dispersion. The graded-index multimode fiber, Figure 12-24(C), is a good compromise. It is discussed further in the next subsection.

Graded-Index Multimode Fiber

Graded-index multimode fiber construction decreases modal dispersion. The paths of light rays in such a fiber are shown in Figure 12-25. Observe that the mode with the longest path is the *higher-order mode*. You may think that since this mode travels the longest distance it would arrive much later than the *lower-order modes*. This is not so because the higher-order mode spends much of its time traveling in that part of the optical fiber

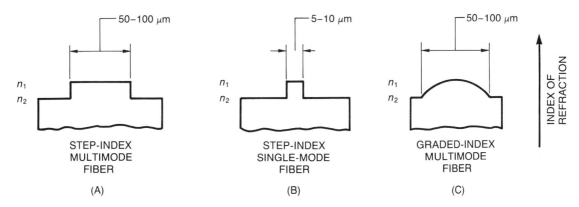

Figure 12-24 Refractive index profiles.

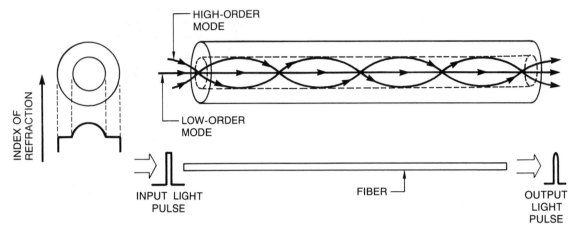

Figure 12-25 Graded-index multimode fiber.

that has the lowest index of refraction, and this is where the light travels faster. The lower modes, with shorter travel distances, spend most of their time in the area of the optical fiber that has the largest index of refraction, thus producing a much slower speed for these light rays. The net result is that the rays all reach the end of the optical fiber at about the same time, so modal dispersion is reduced.

Bandwidth

The bandwidth of optical fibers is an important consideration. The wider the bandwidth, the greater the amount of information that can be transmitted along the optical fiber. Bandwidth is limited by *modal dispersion* and *material dispersion* (sometimes called *spectral dispersion*). Modal dispersion has already been described. Material dispersion results from the variation in the velocity of light through the fiber with the wavelength of the light.

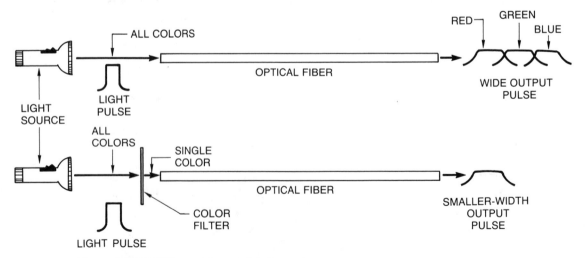

Figure 12-26 Effects of material dispersion.

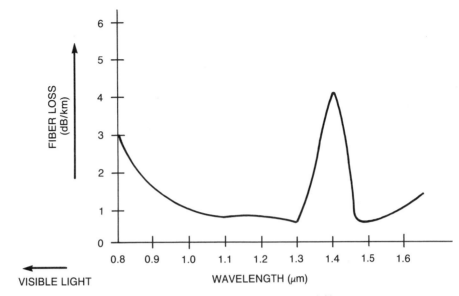

Figure 12-27 Net spectral loss curve for glass core optical fiber.

That is, if the light source emits more than one wavelength of light, the different wavelengths will travel at different velocities along the optical fiber and thus arrive at the output at different times. This dispersion can be reduced if a light source (such as a laser) with a single wavelength is used. The effect of material dispersion is shown in Figure 12-26.

There is a relationship between the net loss in an optical fiber versus the wavelength of the transmitting light. The curve is shown in Figure 12-27. At two wavelengths the loss is minimal (near 1.3 and 1.5 microns). Most of the state-of-the-art work is making light sources and light detectors, as well as optical fibers, that operate efficiently at 1.3 and 1.5 microns.

Conclusion

This section introduced the construction considerations of practical optical fibers. You also saw that the wavelength of the transmitted light affected the information-carrying characteristics of the optical fiber system. The next section will present fiber optics subsystems and components.

12-4 Review Questions

1. Describe how a damaged surface of an optical fiber could affect its transmission capabilities.
2. Explain the difference between a single-strand optical fiber and a multistrand optical fiber.
3. Describe the phenomenon of optical dispersion as it pertains to light transmission in optical fibers.
4. State several ways to reduce optical dispersion.
5. What affects the bandwidth of an optical fiber?
6. What wavelengths exhibit the least loss for transmission in glass?

12-5 CHARACTERISTICS OF OPTICAL FIBERS

Discussion

This section presents the practical considerations of optical fibers. You will see what is involved in putting optical fibers to work. Optical linking, sources, detectors, and causes of signal reduction will be covered.

Typical Optical Fiber Link

A typical optical fiber is shown in Figure 12-28. The transmitter consists of a *driver* circuit. The driver amplitude modulates the *light emitting diode* (LED). An LED is a solid-state device that emits light when forward biased. The more current delivered to the LED by the driver, the brighter the LED will be. Hence, the driver circuit intensity modulates the light entering the optical fiber.

Figure 12-28 also shows an *optical fiber coupler*. An optical fiber coupler is a mechanical device that physically connects two optical fibers. They are needed when optical fibers are used to transmit over long distances.

The receiver of Figure 12-28 consists of a solid-state light detector, such as the *photodiode*. The resistance of a photodiode varies according to the intensity of the light striking its surface. Thus the changes in light intensity arriving at the receiving end cause the current into the receiving amplifier to change, thus reproducing the original input signal.

Practical Optical Links

A practical fiber-optic link is shown in Figure 12-29. This system was installed by Pacific Telesis between San Francisco and San Diego. The total system contains over 78 000 fiber-kilometers of lightwave circuits and was placed in service in 1983. The system is capable of handling three signals, each of 90 000 000 bits per second, over the same fiber. This gives more than 240 000 digital channels at 64 000 bits per second in a cable with 144 optical fibers. A similar system installed by AT&T called the Northeast Corridor System connects Boston to New York, Washington, DC, and Richmond, Virginia. An undersea cable system using fiber optics is now more economical (and more private) than an equivalent satellite communication system.

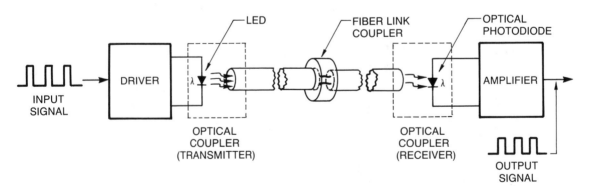

Figure 12-28 Typical fiber-optic link.

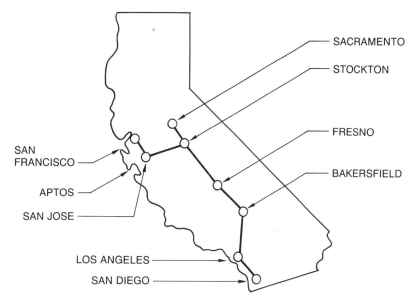

Figure 12-29 Practical fiber-optic link.

Numerical Aperture

The *numerical aperture* is a measure of the optical fiber's light-gathering capabilities. This is an important measurement when a light source is coupled to an optical fiber or when two optical fibers are coupled. The numerical aperture is shown in Figure 12-30. It is equal to the sine of the angle formed by the outer wall of a *light cone* and the axis of the optical fiber.

$$NA = \sin \phi \qquad \text{(Equation 12-5)}$$

where   NA = Numerical aperture
ϕ = Maximum acceptance angle for total internal reflection

The light cone represents the limits of the acceptance from outside light into the fiber. Light entering from outside this cone will strike at less than the critical angle and not be transmitted down the fiber.

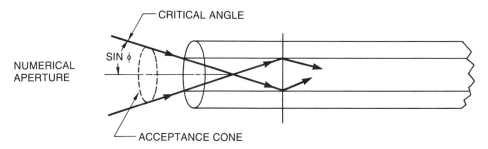

Figure 12-30 Concept of numerical aperture.

Surprisingly, unlike a camera, the numerical aperture does not depend on the physical dimensions of the optical fiber. The numerical aperture is determined by

$$\sin \phi = \frac{\sqrt{n_1^2 - n_2^2}}{n_0}$$ **(Equation 12-6)**

where ϕ = Maximum acceptance angle for total internal reflection
n_0 = Index of refraction of the input medium
n_1 = Index of refraction of the fiber core
n_2 = Index of refraction of the fiber cladding

Example 1

Figure 12-31 shows two optical fibers with their corresponding indexes of refraction. Determine which of the two has the greater numerical aperture.

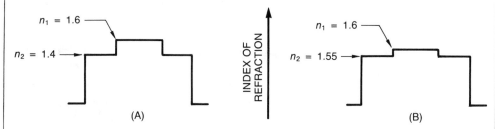

(A)   (B)

Figure 12-31

Solution

$$\sin \phi = \frac{\sqrt{n_1^2 - n_2^2}}{n_0}$$ **(Equation 12-6)**

Since the input medium for both fibers is air, $n_0 \approx 1$ and the formula reduces to

$$\sin \phi = \sqrt{n_1^2 - n_2^2}$$

For the optical fiber in Figure 12-31(A),

$$\sin \phi = \sqrt{1.6^2 - 1.4^2}$$
$$\sin \phi = \sqrt{2.56 - 1.96} = \sqrt{0.6}$$
$$\sin \phi = 0.774$$
$$\phi = \sin^{-1}(0.774) = 50.8°$$

For the optical fiber in Figure 12-31(B),

$$\sin \phi = \sqrt{1.6^2 - 1.55^2}$$
$$\sin \phi = \sqrt{2.56 - 2.402\ 5} = \sqrt{0.157\ 5}$$
$$\sin \phi = 0.397$$
$$\phi = \sin^{-1}(0.397) = 23.4°$$

Thus, the fiber in Figure 12-31(A) has the greater numerical aperture.

The optical fiber with the greatest numerical aperture will have the greatest light-gathering capability and the greatest amount of modal dispersion. This is why a small difference between the index of refraction of the fiber core and the fiber cladding reduces modal dispersion in optical fibers.

Coupling Mismatch

Coupling mismatch is a source of signal loss in optical fiber systems. This loss is the result of

- Improper alignment of the fibers (lateral misalignment)
- Connecting ends of the fibers not being in parallel with each other
- Separation of the two connecting fibers, resulting in light scattering

Scattering Loss

Scattering loss is due to the slight imperfections contained within the optical fiber. Because of better manufacturing methods, these losses are now very small.

Bending Loss

Bending loss can occur when there is a "kink" in the optical fiber. For commercial single-mode fibers, this loss is very small.

Conclusion

This section concludes the presentation of fiber optics. The next section is an introduction to laser light.

12-5 Review Questions

1. Describe the action of an LED.
2. What is the numerical aperture of an optical fiber?
3. How is the numerical aperture related to the difference between the indexes of refraction of the fiber-optic core and sheathing?
4. State some factors that contribute to a coupling mismatch between two optical fibers.
5. Define scattering and bending losses in optical fibers.

12-6	INTRODUCTION TO LASER LIGHT

Discussion

Laser light is a very unique form of light. Ordinary light waves are out of phase with each other. Laser light is a single frequency and in phase.

Laser light can be obtained without using any electrical device, as shown in Figure 12-32. You can use a color filter to produce one wavelength. However, the resulting waves are not in phase. Figure 12-32(B) shows light being projected through a tiny *pinhole*. The resulting light waves from this are in phase. If you put the pinhole and filter together along with a lens to focus the beam, Figure 12-32(C), you will produce a weak *coherent* beam of light. This system is inefficient because the amount of output light is less than one billionth of that produced by a very weak laser.

In this section, you will discover many ways to measure and analyze the properties of light. In doing this, you will be preparing a solid foundation in laser technology.

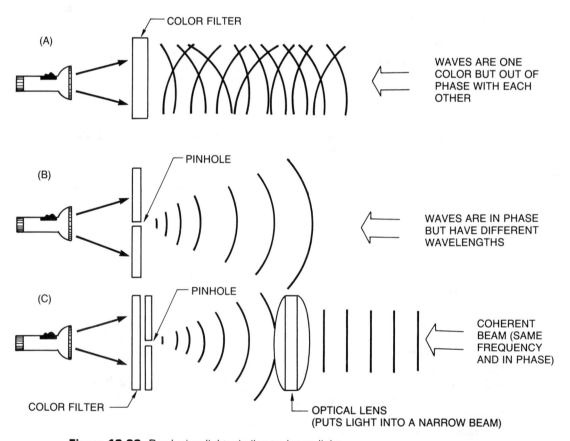

Figure 12-32 Producing light similar to laser light.

Laser Characteristics

Table 12-3 summarizes the characteristics of laser light.

Table 12-3	CHARACTERISTICS OF LASER LIGHT	
Term	**Meaning**	**Figure**
Coherence	Light waves in phase with each other	12-33(A)
Divergence	The spreading of light as it leaves the light source	12-33(B)
Monochromatic	Light that consists of only one wavelength (single color)	12-33(C)

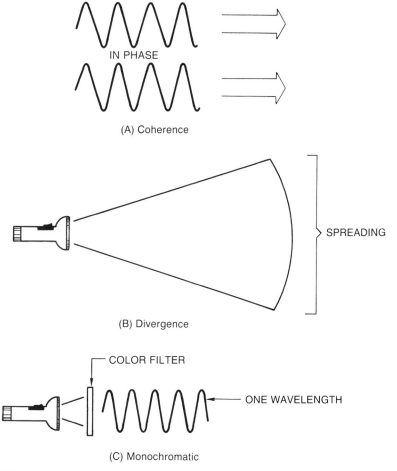

(A) Coherence

(B) Divergence

(C) Monochromatic

Figure 12-33

Radiometric Measurements

Radiometric measurements are measurements of the full spectrum of electromagnetic energy within the *optical spectrum. Photometric measurements,* on the other hand, deal only with visible light. Recall from Section 12-2 that visible light is a small part of the total optical spectrum. To give the broadest application to laser communications, radiometric measurements will be presented.

Energy and Power

Energy is defined as the ability to do work. One of the most common units of measurement for energy is the *joule* (J). One joule is the amount of energy required to lift 0.738 pound a distance of one foot.

Power is a measurement of how quickly energy is delivered. It is measured in watts (W). Stating power in terms of energy and time yields

$$P = \frac{E}{t}$$ **(Equation 12-7)**

where P = Power in watts
  E = Energy in joules
  t = Time in seconds

Example 1

A laser emits 10 mJ for a period of 2 ns. Calculate the power delivered by the laser.

Solution

$$P = \frac{E}{t}$$ **(Equation 12-7)**

$$P = \frac{10 \times 10^{-3}}{2 \times 10^{-9}}$$

$$P = 5 \times 10^{-3} \times 10^{9} = 5 \times 10^{6}$$

$$P = 5\ 000\ 000\ \text{W} = 5\ \text{MW}$$

So what appears to be a very small amount of energy output turns out to be a tremendous amount of power! This is so because the output time was so small. As you will see, the output power of a laser is one of its most important measurements.

Irradiance

Irradiance is the power per unit area. Mathematically it is

$$\text{IRD} = \frac{P}{A}$$ **(Equation 12-8)**

where   IRD = Irradiance in watts per unit area
  P = Power in watts
  A = Area in square units

Example 2

Determine the irradiance of the laser of Example 1 if the beam area on the target is 2 cm^2.

Solution

$$\text{IRD} = \frac{P}{A} \qquad \text{(Equation 12-8)}$$

$$\text{IRD} = \frac{5 \times 10^6 \text{ W}}{2 \text{ cm}^2}$$

$$\text{IRD} = 2\ 500\ 000 \text{ W/cm}^2$$

Thus, if the laser beam has a small divergence, a tremendous amount of power can be concentrated within a very short time. A laser beam traveling at the speed of light can deliver large amounts of energy faster than any other controlled device.

Radiance

Radiance is a measurement of the strength of the light source and is another important measurement in lasers. Radiance is expressed mathematically as

$$\text{RA} = \frac{P}{\omega A_{\text{ap}}} \qquad \text{(Equation 12-9)}$$

> *where* RA = Radiance in watts per steradian-square centimeters
> P = Power in watts
> ω = Solid angle in steradians
> A_{ap} = Aperture area of laser output in square centimeters

To solve problems in *radiance,* it is necessary to have a working knowledge of solid angles.

Solid Angles

The concept of a *solid angle* is shown in Figure 12-34. The *steradian* (st) is a measurement of a solid angle. A solid angle of 1 steradian is equal to a surface area of the square of the radius. The equation for determining a solid angle is

$$\omega = \frac{A}{r^2} \qquad \text{(Equation 12-10)}$$

> *where* ω = Solid angle in steradians
> A = Area of the curved surface
> r = Radius of the sphere

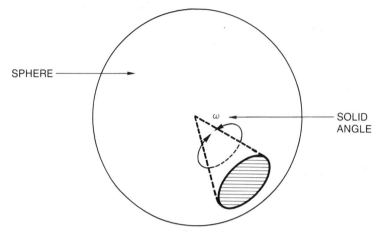

Figure 12-34 Concept of a steradian.

Example 3

Find the number of steradians in a complete sphere.

Solution

$$\omega = \frac{A}{r^2}$$ **(Equation 12-10)**

First, determine the area of a sphere. The formula for the surface area of a sphere is

$$A = 4\pi r^2$$

Substituting this into Equation 12-10 gives

$$\omega = \frac{4\pi r^2}{r^2}$$

$$\omega = 4\pi \text{ steradians}$$

Since $\pi \approx 3.141\,59$,

$$\omega = 12.566\,371 \text{ steradians}$$

Example 4

Calculate the radiance of a 100-W light bulb that is a perfect sphere with a radius of 6 cm.

Solution

$$RA = \frac{P}{\omega A_{ap}}$$
(Equation 12-9)

Example 12-7 showed that the solid angle of a perfect sphere is 12.56 st. The aperture area is equal to the total surface area of the light source (assume a perfect sphere).

$A_{ap} = 4\pi r^2$
$A_{ap} = 4 \times 3.14 \times 6 \text{ cm}^2$
$A_{ap} = 452.4 \text{ cm}^2$

Hence, the radiance is

$$RA = \frac{100 \text{ W}}{12.56 \text{ st} \times 452.4 \text{ cm}^2}$$
$$RA = \frac{100}{5.68 \times 10^3}$$
$RA = 17.6 \times 10^{-3}$
$RA = 0.017 \, 6 \text{ W/st-cm}^2$

You can better appreciate the power of the laser with its small aperture and small divergence by comparing the radiance of the light bulb in Example 4 to the radiance of a laser in Example 5.

Example 5

A small 2-W laser has a 0.5-mrad beam divergence with an aperture radius of 0.2 cm. Determine the radiance of the laser.

Solution

$$RA = \frac{P}{\omega A_{ap}}$$
(Equation 12-9)

First determine the aperture area by using the formula for the area of a circle:

$$A_{ap} = \pi r^2$$
$$A_{ap} = 3.14 \times (0.2 \text{ cm})^2 = 3.14 \times 0.04$$
$$A_{ap} = 0.125\ 6 \text{ cm}^2$$

The next part of the solution is to calculate the solid angle in steradians. To do this, assume that the laser beam creates a perfect circle. Since a radian is defined as the angle that subtends an arc equal to the length of the radius, then 0.5 mrad will subtend an arc equal to 0.000 5 times the radius.

The relationship between a radian and a steradian is

$$\omega = \pi \left(\frac{\theta}{2}\right)^2$$

where ω = Steradian angle
π = 3.141 59 . . .
θ = Radian angle

Hence

$$\omega = 3.14 \left(\frac{0.5 \times 10^{-3}}{2}\right)^2$$
$$\omega = 3.14(0.25 \times 10^{-3})^2$$
$$\omega = 3.14(0.062\ 5 \times 10^{-6}) = 0.196 \times 10^{-6} \text{ st}$$
$$\omega = 1.96 \times 10^{-7} \text{ st}$$

Calculating radiance

$$RA = \frac{2 \text{ W}}{1.96 \text{ st} \times 0.125\ 6 \text{ cm}^2}$$
$$RA = \frac{2}{0.246 \times 10^{-7}}$$
$$RA = 8.124 \times 10^7$$
$$RA = 81\ 240\ 000 \text{ W/st-cm}^2$$

This example demonstrates the enormous amount of radiance from even a 2-W laser. The radiance given off by this laser is much more powerful than the sun! Looking directly at this laser beam is much more dangerous than looking directly at the sun. It will cause irreparable damage to the eye. Laser safety is a very important topic. As demonstrated, even a very-low-power laser can be dangerous to the eyes as well as the skin.

Conclusion

You have completed your study of the fundamental properties of laser light. In the next section you will see how a laser is constructed and what makes it "laser."

12-6 Review Questions

1. Describe some of the unique properties of laser light.
2. What is the difference between radiometric and photometric measurements?
3. State the relationship between energy and power.
4. Define the term irradiance.
5. Describe a solid angle. State the units of measurement.
6. Define radiance.

12-7 | CREATING LASER LIGHT

Discussion

All lasers operate on the principle of *simulated emission*. This section describes the physics of simulated emission.

Atomic Energy Levels

To understand laser action, you must have a mental picture of the relationship of electrons to the nucleus of an atom. When electrons are removed from an atom, the atom is a positively charged *ion*. Electron removal occurs when energy is applied from some external source to the atom, which can be electrical, chemical, heat, or light.

Figure 12-35 shows the *Bohr model* of an atom. Figure 12-35(A) is a physical representation of an atom with positively charged *protons* in the nucleus around which negatively charged *electrons* are orbiting. This model is familiar to electronics students. Figure 12-35(B) shows the *energy level* model of the same atom.

You can imagine that each electron orbits a definite distance from the nucleus. Each distance is called a *shell*. The field of physics called quantum mechanics states that electrons are most likely to be found orbiting the nucleus in one of these shells and not anywhere in between.

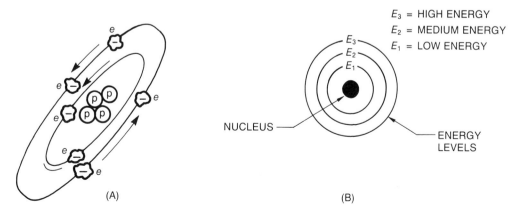

Figure 12-35 Bohr model of an atom.

An electron in the shell *closest* to the nucleus has the *lowest* energy level, because it requires more external energy to remove this electron than an electron in a shell further from the nucleus.

Thus, each shell of an atom can represent the energy level of an electron: the further away the shell, the greater the energy level of the electron in that shell.

Quantized Energy

Figure 12-36(A) shows an external source of energy, a photon, hitting an electron and causing it to go from a lower energy level to a higher energy level. Figure 12-36(B) shows an *energy level diagram* of the resulting action.

Quantum mechanics states that when an electron goes from a lower energy shell to a higher energy shell, it will take an exact predictable amount of energy. In other words, the energy gained by the electron is *quantized*. This quantization of energy gained by the electron is shown in Figure 12-36(B).

The electron will stay at this new energy level only about a microsecond, then it drops back to its old energy level. It must give up the energy that it previously acquired. This energy is usually given up in the form of a photon. When an electron does this, it is said to be returning to its *ground state*.

Energy and Wavelength

Energy and wavelength are related mathematically by

$$E = hf \qquad \text{(Equation 12-11)}$$

where E = Energy emitted
   h = Planck's constant (6.63×10^{-34}/J-s)
   f = Frequency of the emitted photon in hertz

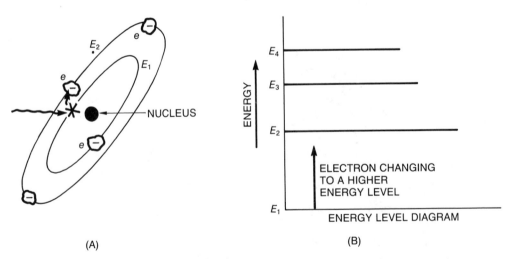

(A)   (B)

Figure 12-36 Increasing an energy level.

Example 1

Determine the frequency of an emitted photon that releases 5×10^{-19} joules of energy as it returns to its ground state.

Solution

$$E = hf \qquad \text{(Equation 12-11)}$$
$$5 \times 10^{-19} \text{ J} = (6.63 \times 10^{-34} \text{ J-s})f$$

Solving for f, we get

$$f = \frac{5 \times 10^{-19}}{6.63 \times 10^{-34}}$$
$$f = 0.754 \times 10^{-19} \times 10^{34}$$
$$f = 0.754 \times 10^{15} \text{ Hz}$$
$$f = 754 \text{ GHz}$$

The energy given off from the electron in Example 1 is near the blue part of the visible light spectrum.

Emission

An atom that receives energy will emit a photon. When it will do this cannot be precisely predicted. When an atom receives energy and one or more of its electrons increase their energy levels, then the atom is said to be *excited*. These atoms will soon return to their ground state, releasing energy in return. However, exactly when these atoms will return to their ground state is not predictable; this kind of emission is called *spontaneous*. Spontaneous emission is shown in Figure 12-37.

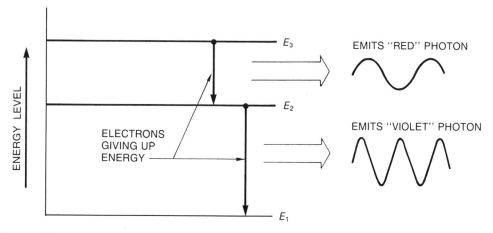

Figure 12-37 Process of spontaneous emission.

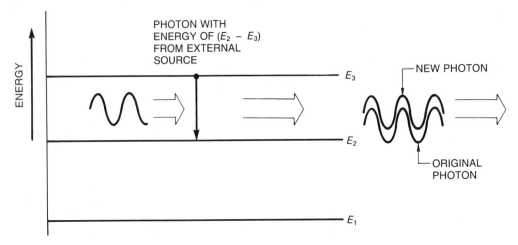

Figure 12-38 Process of stimulated emission.

In lasers, a process called *stimulated emission* occurs when a photon that has a specific frequency enters an excited atom (an atom that will soon release energy and return to its ground state).

If the excited atom has an energy structure that will cause it to release exactly the same energy when it goes to its ground state as the energy of the incoming photon, then stimulated emission will occur. See Figure 12-38.

Lasing

Many different kinds of materials will laser, including solids (ruby), gases (helium–neon), and semiconductors (gallium arsenide); even Jello® has been made to laser. The basic process for lasing requires a medium, such as a ruby rod with an internally reflecting mirror at each end. One of the reflecting mirrors allows a small percentage of light to pass to the outside medium. The lasing process is shown schematically in Figure 12-39.

Figure 12-39(A) shows the rod with some of the atoms in the excited state because the rod is at room temperature. However, most of the atoms are at the ground state.

When an external source of energy, such as a strong light source, is applied to the rod, most of the atoms become excited. This excited condition is shown in Figure 12-39(B).

As shown in Figure 12-39(C), stimulated emission takes place as more light is *pumped* into the rod. Some of this stimulated emission escapes the rod and goes into the surrounding medium. Some of the other stimulated emission causes more stimulated emission in the surrounding atoms.

Figure 12-39(D) shows that the stimulated emission occurring parallel to the axis of the rod is reflected from the mirrored surfaces. Because most of this energy gets reflected back and forth along this axis, causing more stimulated emission, light *amplification* is said to take place.

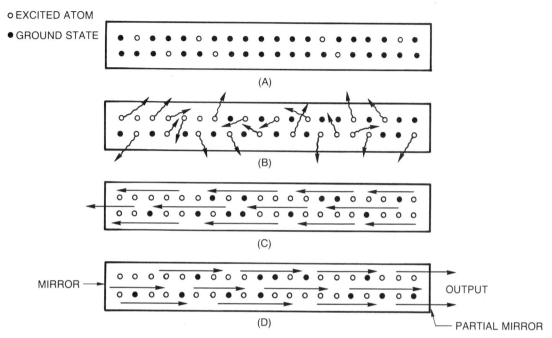

Figure 12-39 The lasing process.

This process continues and, as shown in Figure 12-39(D), *light amplification by stimulated emission radiation* (laser) takes place. The output end of the rod has the partial mirror, where a specific amount of coherent light radiates from the laser.

Conclusion

In this section, you learned about a model to explain the lasing process. In the next section, you will see what it takes to construct a laser.

12-7 Review Questions

1. Describe the Bohr model of the atom.
2. Explain the concept of shells and energy levels.
3. Describe the relationship between energy level and wavelength.
4. Discuss the difference between spontaneous emission and stimulated emission.
5. Explain the lasing process.

12-8 | THE LASER IN COMMUNICATIONS

Discussion

In this section, you will learn the major sections required by all lasers. You will also see how lasers can be used in communication systems. When the laser is used with fiber optics, a new and exciting mode of communications becomes possible.

Major Sections

All lasers must have four major sections:

1. An active medium
2. A method of exciting the medium
3. A feedback mechanism
4. A method of outputting the radiation

Figure 12-40 illustrates the major sections of a laser. The *active medium* is terminated on either side by two mirrors. These mirrors act as the *feedback mechanism*. The light source is used to *excite* the medium so that laser action can take place. The mirror system consists of one highly reflective mirror and one that allows a small percentage of radiation to escape. This process permits outputting of the radiation.

Semiconductor Lasers

One of the most common lasers used in communications is the *semiconductor laser*. The semiconductor laser looks very much like a transistor. See Figure 12-41.

A semiconductor laser is constructed much like a PN junction diode. The most common material used for this construction is *gallium arsenide* (GaAs). For the laser diode, the active medium is the junction of the *forward-biased* diode, Figure 12-42.

The junction of any forward biased PN junction will emit radiation. In particular, this is energy being released from the current in the diode. Since current can be thought of as being "injected" into the diode junction, this type of laser is sometimes referred to as an *injection laser*.

The semiconductor laser does not use mirrors to get the required feedback for light amplification. The reflectivity between the semiconductor material and air (about 35%) is

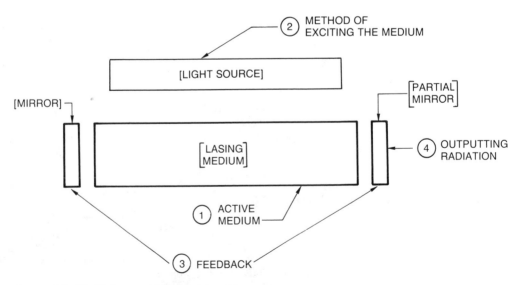

Figure 12-40 Major sections of a laser.

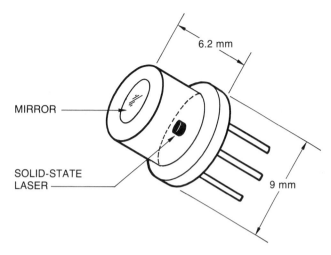

Figure 12-41 Semiconductor laser.

enough to give the necessary feedback for laser action to take place. This means that the laser beam will be radiated from both ends. Usually, one end of the semiconductor laser is coated with a reflective material to increase the output from one end.

The semiconductor laser is very temperature sensitive, which means that its output will decrease as its internal temperature increases. Therefore most of these devices, at room temperature, are not operated continuously but are "pulsed" with a duty cycle of about 1%.

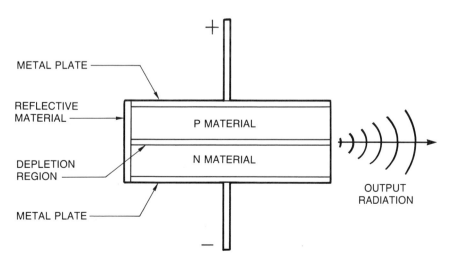

Figure 12-42 Construction of semiconductor laser.

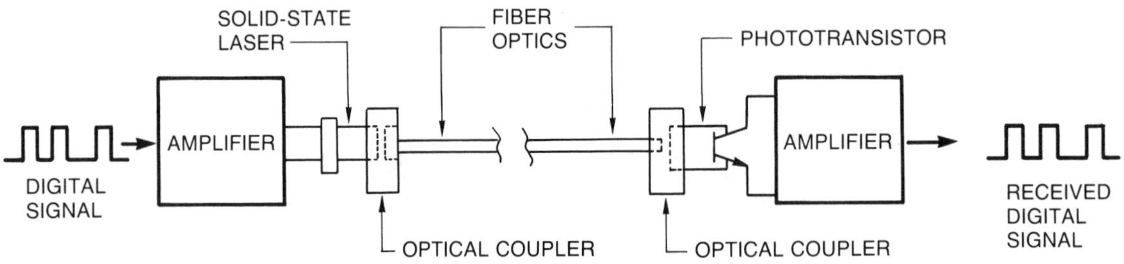

SOLID-STATE
LASER

FIBER
OPTICS

PHOTOTRANSISTOR

AMPLIFIER

DIGITAL
SIGNAL

AMPLIFIER

RECEIVED
DIGITAL
SIGNAL

OPTICAL COUPLER

OPTICAL COUPLER

Figure 12-43 Laser fiber-optic link.

Communications Applications

In communications applications, the semiconductor laser is usually connected to a fiber-optic cable. This arrangement provides a small, low-cost communications link for transmitting digital and analog data. Figure 12-43 shows such a system.

Pulse modulation is the easiest type to use in a laser communication system. It offers a greatly improved method of transmitting confidential information.

Conclusion

This section showed the necessary elements required of all lasers. The semiconductor laser was also introduced in this section along with a laser–fiber-optic communication system. This section concludes our study of laser theory. Laser safety is presented in the next section.

12-8 Review Questions

1. Name the four major sections required by lasers.
2. Give an example of each section in question 1.
3. Describe the action of a semiconductor laser.
4. What is the feedback mechanism for a semiconductor laser?
5. Describe a fiber-optic link using a solid-state laser.

TROUBLESHOOTING AND INSTRUMENTATION

Laser Safety

Laser safety is such an important topic that it warrants inclusion in the troubleshooting and instrumentation section of this chapter. Lasers produce an extremely intense light that can cause permanent eye and skin damage if proper precautions are not taken. Laser man-

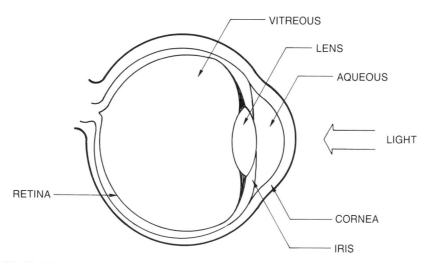

Figure 12-44 Schematic representation of the human eye.

ufacturers must certify that their laser products are in compliance with federal safety standards. This section presents the potential hazards of working with lasers and precautions to take.

The Human Eye

The energy absorbed by your eye depends on the wavelength of the energy. Wavelengths from 400 nm to 1 400 nm are a potential danger to the retina of the eye because these wavelengths are absorbed by the eye. Figure 12-44 is a schematic representation of the human eye. The eye consists of the *cornea, aqueous, lens, vitreous,* and *retina*.

Ultraviolet Damage

Ultraviolet wavelengths of 280 nm or less are absorbed by the cornea of the eye. These are frequencies that you and I cannot see, but can severely burn the cornea of the eye. Damage done to the eye in this manner produces a reddish color of the cornea. The effect is similar to snow blindness or "welder's flash," which occurs from watching arc welding. The eye hurts and feels as if sand were in it. If the damage is not severe, recovery takes a few hours to a couple of days.

Other damage to the cornea can be caused by a photochemical reaction due to ultraviolet radiation. This reaction is a delayed effect and usually occurs several hours after exposure. The cornea takes on a milky hue. This damage can result from eye exposure to a short, intense (3 mJ and above) burst of ultraviolet radiation.

Medical science has strong evidence that cataracts are developed from long-term exposure to ultraviolet radiation. Statistically, people who live near the equator have a higher incidence of cataracts than those living further north. Thus the ultraviolet radiation from a brighter sun at regions near the equator could lead to cataracts.

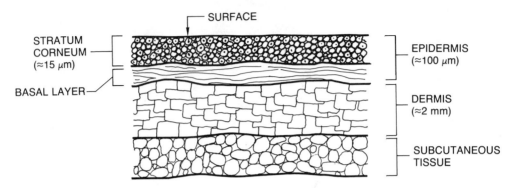

Figure 12-45 Schematic representation of human skin.

Visible Radiation Damage

Since the retina absorbs visible radiation (that's what makes the radiation visible), it can be damaged from overexposure. Laser beams whose wavelengths are 400 nm to 1 400 nm can cause this damage. An extra hazard is caused by the eye's amplification of the light by the lens. Thus, a very weak radiation can be magnified thousands of times and cause even more damage.

Skin Damage

The skin is the largest organ of the body. A section of human skin is shown schematically in Figure 12-45. The skin consists of the *epidermis, dermis,* and *subcutaneous tissue.* The epidermis is further divided into the *stratum corneum* and the *basal layer.* The outermost layer of the epidermis consists mostly of dead skin cells that have moved up from deeper parts of the skin. This layer of dead cells protects living cells from water loss, abrasion, dust, air, and radiant energy.

Skin damage from optical exposure depends on many variables. Usually, the lowest detectable level of damage to the skin is called *erythema,* which causes a reddening of the skin. This photochemical change also takes place during a *sunburn.* The problem is that the effect of erythema is usually not noticed until several hours after the exposure. There is some medical evidence that long-term exposure to this kind of radiation can cause skin cancer.

Laser Safety Standards

The two major safety standards for lasers are (1) standard 1040 published by the federal government and (2) standard Z136 published by the American National Standards Institute (ANSI). The main difference between these standards is that the federal standard classifies lasers according to the maximum possible output from the device during *normal* operation. The ANSI standard classifies the laser in terms of the potential hazard according to the maximum exposure a human could receive.

Generally, lasers are classified for safety purposes as shown in Table 12-4.

Table 12-4	LASER CLASSIFICATION
Class of Laser	**Meaning**
I	Not known to have caused human injury
II	Only visible lasers allowed in this class. Can cause eye damage if laser beam strikes the eye directly for 0.25 s or longer. (Note: It takes about 0.25 s for the eye to blink.)
III	Will produce eye injury if laser beam strikes the eye directly. (This is quicker than the eye can blink!)
IV	Exposure to skin is hazardous. Beam is a fire hazard. Even diffused reflections as well as direct eye contact can cause severe eye injuries.

Typical warning labels for the different classes of lasers are shown in Figure 12-46.

Conclusion

This section should be considered the most important section on lasers. Hopefully, you now have a healthy respect for any laser. Observing the safety precautions that apply to lasers can result in productive applications of the device.

The next section discusses the structured BASIC program for this chapter that simulates a laser.

12-9 Review Questions

1. What parts of the body are most sensitive to laser damage?
2. Describe the most common type of damage to the human eye by laser radiation. Is this radiation always visible?
3. Describe the most common type of damage to the skin by laser radiation. How long does it usually take this kind of damage to develop?
4. Name the two major standards for lasers.
5. Name the different classes of laser radiation.
6. Describe the meaning of each class of laser radiation.

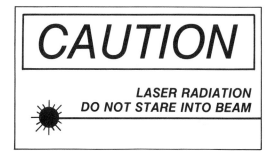

Figure 12-46 Different warning labels for lasers.

MICROCOMPUTER SIMULATION

Troubleshooting simulation number twelve presents a superhetrodyne receiver system. This is a troubleshooting simulation that allows you to use a variety of simulated test equipment to determine if the various parts of the receiver are within the stated specifications. This simulation gives you the opportunity to develop marketable troubleshooting skills required by the electronics industry.

CHAPTER PROBLEMS

(Answers to odd-numbered problems appear at the end of the text.)

1. What does the word "laser" stand for?
2. Describe what is included in the study of fiber optics.
3. Describe coherent light.
4. What was the first recorded attempt at transmitting voice with light as a carrier. Briefly describe how the device worked. What were some of its problems?
5. Who made the first documented prediction that laser light was possible? Who first actually produced laser light?
6. State the advantages of laser light.
7. Name five applications of lasers.
8. Explain the difference between the wave and particle theories of light. Give an example of when each theory would be used.
9. Sketch a diagram of the optical spectrum. What is its frequency range?
10. Sketch a diagram of the visible spectrum. How does visible light caused by various wavelengths differ?
11. State the difference between a light ray and a light beam.
12. Sketch a light ray that reflects from a mirrored surface. Identify the incident and reflected rays. What is the normal? What is the relationship of the angles made by the incident and reflected rays with the normal?
13. State the relationship between the angle of incidence and the angle of reflection.
14. A light ray strikes a mirror with an angle of 15° with respect to the normal. What is the angle of reflection?
15. Describe the difference between reflection and refraction.
16. Why does refraction take place? Use a diagram to show how refraction can cause an object in water to appear in a "phantom" location to a viewer above the surface.
17. What is Snell's law?
18. If one medium allows light to travel at 245 000 m/s and another at 220 000 m/s, what is the angle of refraction if the angle of incidence is 35°?
19. In problem 18, solve for the angle of refraction if the speed of light in each medium is reduced by 10%.
20. Explain the concept of a critical angle.
21. Using the values of Table 12-1, find the critical angle for a glass-water interface.
22. Determine the critical angle for a diamond-air interface.

23. Find the critical angles for the following interfaces: (A) water-diamond, (B) glass-diamond.
24. Sketch a ray diagram of light transmission in an optical fiber.
25. Describe the properties of the following light rays inside an optical fiber: (A) axial ray, (B) reflected ray, (C) surface ray, (D) refracted ray.
26. Explain the difference between a step-index optical fiber and a graded-index optical fiber.
27. Construct a ray diagram of light being transmitted in a multilayered glass fiber consisting of four layers, where the index of refraction is largest on the outer layer and smallest in the inner layer.
28. Sketch the ray diagram of light being transmitted in a multilayered glass fiber consisting of four layers, where the index of refraction is smallest on the outer layer and largest in the inner layer.
29. What is a mode in optical fiber transmission? Describe a multimode step-index fiber.
30. Describe modal dispersion. What causes it?
31. What factor can reduce model dispersion?
32. Describe a single-mode step-index optical fiber.
33. Sketch the refractive index profiles of a: (A) graded-index multimode fiber, (B) step-index multimode fiber, and (C) step-index single mode fiber.
34. Describe how a graded-index multimode fiber reduces modal dispersion.
35. What affects the bandwidth of an optical fiber?
36. What wavelengths produce the least amount of loss in glass fibers?
37. Sketch an optical fiber link. Describe the purpose of each section.
38. What is the numerical aperture of an optical fiber with an index of refraction of 1.7 in the core and 1.5 in the cladding?
39. Determine the numerical aperture of the optical fiber in problem 38 if the index of refraction of each material was decreased by 10%.
40. Describe what constitutes coupling mismatch in fiber optics.
41. Describe the properties of laser light.
42. Sketch and explain the properties of (A) coherence, (B) divergence, and (C) monochromatic light.
43. Explain the difference between radiometric and photometric measurements.
44. Compute the power delivered by the following lasers: (A) energy output = 3 mJ, time = 5 μs; (B) energy output = 12 mJ, time = 6 ns.
45. Calculate the power delivered by the following lasers: (A) energy output = 10 J, time = 2 s; (B) energy output = 120 μJ, time = 0.5 ns.
46. Compute the irradiance of each laser in problem 44 if the beam area on the target is 5 cm^2.
47. Determine the irradiance of each laser in problem 45 if the beam area on the target is 8 cm^2.
48. Calculate the radiance of a 12-W flashlight if it projects $\frac{1}{4}$ of a sphere with a radius of 100 m.
49. Solve for the radiance of the flashlight in problem 48 if the beam is now focused into $\frac{1}{8}$ of a sphere for the same distance.

50. Calculate the radiance of the following lasers: (A) power output = 5 W, beam divergence = 0.24 mrad, aperture radius = 0.1 cm; (B) energy output = 10 mJ, time = 5 μs, beam divergence = 0.5 mrad, aperture radius = 0.4 cm.

51. Calculate the radiance of each laser in problem 50 if each aperture radius is increased by 50%.

52. Determine the frequency of the following photons: (A) energy emitted = 3×10^{-19} J, (B) energy emitted = 5×10^{-18} J.

53. What part of the electromagnetic spectrum is the emitted energy from problem 52?

54. Explain the process of stimulated emission.

55. Describe the major sections of a laser.

56. Describe the construction and operation of a solid-state laser.

57. Describe the major parts of a fiber-optic link.

58. Summarize the safety precautions you should take when working with laser equipment. Include laser classification and the kinds of physical damage that lasers can cause.

SECTION TWO
COMMUNICATION CIRCUITS

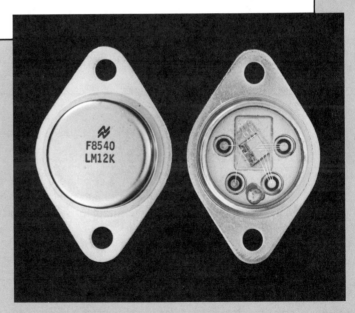

Courtesy of National Semiconductor Corporation.

CHAPTER 13

Phase-Locked Loops

OBJECTIVES

In this chapter, you will learn:

- [ ] The principles of phase-locked loops (PLL).
- [ ] Some basic concepts of analog and digital phase detectors.
- [ ] The theory of operation of the 555 timer and its application as a voltage controlled oscillator.
- [ ] An introduction to the operational amplifier and its use as a noninverting or inverting DC amplifier.
- [ ] Some major applications of phase-locked loops in communications.
- [ ] The theory of operation of a Touch-Tone® decoder system.
- [ ] The operating principles of frequency synthesizes using phase-locked loops.

INTRODUCTION

Phase-locked loops have become a standard in electronic communications. Thanks to the economics of complex circuits brought about by massed-produced integrated circuits, the PLL can be used in home entertainment systems and sophisticated satellite communications systems.

13-1 INTRODUCTION TO PLL

Discussion

The *phase-locked loop* (PLL) is an electronic system consisting of many different circuits. It finds a wide range of applications in communications systems because its integrated circuits (ICs) are inexpensive. Today, a complete PLL system comes in a single IC package and costs less than a reasonable lunch. Figure 13-1 illustrates a PLL in a 14-pin chip.

Figure 13-1 Typical PLL.

Phase-locked loops are used in the following areas:

- Radio telemetry from satellites that have signals *buried* in noise (a signal-to-noise ratio less than 1)
- AM and FM demodulators
- Frequency shift keying (FSK) decoders
- Frequency synthesizer
- Television sync
- Motor speed controls
- Touch-Tone decoders
- Light-coupled analog isolators
- Doppler detector (can be used to solve *Doppler* problems for fast-moving systems or in earth satellites)

Basic Idea

The basic idea of a PLL is shown in Figure 13-2. A typical PLL has a *phase detector (or a comparator)*, a *lowpass filter*, an *amplifier*, and a *voltage controlled oscillator* (VCO).

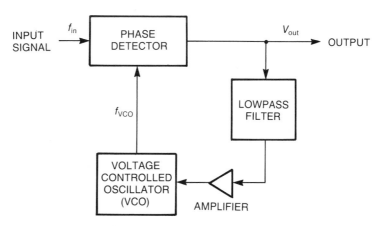

Figure 13-2 Basic idea of a PLL.

There is a single *input* and a single *output*. Part of the output is fed back to the phase detector. The purpose of a PLL is to make the output voltage V_{out} proportional to the phase difference between the VCO (f_{VCO}) and the input signal (f_{in}). This is done by making f_{VCO} = f_{in}.

Table 13-1 describes the four major sections of the PLL.

Table 13-1	SECTIONS OF A PLL
Circuit	**Function**
Phase det.	Compares the phase of the input signal to that of the VCO signal. Its output is a DC voltage whose value is proportional to the phase difference between these two signals.
Lowpass filter	The filter determines many of the PLL *dynamic* characteristics. One is the frequency range over which the loop will acquire and hold its phase lock. Another is how quickly the loop will respond to input frequency variations.
Amplifier	Amplifies the DC offset voltage caused by the frequency or phase difference between the input frequency and the VCO. The amplifier serves to increase the sensitivity of the PLL.
VCO	An oscillator with an output frequency controlled by the value of a DC voltage.

PLL States

Figure 13-3 illustrates the three *states* of a PLL:

1. *Free-running:* [Figure 13-3(A)]
   Free-running occurs when the incoming signal has a frequency different from the VCO signal. The phase detector (ϕ DET) will have a DC output voltage (called the *error voltage*) fed to the VCO, causing the VCO frequency to change.
2. *Capture:* [Figure 13-3(B)]
   Capture starts when the VCO begins to change frequency in order to reduce the frequency difference between itself and the incoming signal.
3. *Phase-lock:* [Figure 13-3(C)]
   When the input frequency is the same as the VCO frequency, the PLL is said to be *phase-locked*. At this time, the VCO phase is slightly different from the phase of the incoming signal. This phase difference produces the error voltage required to maintain the correct VCO frequency.

The important thing about the PLL is that once it is locked, only *phase* is different between the input signal and VCO; the *frequency* is exactly the same.

(A) FREE RUNNING

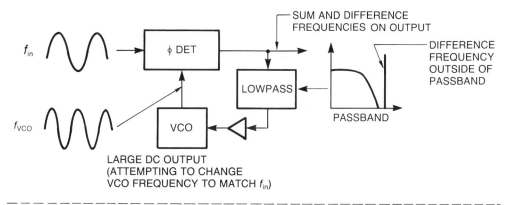

(B) CAPTURE

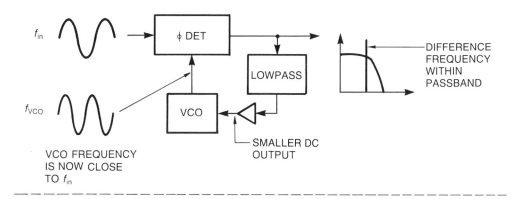

(C) PHASE LOCKED

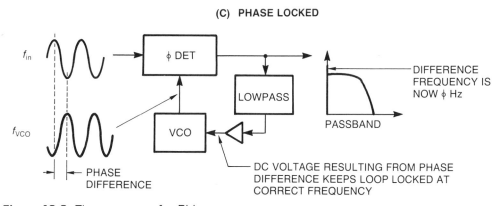

Figure 13-3 Three stages of a PLL.

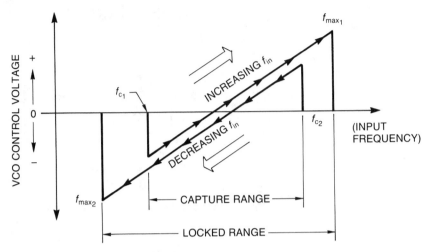

Figure 13-4 Capture and lock range of a PLL.

Capture and Lock Ranges

Figure 13-4 shows the *capture* and *lock* ranges of a typical PLL. As the frequency of the input is increased, the VCO control voltage stays at zero volts until the incoming frequency equals f_{c_1}. At this point, increasing the input frequency increases the VCO input voltage positively, causing its output frequency to follow the frequency of the input. At f_{max_1}, the input frequency has gone above the tracking capabilities of the loop, and the VCO control voltage again falls to zero. If the input frequency is decreased, at f_{c_2} the VCO control voltage starts decreasing negatively and causes the VCO output frequency to stay equal to the incoming frequency. If the incoming frequency is reduced below f_{max_2}, the loop will again lose lock and the VCO control voltage will again be zero volts.

Conclusion

This section presented the basic concepts of phase-locked loops. In the following sections, you will have the opportunity to learn how each part of the VCO works, and how to use a VCO in communications systems.

13-1 Review Questions

1. Why is the PLL so popular in electronic circuits?
2. Name three applications of a PLL.
3. Name the four circuits used in a phase-locked loop.
4. Describe the purpose of each circuit used in a phase-locked loop.
5. Explain the difference between free-running, capture, and phase-locked in the operation of a PLL.

| 13-2 | PHASE DETECTORS |

Discussion

A *phase detector* produces a DC output voltage proportional to the phase difference of two changing signals. This section presents the theory of operation of the two major types of phase detectors: analog and digital. First, you will review the analysis of the phase relations of two signals. Then you will see how these principles are applied in the construction and operation of phase detectors.

Phase Differences

Consider the signals in Figure 13-5. The signals in Figure 13-5(A) are 45° out of phase with each other. In Figure 13-5(B), the two signals, though not sine waves, still have a phase relation—in this case, 90° out of phase.

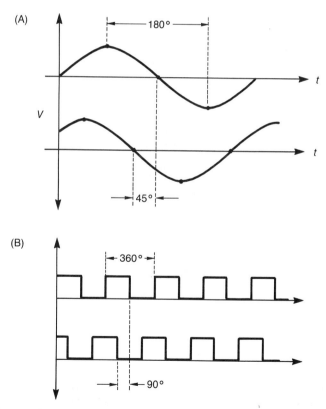

Figure 13-5 Phase relationship of signals.

Often, it is convenient to express the phase relationship of two signals in terms of *radians*. The relationship between radians and degrees is

$$\theta_{rad} = \left(\frac{\pi}{180°}\right)\theta \qquad \text{(Equation 13-1)}$$

where θ_{rad} = Angle in radians

θ = Angle in degrees

Example 1

Determine the phase relationships of the two signals in Figure 13-5 in radians.

Solution
In Figure 13-5(A),

$$\theta_{rad} = \left(\frac{\pi}{180°}\right)\theta \qquad \text{(Equation 13-1)}$$

$$\theta_{rad} = \left(\frac{3.14}{180°}\right)45°$$

$$\theta_{rad} = 0.017\ 4 \times 45 = 0.78 \text{ rad}$$

In Figure 13-5(B),

$$\theta_{rad} = \left(\frac{3.14}{180°}\right)90°$$

$$\theta_{rad} = 0.017\ 4 \times 90 = 1.57 \text{ rad}$$

Analog Phase Detector

Figure 13-6 is a block diagram of an *analog phase detector*. The output voltage of the phase detector is proportional to the phase error of the two input signals. When the signals are in phase, the output voltage is a maximum; when the input signals are 180° out of phase, the output voltage is a minimum. Various kinds of phase detectors are available that produce different output voltages for a given phase error. All phase detectors give a predictable output voltage for a given input phase difference.

Digital Phase Detector

The simplest kind of digital phase detector is an *exclusive OR* (EXOR) gate as shown in Figure 13-7. Recall that the output of an EXOR gate is high (true) only when its inputs are *different*.

The way the circuit works as a phase detector is shown in Figure 13-8. The on time of the output signal increases as the phase relations of the two signals increase toward 180°. Above 180°, the DC output voltage decreases and becomes zero when the two signals are back in phase. For this system to function correctly, the two input signals must have a 50% duty cycle (the on time must equal the off time).

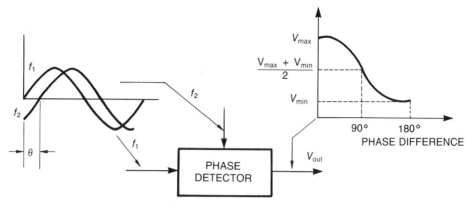

Figure 13-6 Basic analog phase detector.

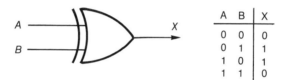

Figure 13-7 Exclusive OR gate.

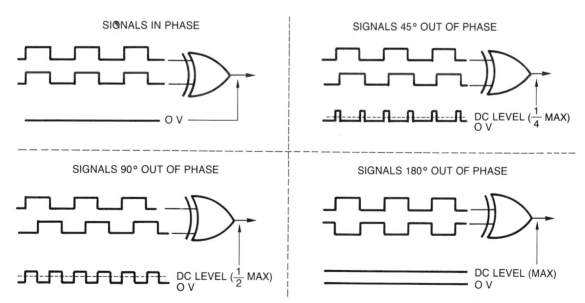

Figure 13-8 EXOR as a phase detector.

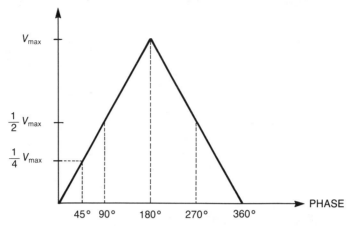

Figure 13-9 Characteristics of EXOR phase detector.

The characteristics of an EXOR phase detector are shown in Figure 13-9. The maximum average DC output is at 180°. Thus, as with the analog phase detector, the average DC output voltage is proportional to the phase difference of the input signals.

Conclusion

This section presented the concepts of analog and digital phase detectors. In both cases, the DC output voltage was a function of the phase difference between the input signals. Most VCOs use a digital phase detector because its output characteristics are linear functions of the input phase differences. The next section presents the concepts of the operation of a VCO.

13-2 Review Questions

1. Describe the purpose of a phase detector.
2. Name the two basic types of phase detectors.
3. Describe how two signals are out of phase. Do this for sine waves and for square waves.
4. Explain the operation of a digital phase detector using an exclusive OR gate.
5. Discuss the relationship between the output voltage of a digital phase detector and the phase difference of the input signals.

13-3 | THE 555 TIMER

Discussion

The *555 timer* is such a widely used IC with so many applications that it is now an industry standard. This tiny, inexpensive IC can be used as a VCO as part of a PLL.

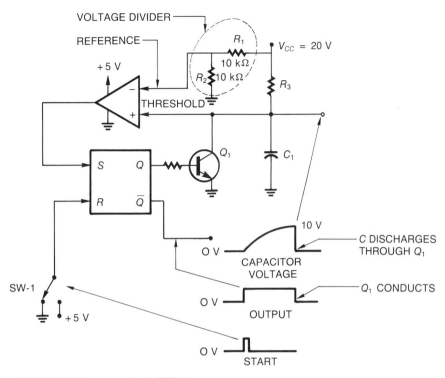

Figure 13-10 Basic concept of 555 timer.

Basic Concept

The basic concept behind the 555 timer is shown in Figure 13-10. The circuit consists of a *Set-Reset* flip-flop. This flip-flop works exactly the same way as the *Set-Reset* flip-flop in Appendix A. When the S input is high ($+5$ V), Q becomes true (causing $\bar{Q}$ to become false or zero). When the C input is high ($+5$ V), the Q output will become false (0 V) and $\bar{Q}$ will become true ($+5$ V). Here is how the system works.

In the condition shown, with SW-1 at the ground position, the Q output of the flip-flop is $+5$ V and the NPN transistor is saturated (looks like a short across capacitor C_1). Consider switch SW-1 as the starting switch. When it is momentarily switched to its $+5$ V position and then switched back to ground, $+5$ V is applied to the *Reset* input of the flip-flop, causing the Q output to go to 0 V, which cuts off the transistor (makes it look like an open). At the same time, the $\bar{Q}$ output goes true ($+5$ V), as shown in Figure 13-10. Since the transistor is now cut off, the capacitor C_1 begins to charge toward V_{CC}.

Note that the voltage across C_1 is applied to the noninverting ($+$), called the *threshold,* input of a comparator. The inverting ($-$) input is *referenced* to $+10$ V from the voltage divider formed by R_1 and R_2. As long as the threshold input to the comparator is less than the $+10$-V reference input, the output of the comparator will stay at 0 V. As soon as the capacitor voltage charges *just slightly above* $+10$ V, the value of the *reference voltage* will be exceeded and the comparator output will become $+5$ V. When this happens, the

R-S flip-flop becomes set, causing Q to be true (+5 V), and once again the NPN transistor will saturate, discharging C_1.

The preceding explanation is the fundamental principle of the 555 timer IC. This IC has so many applications that it may be helpful to read through the explanation again.

Complete Timer

A block diagram of the 555 timer is shown in Figure 13-11. The 555 timer actually has two comparators. The figure shows all of the control and output pins of the 555 timer with their descriptive names. To use the 555 timer as an oscillator, connect external components to it as in Figure 13-12.

Observe from Figure 13-12 that pin 7 (the discharge output) of the 555 is connected to a voltage divider (R_1 and R_2) with an external capacitor. The *trigger* and *threshold* inputs (pins 6 and 2) are connected together and to the capacitor. The other inputs of the two comparators are internally connected to a voltage divider of three equal resistors.

Assume that the capacitor starts at zero volts. This means that the threshold input to comparator 1 will be zero volts while its reference input is $\frac{2}{3}$ V_{CC}. Hence, its output will be zero. The reference input to comparator 2 will be zero volts while its threshold input is $\frac{1}{3}$ V_{CC}. Therefore its output will be +5 V, which is fed to the Reset input of the *R-S* flip-flop. This will cause the Q output to become zero while the $\bar{Q}$ output will be +5 V. Since the Q output is zero, the transistor will be well into cutoff, allowing the capacitor to charge through the two external resistors (R_1 and R_2).

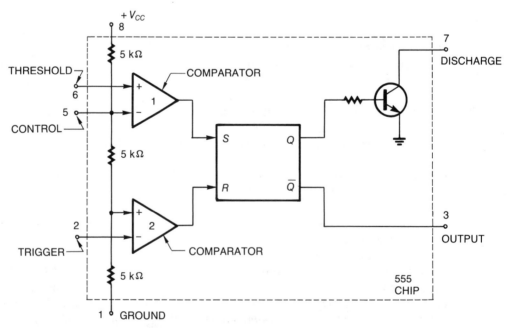

Figure 13-11 Block diagram of 555 timer.

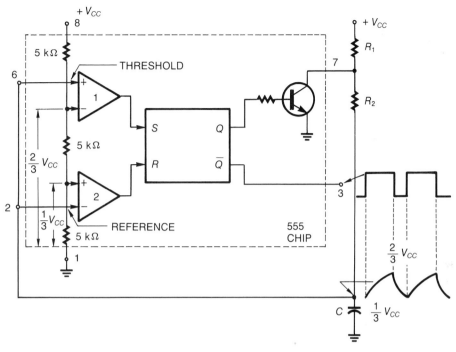

Figure 13-12 Connecting the 555 as an oscillator.

This charging process continues and will eventually reach $\frac{2}{3} V_{CC}$. When this happens, the reference input of comparator 1 will be more positive than its control input, and its output will become +5 V. At the same time, the control input of comparator 2 is at a higher positive voltage than its reference input, and the output of comparator 2 will become zero.

This condition will cause the Q output of the R-S flip-flop to go true, thus placing a +5 V on the base of the transistor. This will cause the transistor to saturate and allow the capacitor to discharge through R_2 of the external voltage divider. The capacitor will continue to discharge until its output voltage is $\frac{1}{3} V_{CC}$. When this happens, the reference input of comparator 2 will now be less positive than its control input, and the output of comparator 2 will go to +5 V, causing the Q output of the R-S flip-flop to return to zero. This in turn will apply zero volts to the base of the transistor, again causing it to go into cutoff.

The output frequency of a 555 may be determined from the following relationship:

$$f = \frac{1.44}{(R_1 + 2R_2)C}$$

(Equation 13-2)

where f = Frequency of oscillation in hertz
R_1, R_2 = Values of external resistors in ohms
C = Value of external capacitor in farads

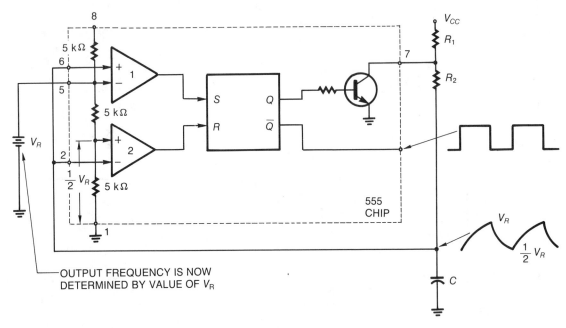

Figure 13-13 The 555 connected for VCO operation.

The duty cycle of the 555 oscillator can be found from:

$$D = \frac{R_1 + R_2}{R_1 + 2R_2} \times 100\%$$ **(Equation 13-3)**

where D = Duty cycle (ratio of the on time to the period) in percent
   R_1, R_2 = Values of external resistors in ohms

Voltage Controlled Oscillator (VCO) Operation

The 555 timer may be operated as a VCO as shown in Figure 13-13. The variable voltage source V_R controls the DC voltage to the control input of comparator 1. This means that both comparators are now controlled by an external input voltage. Under these conditions, the capacitor will now charge up to the value of the control voltage and discharge down to half the value of the control voltage. If the control voltage decreases, the output frequency will increase. If the input control voltage increases, the output frequency will decrease.

The 555 is not the only IC that can be used as a VCO. It was used here to illustrate a practical application of this versatile device.

Conclusion

This section presented the 555 (sometimes called a "triple nickel") timer. You saw how to connect it as an oscillator and then use an external voltage to actually control the frequency of oscillation. In the next section, you will be introduced to operational amplifiers.

13-3 Review Questions

1. Give a general description of the 555 timer.
2. Name the basic circuits used inside the 555 timer IC.
3. Describe the basic operation of the 555 timer.
4. What determines the output frequency of a 555 timer when it is used as an oscillator?
5. Describe how the 555 timer may be used as a VCO.

| 13-4 | OPERATIONAL AMPLIFIERS |

Discussion

Operational amplifiers (often referred to as *op amps*) are high-gain amplifiers capable of amplifying DC voltages. The op amp is such a versatile amplifier that it serves hundreds of applications in the electronics industry and in electronic communications.

A variety of operational amplifiers are available in IC packages, many of which cost less than a package of chewing gum. The op amp symbol is shown in Figure 13-14. It looks like a comparator because the op amp can be used as a comparator. The op amp has two inputs: the *inverting* input, signified by the (−) sign, and a *noninverting* input, signified by the (+) sign. Most op amps are connected to two power sources—one positive with respect to ground, and the other negative—so that the output voltage can be positive or negative with respect to ground.

Gain Characteristics

There are two basic ways to connect an op amp as a voltage amplifier. One way is in the inverting mode, where the output signal is 180° out of phase with the input signal. The other is the noninverting mode, where the output signal is in phase with the input signal. The inverting mode is illustrated in Figure 13-15. The input signal is applied through R_{in} (called the input resistor) to the inverting (−) input of the amplifier while the noninverting (+) input is connected to ground. Also note that part of the output signal is fed back to the input through R_f (called the feedback resistor). This is called *negative feedback;* it controls the voltage gain and produces a "cleaner" output signal.

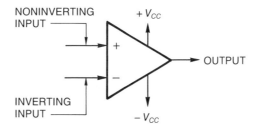

Figure 13-14 Operational amplifier.

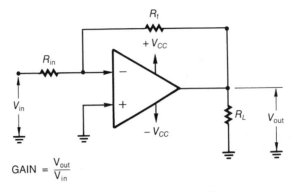

$$\text{GAIN} = \frac{V_{\text{out}}}{V_{\text{in}}}$$

Figure 13-15 Operational amplifier as an inverting amplifier.

The voltage gain of an inverting op amp is

$$A_V = -\frac{R_f}{R_{\text{in}}}$$

(Equation 13-4)

where A_V = Voltage gain of the amplifier
R_f = Value of feedback resistor in ohms
R_{in} = Value of input resistor in ohms

Note: The minus sign indicates that the output signal is 180° out of phase with the input.

Example 1

Determine the voltage gain of the op amp in Figure 13-16 and the amount of output voltage.

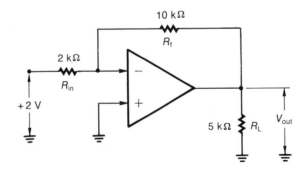

Figure 13-16

Solution

$$A_V = -\frac{R_f}{R_{in}}$$ **(Equation 13-4)**

$$A_V = -\left(\frac{10 \text{ k}\Omega}{2 \text{ k}\Omega}\right)$$

$$A_V = -5$$

This means that the op amp output voltage will be

$$A_V = \frac{V_{out}}{V_{in}}$$

$$-5 = \frac{V_{out}}{2 \text{ V}}$$

Solving for V_{out} gives

$$V_{out} = 2 \times -5 = -10 \text{ V}$$

Hence, the output voltage will be -10 V for an input voltage of $+2$ V.

Noninverting Amplifier

A *noninverting* op amp is illustrated in Figure 13-17. The input signal for a noninverting op amp is applied to the noninverting (+) input. The inverting input is connected to a voltage divider network consisting of R_f and R_{in}. With these arrangements, the input signal and the output signal are in phase with each other.

The relationship of these resistors and the voltage gain of the op amp is

$$A_V = \frac{R_f + R_{in}}{R_{in}} = \frac{R_f}{R_{in}} + 1$$ **(Equation 13-5)**

where A_V = Voltage gain of the amplifier

R_f = Value of feedback resistor in ohms

R_{in} = Value of input resistor in ohms

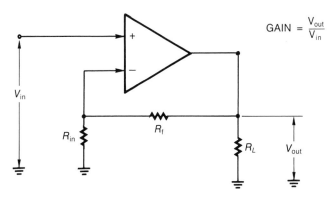

$$\text{GAIN} = \frac{V_{out}}{V_{in}}$$

Figure 13-17 Noninverting operational amplifier.

Example 2

Determine the voltage gain and the value of the output signal of the op amp in Figure 13-18.

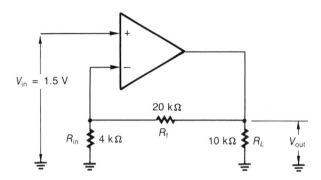

Figure 13-18

Solution

$$A_V = \frac{R_f + R_{in}}{R_{in}} = \frac{R_f}{R_{in}} + 1 \qquad \text{(Equation 13-5)}$$

$$A_V = \frac{20 \text{ k}\Omega}{4 \text{ k}\Omega} + 1 = 5 + 1$$

$$A_V = 6$$

This means that the output signal will be six times larger than the input signal, or

$$V_{out} = 6V_{in}$$
$$V_{out} = 6 \times 1.5 \text{ V} = 9 \text{ V}$$

Thus, the output signal will be 9 V and in phase with the input signal.

Conclusion

Operational amplifiers can be used as a part of the PLL to increase the loop gain. More will be said about that later, but for now think of it as a way of amplifying the DC voltage from the phase detector to make the VCO more sensitive to changes in the input signal. In the next section, you will see how to make some important measurements on the PLL.

13-4 Review Questions

1. State one of the uses of an operational amplifier.
2. Describe the differences between the two inputs of an op amp.
3. Explain how to connect an op amp as an inverting amplifier and as a noninverting amplifier.
4. State why op amps use two power sources.
5. What determines the gain of an op amp?

13-5 | THE COMPLETE PLL

Discussion

This section presents the complete PLL in more detail. Now that you know what each section of the PLL does, you can see how these individual parts are measured and how they affect the whole system. You will also be introduced to the NE565 (Signetics) PLL.

Phase Detector

The output of the *phase detector* is measured as a ratio of the amount of output voltage obtained per degrees of phase shift. This measurement is usually made in terms of radians rather then degrees and is referred to as the *phase detector gain*.

This measurement is

$$A_\phi = \frac{V_\phi}{\theta_{rad}}$$

(Equation 13-6)

where A_ϕ = Phase detector gain in volts per radian
   V_ϕ = Voltage output of phase detector
   θ_{rad} = Phase difference in radians

Example 1

Determine the gain of a phase detector with an output voltage of 0.15 V when the phase difference between its two input signals is 30°.

Solution

First, convert degrees to radians:

$$\theta_{rad} = \left(\frac{\pi}{180°}\right)\theta$$

(Equation 13-1)

$$\theta_{rad} = \left(\frac{3.14}{180°}\right)30° = 0.017\,4 \times 30$$

$$\theta_{rad} = 0.524 \text{ rad}$$

Now

$$A_\phi = \frac{V_\phi}{\theta_{rad}}$$

(Equation 13-6)

$$A_\phi = \frac{0.15 \text{ V}}{0.524}$$

$$A_\phi = 0.29 \text{ V/rad}$$

Effect of Op Amp Gain

The op amp can greatly increase the sensitivity of a PLL, because it amplifies the voltage output of the phase detector and uses this voltage to control the VCO frequency.

If the input phase difference to the phase detector and its gain are known, then its output voltage can be determined:

$$V_\phi = A_\phi \theta_{rad}$$ **(Equation 13-7)**

where V_ϕ = Voltage output of phase detector
   A_ϕ = Gain of phase detector in volts per radian
   θ_{rad} = Phase difference between signals in radians

Example 2

Determine the voltage to the input of the VCO for the PLL in Figure 13-19.

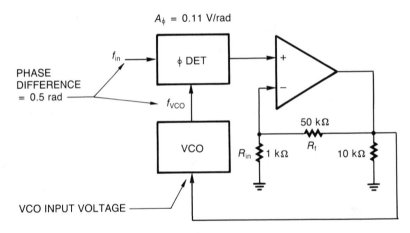

Figure 13-19

Solution

First, determine the output voltage of the phase detector:

$$V_\phi = A_\phi \theta_{rad}$$ **(Equation 13-7)**
$$V_\phi = (0.11 \text{ V/rad})(0.5 \text{ rad})$$
$$V_\phi = 0.055 \text{ V}$$

Calculate the gain of the op amp:

$$A_V = \frac{R_f}{R_{in}} + 1$$ **(Equation 13-5)**

$$A_V = \frac{50 \text{ k}\Omega}{1 \text{ k}\Omega} + 1$$

$$A_V = 51$$

This means that the output voltage of the op amp will be 51 times larger than the input voltage. Hence the input voltage to the VCO will be

$V_{\text{VCO}} = 51 \times 0.055$

$V_{\text{VCO}} = 2.8 \text{ V}$

VCO Measurements

The VCO is measured in terms of the frequency output per amount of control voltage input:

$$V_f = \frac{f_2 - f_1}{V_2 - V_1}$$ **(Equation 13-8)**

where V_f = Voltage-to-frequency conversion of VCO in hertz per volt

$f_2 - f_1$ = Change in VCO output frequency

$V_2 - V_1$ = Change in VCO control voltage

Example 3

Determine the voltage-to-frequency conversion of a VCO with an output frequency of 10 kHz at an input voltage of 2 V and an output frequency of 15 kHz with an input voltage of 3.2 V.

Solution

Using the relationship

$$V_f = \frac{f_2 - f_1}{V_2 - V_1}$$ **(Equation 13-8)**

$$V_f = \frac{15 \text{ kHz} - 10 \text{ kHz}}{3.2 \text{ V} - 2 \text{ V}}$$

$$V_f = \frac{5 \times 10^3}{1.2}$$

$$V_f = 4.17 \text{ kHz/V}$$

Typical IC Op Amp

A typical IC PLL is shown in Figure 13-20. The PLL consists of a phase detector, amplifier, and VCO. The values of the external resistor (R_1) and capacitor (C_1) are used to determine the *free-running* frequency of the VCO. For example, if you want the PLL to lock to an input frequency between 20 kHz and 30 kHz, then the free-running frequency

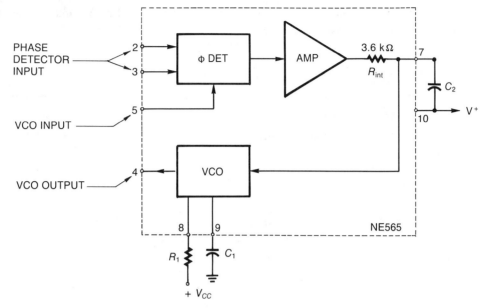

Figure 13-20 Typical IC PLL.

of the VCO should be 25 kHz. The lowpass filter for the PLL is determined by the value of the 3.6-kΩ internal resistor and C_2. The cutoff frequency for this filter is

$$f_c = \frac{1}{2\pi R_{int} C_2}$$

(Equation 13-9)

where f_c = Cutoff frequency in hertz

R_{int} = Value of internal resistor in ohms

C_2 = Value of external filter in farads

The microcomputer simulation program for this chapter presents a structured BASIC program that simulates the action of a PLL.

Conclusion

This section presented some of the finer details of a PLL. In the next section, you will see some of the major applications of the PLL.

13-5 Review Questions

1. Explain the meaning of phase detector gain.
2. Describe what effect the gain of an op amp used in a PLL has on the VCO control voltage.
3. What are the units of measurement for the VCO?
4. Describe the internal construction of a typical IC PLL.
5. How can you adjust the free-running frequency of a VCO when using an IC PLL?

| 13-6 | APPLICATIONS OF PLL |

Discussion

This section presents some of the applications of PLL in electronic communications. This versatile system, now available in an inexpensive IC chip, has found its way into many communications systems.

FM Demodulator

Figure 13-21 is a block diagram of a PLL used as an *FM detector*. Any changes in the input frequency will produce a corresponding voltage at the output of the op amp. Thus, frequency deviations at the input will be converted to voltage amplitude changes at the output.

This system may also be used as a phase detector. Changes of phase in the input signal will also cause a similar change in the output voltage of the PLL system. Thus phase modulation can also be converted into amplitude changes.

Phase-Locked Receiver

Figure 13-22 is a block diagram of a *phase-locked receiver* that is used to receive *deep-space* signals that have a very low signal-to-noise ratio. This receiver is also known as a *Doppler detector* because it is useful in detecting the change in frequency due to the Doppler effect caused by an object, such as a satellite, moving at a very high velocity relative to the receiver. The basic operation of the system is such that it keeps the VCO, which is now the local oscillator, tuned to the correct frequency. Thus, if the incoming signal is buried in noise, the receiver will remain tuned to the proper incoming frequency, because noise, by its very nature, is random and the signal carrier is not.

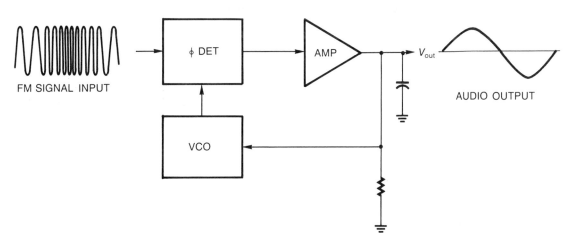

Figure 13-21 PLL used as an FM detector.

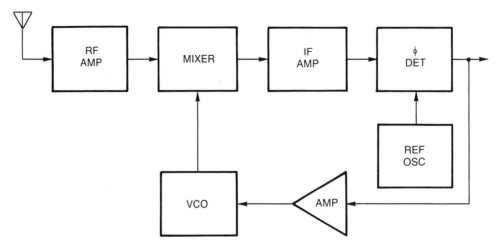

Figure 13-22 Phase-locked receiver.

Television Horizontal AFC

Figure 13-23 is the block diagram of a *TV horizontal automatic frequency control (AFC)*. The TV AFC compares the horizontal sync pulse to the VCO frequency. The VCO is determining the frequency of the horizontal oscillator for the TV horizontal sweep. The offset voltage produced by the frequency difference between the VCO and the horizontal sync adjusts the VCO frequency for a correct horizontal sweep frequency.

FSK Demodulator

Figure 13-24 is a block diagram and waveform of a PLL used as a *frequency shift keying* (FSK) demodulator. Recall that FSK is used to transmit digital data. As shown in Figure 13-24, the change in frequency is being converted to a corresponding change in amplitude. Thus, the original digital information is being restored at the receiver. Such a system can be used to *modulate* or *demodulate* FSK information. When used in this manner, it is called an FSK modem.

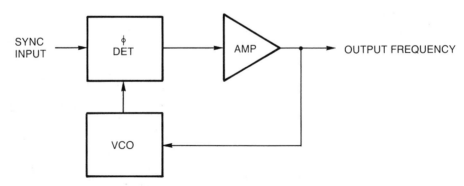

Figure 13-23 TV horizontal AFC using PLL.

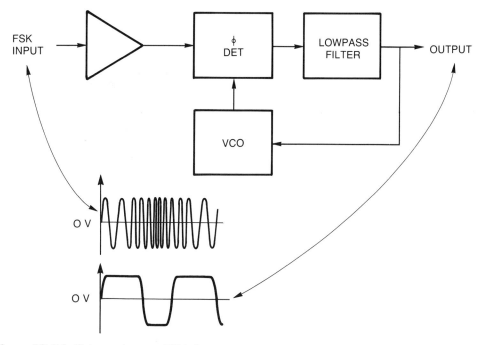

Figure 13-24 PLL used as an FSK demodulator.

Conclusion

This section presented some common applications of the PLL. The next section contains a detailed discussion of the PLL used as a *Touch-Tone decoder*. This versatile device is used in many communication systems.

13-6 Review Questions

1. Explain how the PLL can be used as an FM detector.
2. What is a phase-locked receiver? Describe its basic parts.
3. Describe the circuit used in TV horizontal AFC.
4. Describe the main parts of an FSK demodulator.
5. Describe an FSK modem.

13-7   TOUCH-TONE DECODER

Discussion

This section presents a unique application of a PLL, the *Touch-Tone decoder*. The circuit used is 567 IC as shown in Figure 13-25.

Figure 13-25 Tone decoder phase-locked loop.

Tone Decoder

A *tone decoder* is a circuit with a predictable output at a specific frequency. Specifically, the 567 is a PLL designed to respond to a given tone of frequencies. Its frequency range has an upper end of 500 kHz.

The basic connection of the 567 as a tone decoder is shown in Figure 13-26. The selected frequency of the tone decoder is determined by the value of R_1 and C_1. When the loop inside the 567 is locked, the output voltage from pin 8 will be zero.

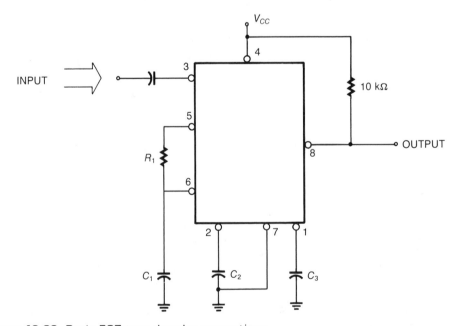

Figure 13-26 Basic 567 tone decoder connections.

Table 13-2	**TOUCH-TONE FREQUENCIES**		
	High-Tone Group		
Low-Tone Group (Hz)	**1 209 Hz**	**1 336 Hz**	**1 477 Hz**
697	1	2	3
770	4	5	6
852	7	8	9
941	*	0	#

Touch-Tone Frequencies

The frequencies used for Touch-Tone are listed in Table 13-2. Seven different frequencies, divided into the *low-tone group* and the *high-tone group* are used to represent 12 different signals. Thus, the digits 0–9 and the * and # symbols are each represented by two frequencies. For example, the number 1 is represented by 697 Hz and 1 209 Hz; the # is represented by 941 Hz and 1 477 Hz.

Frequency Decoding

Since the output of the 567 tone decoder is zero when its loop is locked, two of these devices can be used together to decode a specific number. Figure 13-27 shows the connection to decode the number 1.

Recall that the output of a NOR gate is true when both inputs are false. In the circuit in Figure 13-27, when both decoders are locked on to their respective frequencies, their outputs will be zero volts. At this point, the output of the NOR gate will be true, indicating the presence of the two frequencies that indicated the number 1.

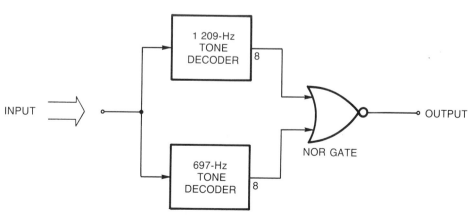

Figure 13-27 Decoding the number 1.

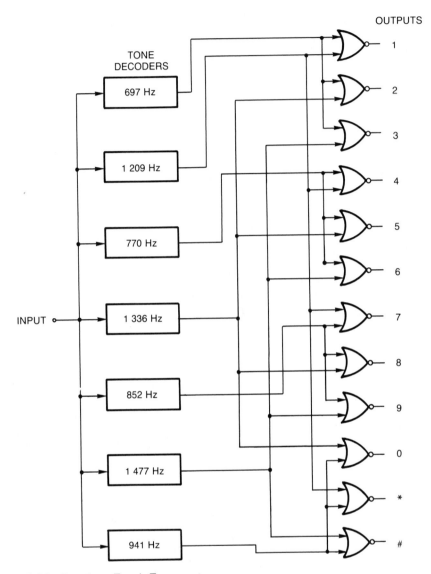

Figure 13-28 Complete Touch-Tone system.

Complete Decoder

The complete Touch-Tone decoder is shown in Figure 13-28. The input will consist of two separate audio frequencies. If you compare Table 13-2 to the decoder in Figure 13-28, you will see that each digit and the * and # symbols will be selected according to the combinations of input frequencies.

Conclusion

This section presented a specific application of a PLL system that you use everyday. The application of PLL in the communications industry is more common now than ever before.

13-7 Review Questions

1. Explain the purpose of a tone decoder.
2. What is the value of the voltage at the output of the 567 tone decoder when the loop is locked?
3. State what determines the selected frequency of the tone decoder.
4. Explain the coding scheme used for the Touch-Tone system.
5. Describe the circuit used to decode a number represented by two frequencies.

TROUBLESHOOTING AND INSTRUMENTATION

Frequency Synthesizers

Frequency synthesizers are much used in communication instrumentation. Frequency synthesis provides stable frequencies that are easily selected by the user. Another advantage of frequency synthesizers is that you can easily select output frequencies by digital means. Thus, different frequencies may easily be selected by a computer or other digital control circuits that have been preprogrammed to respond to specified conditions.

Because of the digital selection of frequency synthesizers, music synthesizers are easily constructed using this basic principle. Frequency synthesizers have also found their way into communication receivers and transmitters. As you may suspect, the backbone of the frequency synthesizer is the PLL.

Basic Idea

Figure 13-29 is the block diagram of a basic frequency synthesizer. Note that the output frequency is *higher* than the input frequency. This is important because the accuracy of the reference frequency is passed on the final output frequency. The whole system consists

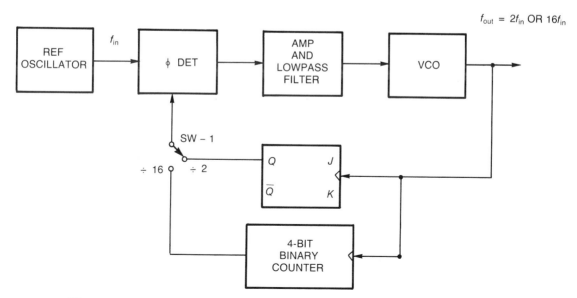

Figure 13-29 Basic frequency synthesizer.

of a PLL with a counter in the feedback loop. In Figure 13-29, the counter is simply a *J-K* flip-flop, which changes states at the leading edge of each input pulse, thus producing an output frequency that is half of the input frequency.

If the switch SW-1 were changed to position $\div 16$, then a 4-bit binary counter is in the feedback circuit. Thus the output frequency would then be 2^4 or 16 times larger than the input frequency.

Example 1

Determine the output frequencies of the frequency synthesizer in Figure 13-30.

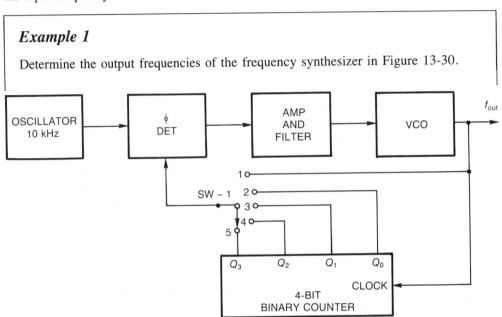

Figure 13-30

Solution

Since the feedback circuit consists of a 4-bit binary counter, the output frequency will be a multiple of the binary count selected by the *frequency selector switch*.

The resultant output frequencies are listed in Table 13-3.

Table 13-3	OUTPUT FREQUENCIES OF FREQUENCY SYNTHESIZER	
Selector Switch Position	**Binary Counter Output**	**Output Frequency (kHz)**
1	0	10
2	2	20
3	4	40
4	8	80
5	16	160

The preceding circuit is a handy troubleshooting device capable of quickly producing five frequency standards that are just as accurate as the source oscillator. However, just one oscillator is enough.

This is a workable scheme for low frequencies. For frequencies in the 30-MHz to 300-MHz range (VHF and UHF), it isn't practical to use this scheme because these frequencies are out of the range of most ICs.

Synthesizing Communication Frequencies

Figure 13-31 illustrates a method of reducing the incoming frequency to a lower frequency for the PLL. The frequency of the crystal-controlled oscillator is reduced by a divide-by-N counter. Thus a crystal frequency of 40 MHz can be reduced to a workable 4 MHz with a divide-by-10 counter.

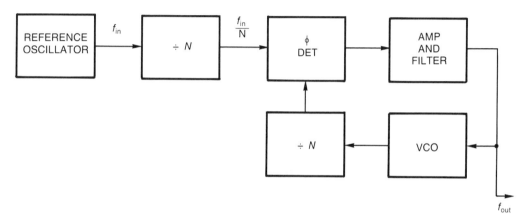

Figure 13-31 Producing a lower frequency for the PLL.

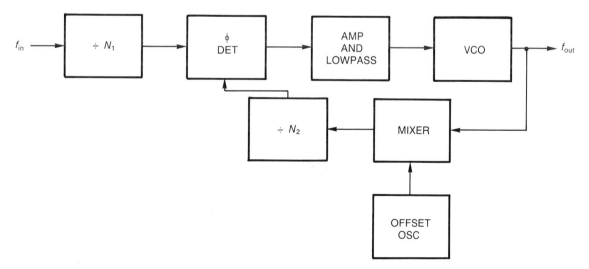

Figure 13-32 Using heterodyning in frequency synthesizers.

Another way to work with higher frequencies in frequency synthesizers is to use the principles of *heterodyning*. Figure 13-32 is a block diagram of a frequency synthesizer using this method. The circuit operation is as follows:

1. The *offset* oscillator is fed into a mixer along with the output of the VCO.
2. The output difference frequency of the mixer is fed into the second divide-by-N_2 counter.
3. The divide-by-N_1 counter output is fed into the phase detector with the lower-frequency output of the second divide-by-N_2 counter.

The advantage of this system is that it can work with a much higher frequency range than other frequency synthesizers.

Conclusion

This section presented the basic concepts of frequency synthesizers. In the programming section, you will work with a program that calculates the bandwidth of a PLL.

13-8 Review Questions

1. Describe some advantages of a frequency synthesizer.
2. Name the major sections of a frequency synthesizer.
3. What section of a frequency synthesizer determines the output frequency?
4. Describe a method of synthesizing communications frequencies.

 MICROCOMPUTER SIMULATION

The thirteenth troubleshooting simulation on your student disk presents the troubleshooting of an RF transmission system. As with the previous troubleshooting simulation, you will again have the opportunity to develop your troubleshooting skills using a variety of simulated equipment.

CHAPTER PROBLEMS

(Answers to odd-numbered problems appear at the end of the text.)

1. Name three applications of a PLL.
2. Describe the circuits needed to form a PLL.
3. Explain the purpose of each of the following circuits, as used in a PLL: (A) phase detector, (B) lowpass filter, (C) amplifier, and (D) VCO.
4. Describe the following states of a PLL: (A) free-running, (B) capture, and (C) phase-lock.
5. When the loop in a PLL is locked what is the difference in frequency detected by the phase detector?
6. When the loop in a PLL is locked what is the difference in phase detected by the phase detector?
7. Explain the meaning of capture range as used in a PLL.
8. Explain the meaning of lock range as used in a PLL.
9. Describe the relationship between the VCO control voltage and the output voltage of the phase detector.

10. Convert the following to radians: (A) 22°, (B) 45°, (C) 180°.
11. Convert the following to degrees: (A) 0.25 rad, (B) 2π rad, (C) π rad.
12. Determine the phase relations of the signals in Figure 13-33(A).
13. Determine the phase relations of the signals in Figure 13-33(B).

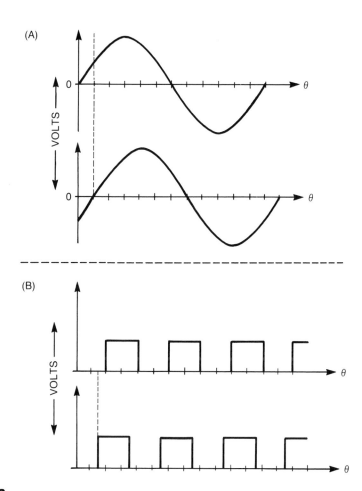

Figure 13-33

14. Find the output voltage of the phase detector in Figure 13-34(A).
15. For the phase detector in Figure 13-34(B), determine the phase error if its output voltage is 1 volt.

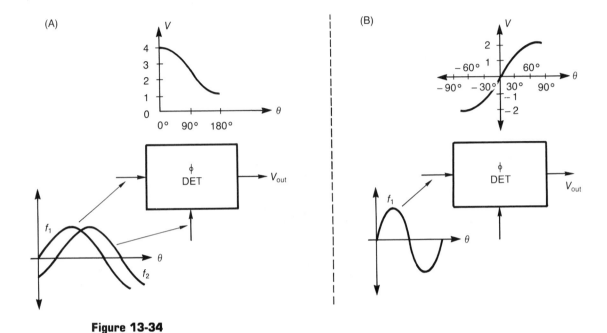

Figure 13-34

16. Construct the truth table for an EXOR gate. What voltage levels do the values in the truth table indicate?
17. Sketch the output waveform and indicate the average DC level for the EXOR phase detector in Figure 13-35(A).
18. Sketch the output waveform and indicate the average DC level for the EXOR phase detector in Figure 13-35(B).

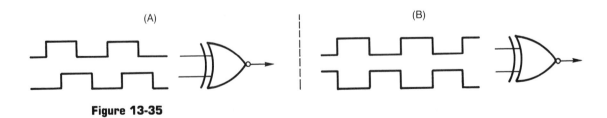

Figure 13-35

19. The graph in Figure 13-36(A) represents the output characteristics of an EXOR phase detector. Determine the output voltage at a phase error of (A) 90° and (B) 270°.
20. For the graph of the output characteristics of an EXOR phase detector in Figure 13-36(B), determine the output voltage when the phase error is (A) 45° and (B) 135°.

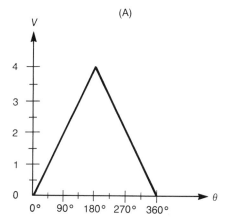

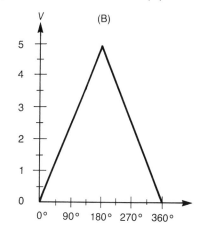

Figure 13-36

21. Construct the output waveform you would expect to see for the 555 timer in Figure 13-37(A). Show values of period, duty cycle, and peak voltage.
22. For the 555 timer in Figure 13-37(B), construct the output waveform you would expect to see. Show values of period, duty cycle, and peak voltage.

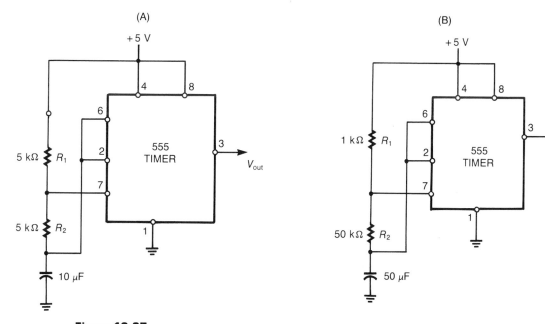

Figure 13-37

23. For the values of the 555 timer in Figure 13-38, show the output waveform you would expect to get across C_1.
24. Show the output waveforms if the reference voltage (V_R) of the 555 timer in Figure 13-38 is doubled.
25. For the 555 timer in Figure 13-38, sketch the output waveform if V_R is greater than V_{CC}.
26. For the 555 timer in Figure 13-38, sketch the output waveform if $V_R = V_{CC}$.

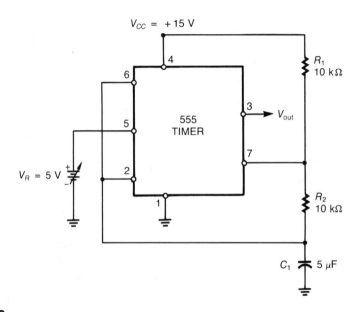

Figure 13-38

27. Determine the gains of the op amp in Figure 13-39.

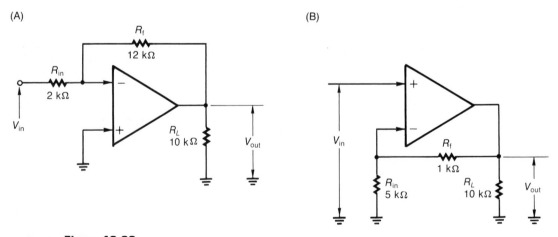

Figure 13-39

28. Find the gains for the op amp in Figure 13-40.

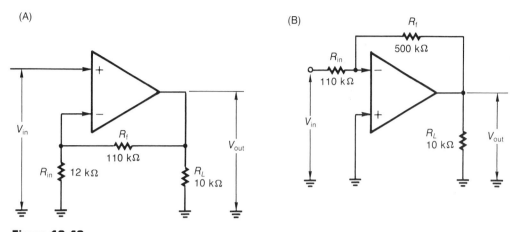

Figure 13-40

29. Determine the output voltage for each op amp in Figure 13-39 if the input voltage is 2.5 volts. Be sure to indicate the correct polarity.
30. Calculate the output voltage for each op amp in Figure 13-40 if the input voltage is -1.75 volts. Be sure to show the correct polarity.
31. Determine the gain of a phase detector with an output voltage of 0.33 volt when the difference between its two input signals is $10°$.
32. A phase detector with a phase difference of $45°$ between its input signals has an output voltage of 1.2 volt. Determine the gain of the phase detector.
33. For the PLL in Figure 13-41, determine the control voltage of the VCO.

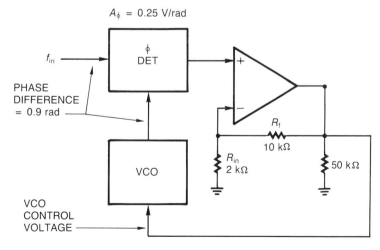

Figure 13-41

34. Determine the gain of the phase detector for the values in the PLL in Figure 13-42.

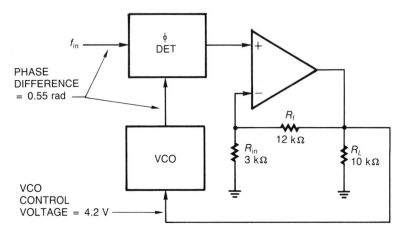

Figure 13-42

35. Determine the free-running frequency of a VCO if its lock range is to be between 100 kHz and 150 kHz.
36. The free-running frequency of a VCO is 30 kHz. Determine the lock range if the lowest frequency it can lock to is 15 kHz.
37. For the NE565 PLL in Figure 13-43, determine the cutoff frequency.
38. Find the cutoff frequency of the PLL in Figure 13-43 if the capacitor C_2 is doubled.

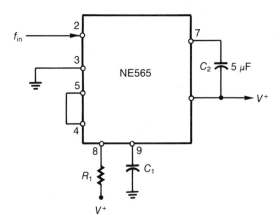

Figure 13-43

39. Sketch the block diagram of a PLL used as an FM detector.
40. Sketch the block diagram of a PLL used as a phase-locked receiver.
41. Sketch the block diagram of a PLL used as an FSK demodulator.

42. Sketch the block diagram of a PLL used as an FSK modulator.
43. Describe the operation of a tone decoder.
44. Describe the operation of the circuit in Figure 13-44.

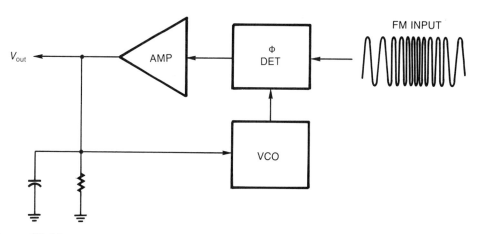

Figure 13-44

45. Determine the output frequency of the frequency synthesizer in Figure 13-45.
46. Find the output frequency of the frequency synthesizer in Figure 13-45 if the switch is in position 3.

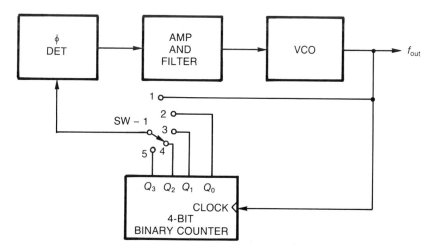

Figure 13-45

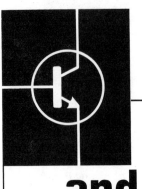

CHAPTER 14

Audio, Video, and RF Amplifiers

OBJECTIVES

In this chapter, you will study:

☐ How amplifiers are classified in terms of the frequencies they amplify.

☐ The various kinds of devices used for amplifiers.

☐ What all amplifiers have in common.

☐ Different classes of amplifier operation.

☐ The difference between small-signal amplifiers and power amplifiers.

☐ How amplifiers are used in communications equipment.

☐ How to analyze a transistor audio amplifier circuit from a troubleshooting standpoint.

INTRODUCTION

Amplifiers are the building blocks of communications equipment. Even though many kinds of amplifiers are used, they all have much in common. This section introduces the basic concept of amplifiers as they apply to communication circuits, and then shows how they are applied to communication systems. The chapter concludes with an introduction to amplifier troubleshooting.

14-1 CLASSIFICATION OF AMPLIFIERS

Discussion

Amplifiers can be classified according to the range of frequencies they can amplify. This section introduces what an amplifier does and how they are classified.

Amplifiers

Amplifiers were first presented in Section 4-4. An *amplifier* is a circuit that increases the *amplitude* of a signal. This definition worked well for the generalized approach given to amplifiers in that chapter. However, since you will now be dealing with the specifics of

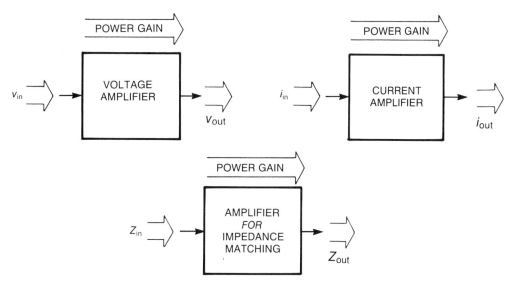

Figure 14-1 Functions of amplifiers.

circuits, a more precise definition will be used: An *amplifier* is a circuit that reproduces a more powerful replication of a signal. Figure 14-1 shows amplifiers that increase the voltage or current gain of a signal. Unlike a transformer, an amplifier delivers a *power gain.* Recall from basic electrical circuits that even though a step-up transformer gives a voltage gain and a stepdown transformer gives a current gain, a transformer can *never give a power gain.*

Frequency Response

The *frequency response* of an amplifier is the range of frequencies from the lowest to the highest that the amplifier is capable of amplifying at a *power gain* that is more than half of its maximum power gain.

Figure 14-2 illustrates the frequency response of a typical amplifier. Points A and B on the frequency-response graph (called a *Bode* plot) represent the *half-power* or 3-dB points of the amplifier. Note that the vertical scale of the Bode plot is *linear* and measured in decibels, but the horizontal scale is *logarithmic* and measured in hertz. To compute the *bandwidth,* use the relationship

$$\text{BW} = f_U - f_L \qquad\qquad \textbf{(Equation 4-19)}$$

As shown in Figure 14-2, all amplifiers have a frequency response. Amplifiers are classified according to the *range of frequencies they are capable of amplifying.*

Amplifier Classification

Figure 14-3 shows the frequency response of the major classifications of amplifiers. Table 14-1 summarizes the classifications.

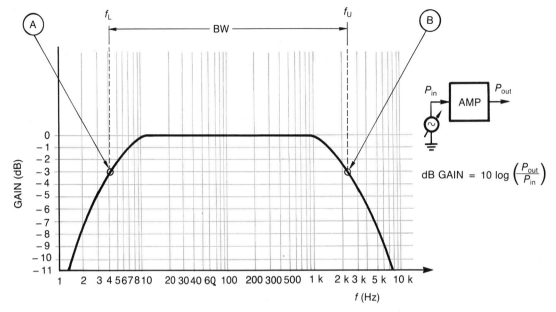

Figure 14-2 Frequency response of typical amplifier.

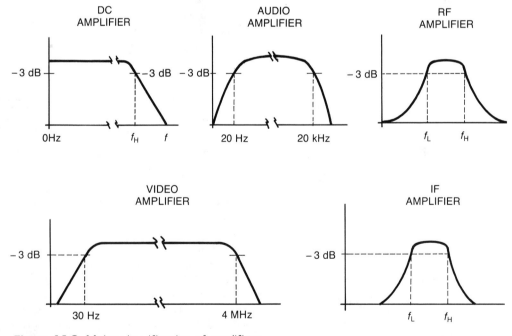

Figure 14-3 Major classification of amplifiers.

Table 14-1	AMPLIFIER CLASSIFICATION ACCORDING TO FREQUENCY RESPONSE	
Frequency Range	**Classification**	**Comments**
DC to some AC frequency	DC amplifier	Used to amplify DC or very low frequencies
20 Hz–20 kHz	Audio amplifier	Used to amplify audio frequencies
30 Hz–4 MHz	Video amplifier	Used to amplify a wide range of frequencies
Communication frequencies	RF amplifier	Used to amplify radio frequencies within the frequency limitations of the desired signals
Intermediate frequencies	IF amplifier	Used to amplify a specific bandwidth of frequencies at the IF frequency

Conclusion

This section presented a classification of amplifiers according to the range of frequencies the amplifier would amplify. In the next section, you will see the different types of devices used to construct an amplifier.

14-1 Review Questions

1. State one method of classifying amplifiers.
2. Explain the difference between an amplifier and a transformer.
3. Define frequency response.
4. Explain the coordinates in a Bode plot.
5. Name the five amplifiers presented in this section.

14-2 DEVICES USED IN AMPLIFIERS

Discussion

This section introduces *devices* used in amplifiers. In this section, a device is considered to be a single electronic component used to create an amplifier, such as a bipolar transistor, field-effect transistor (FET), vacuum tube, or linear IC. A linear IC actually consists of many different components. However, for the purpose of troubleshooting and replacement, it is treated as a single component.

Types of Devices

The major devices used in communication systems are shown in Figure 14-4. The IC device (the LM 3820 AM Radio System from National Semiconductor Corp.) contains all of the active components needed to construct a complete AM radio. Each device in Figure 14-4 will be discussed in detail, starting with the bipolar transistor (simply called the *transistor*).

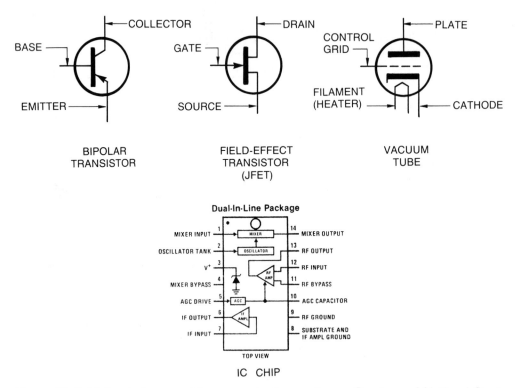

Figure 14-4 Major devices used in communication systems. *Courtesy* of National Semiconductor Corporation

Transistor

The two major types of transistors are NPN and PNP. Both operate under the same *semiconductor* principles. Figure 14-5 shows both types of transistors and their corresponding nomenclature. The significant difference between the two types is the voltage polarities applied to each device.

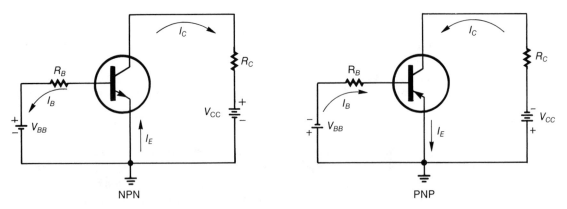

Figure 14-5 NPN and PNP transistors.

A transistor operates effectively as a *current*-controlled device, where a change in the input current causes a corresponding change in the output current, with a resultant increase in signal power. Transistors may be used to give *voltage* and *current gains* along with a corresponding power gain.

Field-Effect Transistor (FET)

The most common types of FETs are illustrated in Figure 14-6. The JFET is a junction field-effect transistor. The MOSFET is a metal-oxide silicon field-effect transistor. The major differences between them are the process and materials used in their manufacture. There are N and P channel devices as well as those that operate in the *enhancement* or *depletion* modes. The actual theory of operation is covered in devices courses. You should recognize these symbols as they appear in communication circuits. The FET is a *voltage* controlled device: a change in the input voltage causes a corresponding change in the output current thus producing a power gain.

Vacuum Tubes

Vacuum tubes are voltage controlled devices in which a change in input voltage causes a corresponding change in output current. They are seldom used in communication equipment. A wide variety of vacuum tubes—for example, the *pentode, tetrode,* and *pentagrid converter*—were used in older equipment. However, for the purpose of this section, only the *triode vacuum tube* will be illustrated because its principles are common to those of all vacuum tubes.

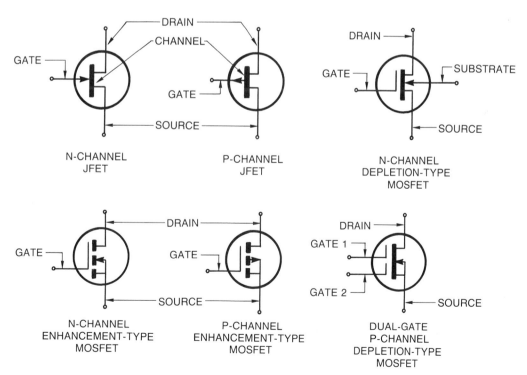

Figure 14-6 Common types of FETs.

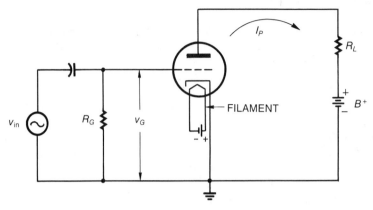

Figure 14-7 Triode vacuum tube.

Connection Diagram

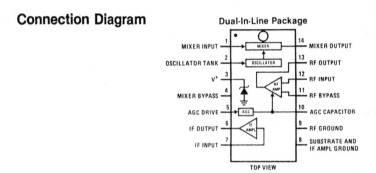

Circuit Schematic

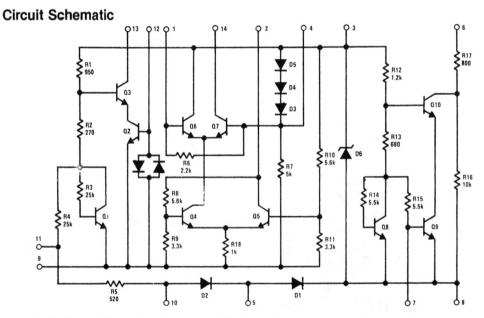

Figure 14-8 Typical linear IC. *Courtesy* of National Semiconductor Corporation

Figure 14-7 illustrates a triode vacuum tube. Unlike the transistor or FET, there is only one kind of voltage polarity connection used in the vacuum tube. The major difficulty with this device was the need for a *filament* to heat the *cathode* in order to free the electrons from the cathode's surface.

Linear ICs

A *linear* IC can be a most complex device. See Figure 14-8. The inside of the IC contains transistors, diodes, and resistors. Even though this is a complex circuit, it is serviced as a single component—that is, it cannot be repaired; it must be replaced.

Such a variety of linear ICs are available for communication systems that it is difficult to make any generalization about them. The remaining circuits section illustrates the major types of ICs, ranging from complex radio circuits to a single op amp.

Conclusion

Different kinds of devices used in communication systems were described. These devices are the focal point of amplifier circuits. Their corresponding circuits are described in the following sections.

14-2 Review Questions

1. Name three devices used in communication systems.
2. State why ICs are considered to be a device.
3. Name two types of bipolar transistors.
4. Describe the basic operating principle of a JFET.
5. Describe the major elements of the vacuum tube.
6. State how an IC can be used in a communication system.

14-3   CLASSES OF OPERATION

Discussion

Amplifiers are classified according to their *class* of operation. These classes of amplifiers are given the alphabetical designations *class A, class B,* and so on. This section presents the most common classes of amplifiers used in communications and shows how the various classes are determined.

General Idea

Figure 14-9 shows three different amplifiers and the relationship between the input and output signals of each amplifier. The class of an amplifier can be determined from the percentage of the input signal that appears on the output. Table 14-2 summarizes the major classes of operation for amplifiers.

Any amplifier device discussed in the last section may be operated as any class. Thus, a transistor may be operated as a class A, class B, or class C amplifier. This is also true of an FET, vacuum tube, or linear IC.

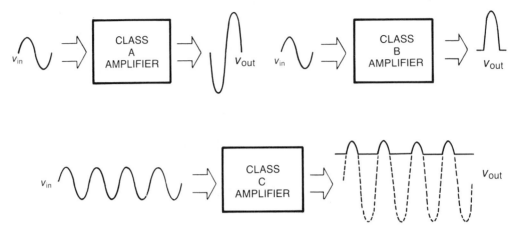

Figure 14-9 Major classes of amplifiers.

Table 14-2	MAJOR CLASSES OF OPERATION FOR AMPLIFIERS		
Class of Operation	Input–Output Relations (degrees)	Approximate Efficiency (%)	Typical Application
A	360	15	Audio frequency
B	180–200	75	Push-pull audio and RF
C	<180	95	Tuned RF

Class A Amplifier

A *class A* amplifier means that the device is conducting at all times. Continuous conduction allows the positive-going and negative-going portions of the signal to be amplified. *Audio frequency* amplifiers must have both halves of the signal equally amplified. Figure 14-10 shows a typical class A amplifier circuit. The circuit elements that actually determine the class of amplifier operation are the voltage-divider biasing resistors R_1 and R_2.

Class B Amplifier

A *class B* amplifier means that the device is conducting only when there is an input signal to the amplifier. A class B amplifier is most frequently used as a *power amplifier* to deliver audio power to the loudspeaker in an audio amplifier. Figure 14-11 illustrates a typical class B *push-pull* audio amplifier.

In the class B push-pull, only one transistor conducts at a time. Transistor Q_1 conducts during the positive portion of the input signal, transistor Q_2 during the negative portion. The devices that determine the class of operation are resistors R_1 and R_2.

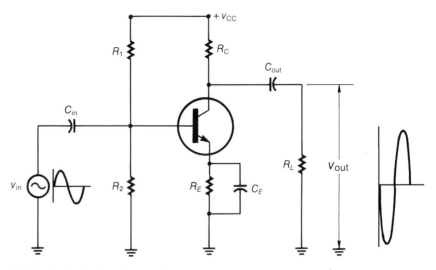

Figure 14-10 Typical class A amplifier.

Class C Amplifier

A *class C* amplifier means that the device is conducting only during a small portion of the input signal. Hence, a class C amplifier must have a tuned circuit to restore the rest of the input signal. This restoration takes place through the action of the *flywheel* effect

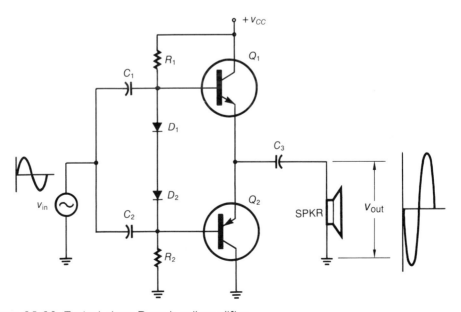

Figure 14-11 Typical class B push-pull amplifier.

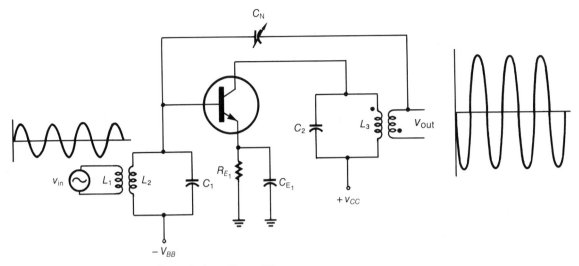

Figure 14-12 Typical class C amplifier.

of the tuned circuit, where the remainder of the incoming RF sine wave is restored by the action of the tuned circuit. A typical class C amplifier is illustrated in Figure 14-12.

Capacitor C_N is a *neutralization capacitor* that compensates for the internal capacitance of the device. Without proper neutralization, the RF amplifier could become an *oscillator,* producing its own frequency and ignoring the incoming frequency. The class of operation of the amplifier is determined by the biasing network indicated by $-V_{BB}$.

Other Classes

Other classes of amplifier operation are listed in Table 14-3.

Table 14-3	OTHER CLASSES OF AMPLIFIER OPERATION
Class of Operation	**Amplifier Characteristics**
AB	Amplifiers are adjusted halfway between class A and class B operation.
D	Two amplifiers used as switches in push-pull operation. Device goes between saturation and cutoff; thus its efficiency is almost 100%.
E	Uses a *high-impedance* load, such as an RF choke, to allow the device to be in saturation for 180° of the input signal signal, thus greatly improving the efficiency of a single stage.
F	Single stage that acts essentially as a *switch,* thus allowing almost 100% efficient operation. Since the output looks like a square wave, two sets of tuned circuits are used: One removes the third harmonic; the other passes on the fundamental.
S	Used in *switching regulators.*

Conclusion

This section presented the classes of operation for amplifiers. You should be able to quickly spot the major class of operation (A, B, or C) of a communication amplifier by observing the effects of the input signal on the conduction of the device. The next section presents the various types of amplifiers.

14-3 Review Questions

1. Name the three main classes of operation for amplifiers.
2. State what determines the amplifier's class of operation.
3. Describe the differences between the major classes of operation.
4. State what kind of circuit is required for a class C amplifier. Why is it necessary?
5. Describe the operation of a class B push-pull amplifier.

14-4 | TYPES OF AMPLIFIERS

Discussion

You have seen that amplifiers are classified into the frequency range of operation and the class of operation. This section shows how amplifiers are classified into *types*. This classification determines if the amplifier will be used primarily to match impedance, or to produce a voltage or current gain along with the corresponding power gain.

General Idea

An amplifier *type* is based upon the part of an amplifier that is common to the input and output signals. Figure 14-13 illustrates amplifiers that are called *common base, common gate,* and *common grid* (better known as *grounded grid*) amplifiers. In each circuit the output signal is 180° out of phase with the input signal. Other characteristics of this type

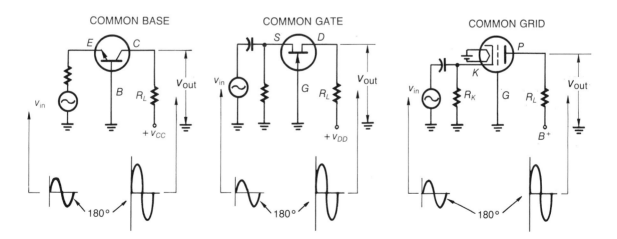

Figure 14-13 Common base, common gate, and common grid amplifiers.

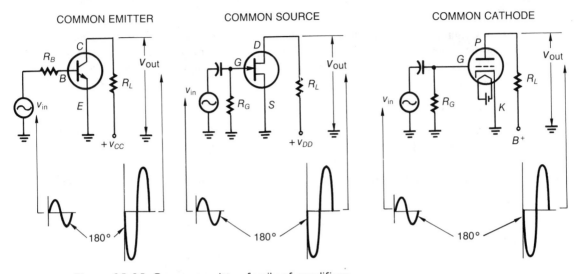

Figure 14-14 Common emitter family of amplifiers.

of amplifier include *low input impedance* and *high output impedance*. The current gain in each of these amplifiers is unity or less, but, because of the large difference in input and output impedances, the amplifier has a large voltage gain and power gain.

Other Types

Two other families of amplifier configurations are shown in Figures 14-14 and 14-15. The *common emitter amplifier* is shown in Figure 14-14. Note that the output signal is 180° out of phase with the input signal. Another characteristic of this amplifier is that it pro-

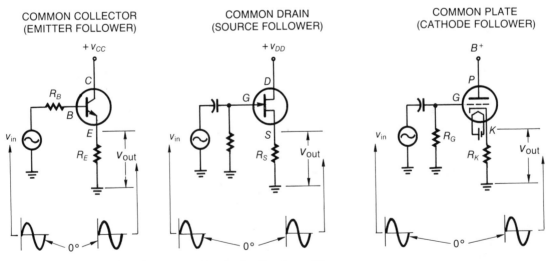

Figure 14-15 Common collector family of amplifiers.

duces a voltage gain and a current gain, thus giving the largest power gain of any of the three types. The common collector (usually referred to as the *emitter follower*) amplifier is illustrated in Figure 14-15. This type of amplifier has an output signal that is *in phase with the input signal*. It also has a *large input impedance* and a *small output impedance*, usually a few hundred ohms or less. The voltage gain is less than 1, but the current gain and corresponding power gain are large.

Summary of Amplifier Types

Table 14-4 summarizes the three types of amplifiers.

Table 14-4		**SUMMARY OF AMPLIFIER TYPES**			
Type of Amplifier			**Input–Output Phase Relation**	**Impedance**	
Transistor	**FET**	**Tube**		**Input**	**Output**
Common Emitter	Common Source	Common Cathode	180°	Medium	Medium
Common Base	Common Gate	Common Grid	0°	Small	Large
Common Collector (emitter follower)	Common Drain (source follower)	Common Plate (cathode follower)	0°	Large	Small

Typical Circuits

A typical application of the common source amplifier family as a class A audio amplifier is shown in Figure 14-16. Figure 14-17 shows a typical application of the common base amplifier family as RF amplifiers. They are used here in the front end of a communications

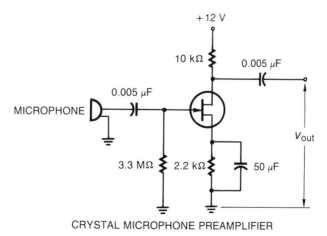

CRYSTAL MICROPHONE PREAMPLIFIER

Figure 14-16 Application of common source family.

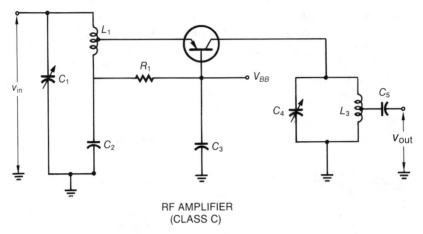

RF AMPLIFIER
(CLASS C)

Figure 14-17 Application of common base family.

receiver where the input impedance matches the low output impedance of the receiving antenna.

The common collector family application is illustrated in Figure 14-18. This circuit is called a *Darlington amplifier*. The Darlington amplifier has high input impedance and low output impedance. The voltage gain of this arrangement is less than 1, but because of the great difference in input and output impedances, the amplifier gives a large power gain.

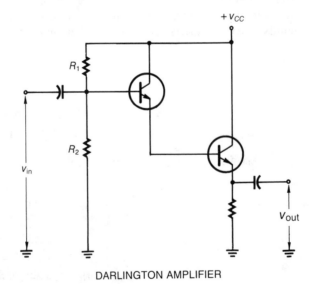

DARLINGTON AMPLIFIER

Figure 14-18 Application of common collector family.

Conclusion

This section presented the types of amplifiers used in communications circuits. In the next section, you will be introduced to small-signal amplifiers and power amplifiers.

14-4 Review Questions

1. Describe the similarities between common base, common gate, and grounded grid amplifiers.
2. Explain the major characteristics of a common emitter amplifier.
3. Give another name for a common drain amplifier.
4. State the phase relationships between the input and output signals for the (A) common base, (B) cathode follower, and (C) common source amplifiers.
5. What type of amplifier is used to give a high input impedance and a low output impedance?

14-5 SMALL-SIGNAL AMPLIFIERS AND POWER AMPLIFIERS

Discussion

Typically, small-signal amplifiers have a power rating of less than half a watt. Power amplifiers have power ratings greater than half a watt.

Small-signal amplifiers are used at the front end of communications systems, where the signals are small. Power amplifiers are used at the end of communications systems.

Examples of small-signal amplifiers are

■   An RF amplifier for a communications receiver
■   An audio amplifier for a communications transmitter

Examples of power amplifiers are

■   Audio power amplifier to operate speakers in communication receivers
■   RF power amplifier to drive an antenna in a communication transmitter

Small Signal

A *small signal* is defined as any input signal to an amplifier that makes the amplifier operate *in the linear portion of its characteristics*. See Figure 14-19. A small input signal produces an *undistorted* reproduction. As the amplitude of the input signal is increased, the amplifier will begin to operate in the *nonlinear* portion of its operational characteristics. When the amplifier operates in this range, the output signal is a *distorted* reproduction of the input signal.

Small-Signal Analysis

You must understand the purpose of the circuit components in small-signal amplifiers. A detailed discussion is given in any standard devices and circuits course. The discussion here is in the context of communication circuits.

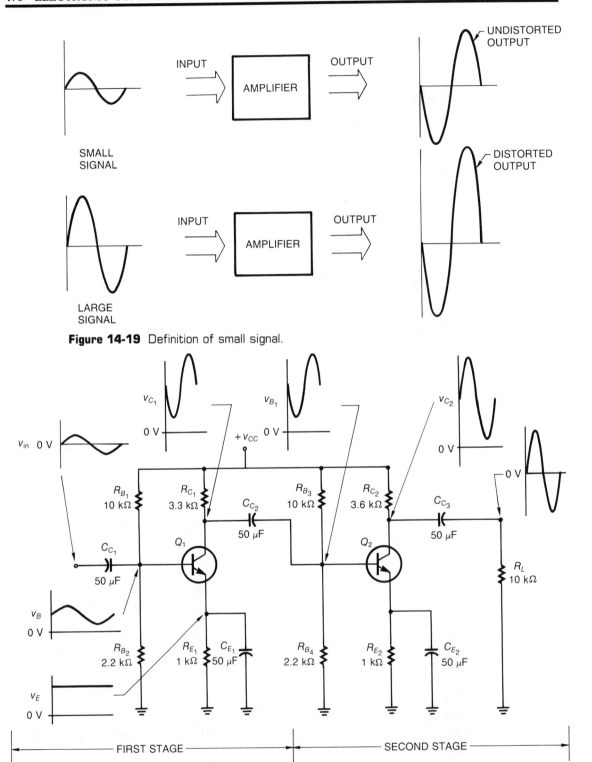

Figure 14-19 Definition of small signal.

Figure 14-20 Two-stage low-power transistor audio amplifier.

Figure 14-20 is a *two-stage, small-signal* audio, transistor amplifier. Both stages are operated as class A amplifiers. Table 14-5 summarizes the purpose of each component of the first stage. The components in the second stage serve the same purpose.

Table 14-5	SUMMARY ANALYSIS OF TRANSISTOR AUDIO AMPLIFIER
Component(s)	**Purpose**
R_{B_1}, R_{B_2}	*Base biasing resistors* Develops the proper biasing voltage V_B to ensure proper operating characteristics
C_{C_1}, C_{C_2}	*Coupling capacitors* Blocks the DC voltage of the previous stage, yet allows the signal to pass from one stage to the next
R_{E_1}	*Emitter resistor* Sets the amount of emitter current, which sets the collector current so that $V_C \approx \frac{1}{2} V_{CC}$
C_{E_1}	*Emitter bypass capacitor* Maintains a constant DC voltage across R_E, which prevents a signal from developing across R_E that would decrease the gain of the stage
R_{C_1}	*Collector resistor* Determines the voltage gain of the stage and develops a changing voltage across Q_1

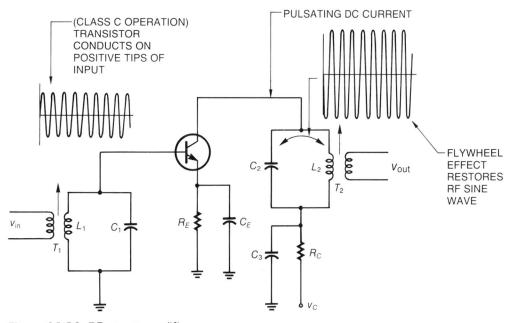

Figure 14-21 RF power amplifier.

Power Amplifier

An RF power amplifier is illustrated in Figure 14-21. It is a *class C* transistor power amplifier used to provide power to an RF carrier.

Table 14-6 summarizes the purpose of each amplifier component.

Table 14-6	SUMMARY ANALYSIS OF TRANSISTOR RF AMPLIFIER
Component(s)	**Purpose**
T_1	*Input RF transformer* Couples the input signal to the RF amplifier (turns ratio selected for an optimum impedance match)
L_1, C_1	*Base tuned circuit* Acts as a tuned circuit for the frequency to be amplified; develops a maximum impedance at resonance; T_1 is "slug" tuned to adjust for correct frequency
R_E	*Emitter resistor* Temperature stabilizes the amplifier and prevents "thermal runaway" (i.e., the transistor heats up, causing it to conduct more and produce more heat)
C_E	*Emitter bypass capacitor* Prevents signal degeneration of the output signal by keeping the voltage across R_E constant
C_2, L_2	*Collector tuned circuit* A parallel resonant circuit used to develop a high collector load impedance at the frequency to be amplified (T_2 is slug tuned to adjust to correct frequency)
R_C, C_3	*Decoupling network* Removes any RF from power source line (V_{CC}); the capacitor presents a lower-impedance path to ground than that offered by the value of R_C
T_2	*Output transformer* Couples the amplified signal to the next stage (turns ratio selected for optimum impedance matching)

Conclusion

This section presented examples of a small-signal amplifier and a power amplifier with typical circuit applications.

The next section presents several amplifier circuits used in communications systems.

14-5 Review Questions

1. What is the difference between a small-signal amplifier and a power amplifier?
2. Cite some examples of small-signal amplifiers and power amplifiers.

3. Define small signal.
4. Why are base biasing resistors used in a transistor amplifier?
5. Summarize the main features of a class C transistor power amplifier.

14-6 | AM, FM, AND TV AMPLIFIERS

Discussion

This section contains some practical applications of the circuits and devices described in the previous sections. Here you will see the application of amplifiers in an AM–FM radio and a television receiver.

AM–FM Radio

Figure 14-22 shows an IC chip that provides a combination AM–FM radio receiver. The block diagram is of the combination receiver. The features of the IC include:

■ DC selection of AM–FM mode
■ Regulated supply
■ Decreased audio amplifier bandwidth in AM mode, reducing amplifier noise in the AM band
■ AM converter AGC (automatic gain control) for excellent overload characteristics

AGC means automatic gain control. This is simply a way to automatically control the radio gain, so as you tune in stations of different strengths it isn't necessary to readjust the volume control.

Figure 14-23 is a schematic of the IC chip. You can see the circuit complexity inside an AM–FM receiver chip. If any component inside the chip breaks down, the whole chip must be replaced. In general, the cost of the replacement time is more than the cost of the chip. In older AM–FM receivers, all of the components, including transistors, were discrete.

TV Amplifiers

Figure 14-24 illustrates a TV video amplifier IC chip. The video amplifier contains a video preamplifier, DC contrast control that can be ganged with the chroma gain control, and beam-current limiting through the contrast control. This chip contains a separate NPN transistor to allow more design flexibility.

Figure 14-25 is the schematic of the TBA970 video amplifier. All internal transistors are DC coupled to improve the low-frequency response of the video amplifier. Again, if any component inside this chip breaks down, the whole chip must be replaced. In older TV sets, the video amplifier contained discrete components, including the transistors themselves.

Block Diagram

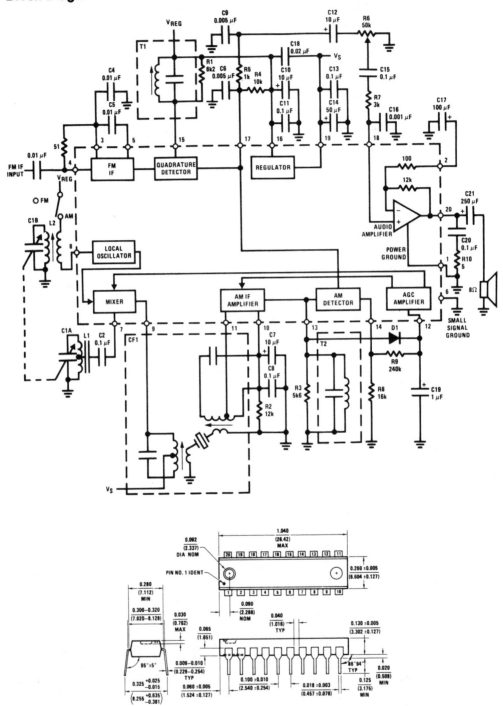

Figure 14-22 Combination AM–FM radio receiver, the LM 1868. *Courtesy* of National Semiconductor Corporation

Equivalent Schematic Diagram

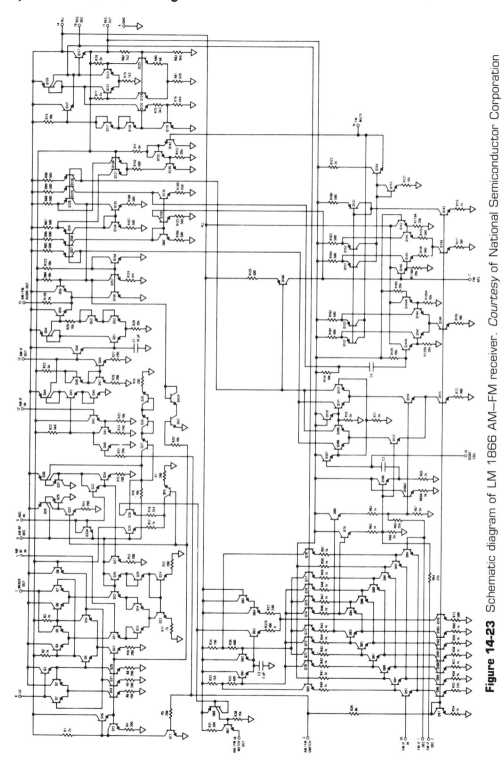

Figure 14-23 Schematic diagram of LM 1866 AM–FM receiver. *Courtesy* of National Semiconductor Corporation

Connection Diagram

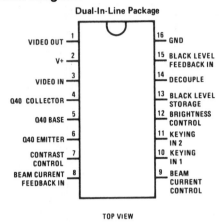

Block Diagram

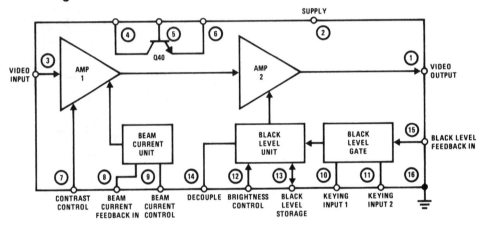

Figure 14-24 Television video amplifier TBA970. *Courtesy* of National Semiconductor Corporation

Equivalent Circuit

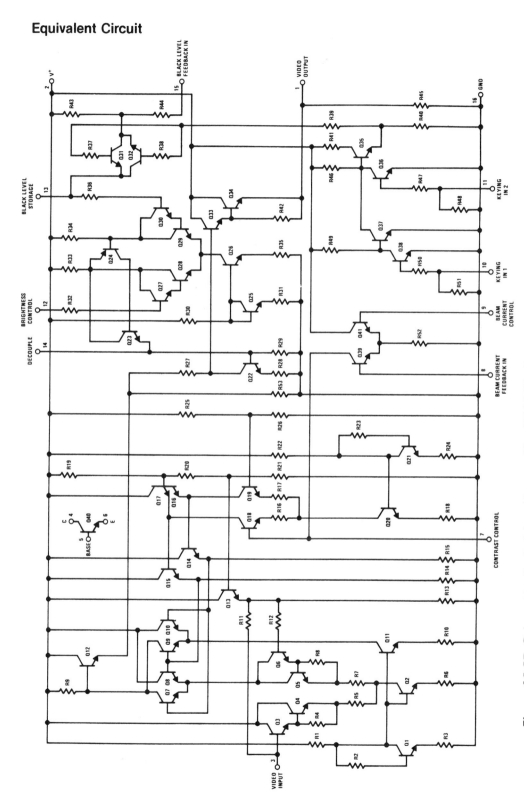

Figure 14-25 Schematic of television video amplifier TBA970. *Courtesy of National Semiconductor Corporation*

Figure 14-26 Miniature TV set. *Courtesy* of Radio Shack

Figure 14-26 shows a portable TV set that weighs less than this textbook. The entire receiver contains fewer discrete components than a simple AM radio of only a few decades ago.

Figure 14-27 is a schematic of the miniature TV set. The TV contains a *back-lighted liquid crystal display* (LCD) digital screen. The TV must be viewed in a mirror in order to achieve the required *contrast*. One advantage of this is that the brighter the surroundings are, the brighter the picture is. Figure 14-28 illustrates a back-lighted LCD display.

Figure 14-29 shows the inside of the miniature TV. There are only a few discrete components; most of the system is constructed from a few ICs.

Conclusion

This section introduced amplifiers used in AM, FM, and TV. In the next section, you will see how to troubleshoot a single transistor amplifier. These can still be found in communication receivers.

14-6 Review Questions

1. Is it possible to have an AM–FM radio contained on a single IC chip? Explain.
2. Explain the purpose of AGC.
3. Describe the servicing procedure for a bad chip.
4. Explain the meaning of LCD.
5. Explain the operation of a back-lighted LCD display.

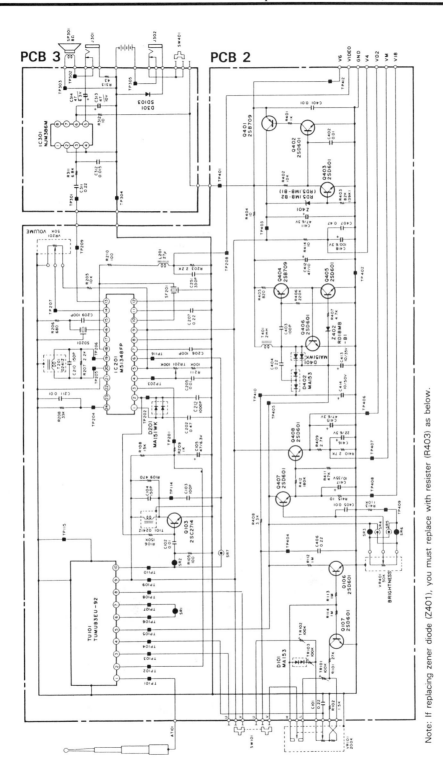

NOTES:
(1) ALL RESISTANCE VALUES ARE INDICATED IN "Ω". (KΩ = 10^3 Ω, MΩ = 10^6 Ω)
(2) ALL CAPACITANCE VALUES ARE INDICATED IN "μF". (pF = 10^{-6} μF)

Note: If replacing zener diode (Z401), you must replace with resister (R403) as below.

	Z401		R403
A73-0130 (RD5.1MB-B2)		W22-8233 (82 kΩ)	
A73-0220 (RD5.1MB-B1)		W22-3933 (39 kΩ)	

Figure 14-27 Schematic of miniature TV set. *Courtesy of* Radio Shack

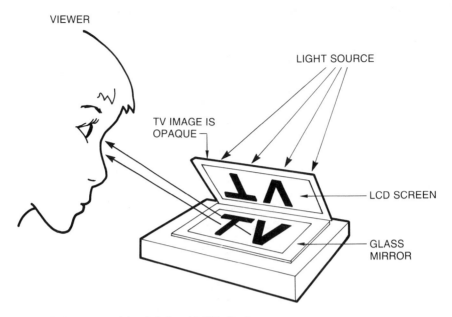

Figure 14-28 Process of back-lighted LCD display.

Figure 14-29 Inside a miniature TV.

TROUBLESHOOTING AND INSTRUMENTATION

Troubleshooting a Transistor Amplifier

This section will take you step-by-step through the analysis and troubleshooting of a transistor amplifier.

Transistor Amplifier

A typical transistor amplifier is shown in Figure 14-30. The troubleshooting analysis of this amplifier assumes that any of the transistor resistors can either *increase* or *decrease* in value.

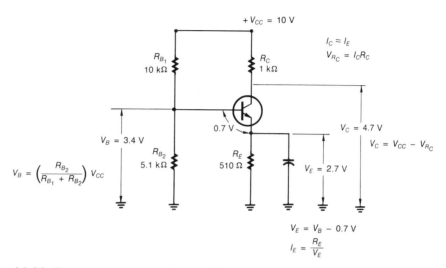

Figure 14-30 Typical transistor audio amplifier.

Table 14-7		TRANSISTOR AMPLIFIER TROUBLESHOOTING		
		Effect on		
Component	Problem	V_B	V_E	V_C
R_{B_1} (refer to Fig. 14-31)	Increase Decrease	Decrease Increase	Decrease Increase	Increase Decrease
R_{B_2} (refer to Fig. 14-32)	Increase Decrease	Increase Decrease	Increase Decrease	Decrease Increase
R_E (refer to Fig. 14-33)	Increase Decrease	None None	None None	Increase Decrease
R_C (refer to Fig. 14-34)	Increase Decrease	None None	None None	Decrease Increase

Table 14-7 is a troubleshooting summary for the circuit of Figure 14-30. To encourage circuit analyzing, the problems caused by the resistors will be viewed in terms of the circuit voltage readings with respect to ground.

Note: Readings are for a percentage change of less than 100% in resistance values, not a drastic change approaching an open or a short.

The following figures explain the reason for the readings in Table 14-7.

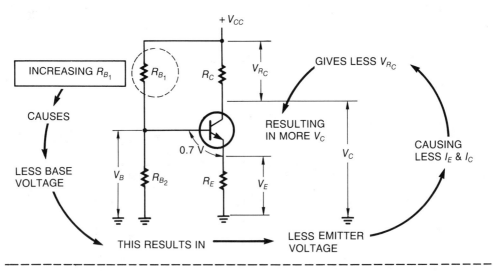

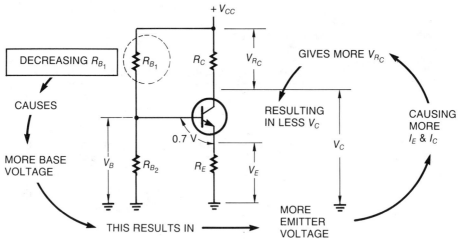

Figure 14-31 Effects of changes in R_{B_1}.

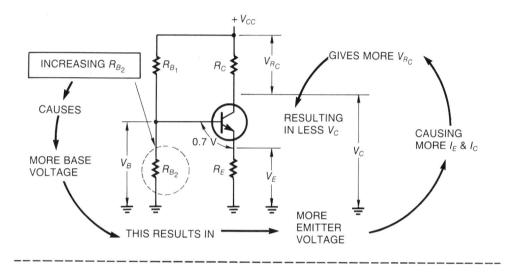

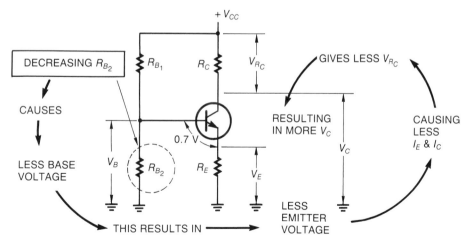

Figure 14-32 Effects of changes in R_{B_2}.

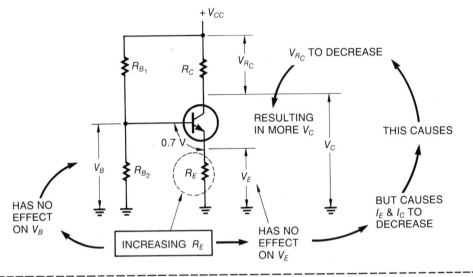

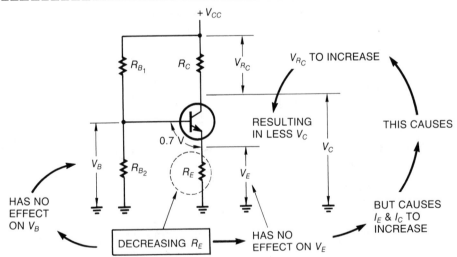

Figure 14-33 Effects of changes in R_E.

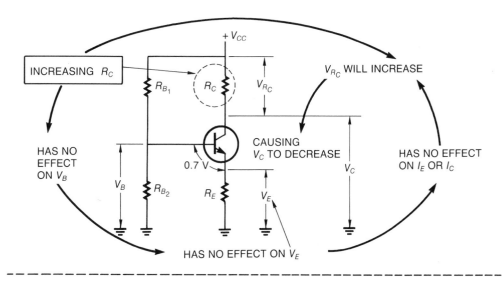

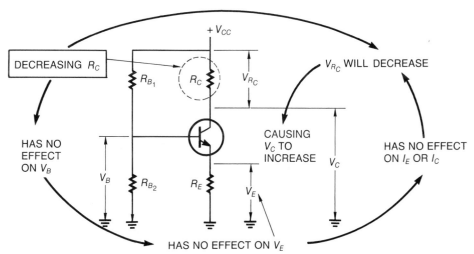

Figure 14-34 Effects of changes in R_C.

Conclusion

This section analyzed in detail the effects of resistor value on a transistor amplifier from a troubleshooting standpoint. The microcomputer simulation section contains a structured BASIC program that allows you to practice your troubleshooting skills.

14-7 Review Questions

Questions 1–3 refer to the transistor amplifier given in this section.
1. Why are the voltage divider resistors R_{B_1} and R_{B_2} used?
2. How do the values of R_{B_1} and R_{B_2} affect the base voltage V_B?
3. How does the value of R_E affect the base voltage V_B?
4. How does the value of R_E affect the transistor collector voltage V_C?
5. Describe the relationship between the values of R_C and V_C.

MICROCOMPUTER SIMULATION

Troubleshooting simulation fourteen is your first opportunity to develop your troubleshooting skills on a circuit. This simulation shows a solid state amplifier and you are shown how to make a DC analysis using a simulated DC voltmeter. In the demonstration and test mode, problems are introduced into the circuit by the computer for you to find using DC troubleshooting methods. This is an excellent troubleshooting program for building your circuit skills. It will also help you in future troubleshooting simulation programs.

CHAPTER PROBLEMS

(Answers to odd-numbered problems appear at the end of the text.)

TRUE/FALSE
Answer the following questions true or false.
1. All amplifiers have a power gain.
2. The difference between transformers and amplifiers is that a transformer cannot have a current gain.
3. The main difference between a video amplifier and an RF amplifier is the range of frequencies they are capable of amplifying.
4. An integrated circuit is not considered as a device in this text because it has more than one component in it.
5. The main difference between a PNP transistor and an NPN transistor in communication circuits is that one is used for low frequencies and the other for high frequencies.

MULTIPLE CHOICE

Answer the following questions by selecting the most correct answer.

6. The major classification of amplifiers is determined by:
   (A) the frequency response.
   (B) the type of device used.
   (C) the circuit in which the device is used.
   (D) none of the above.

7. The classes of amplifiers are determined by:
   (A) the circuit in which the device is used.
   (B) the frequency response.
   (C) how much of the input signal causes the device to conduct.
   (D) how much of the output signal looks like the input signal.

8. The amplifier with no voltage gain and with input and output signals in phase is
   (A) an emitter follower.
   (B) a source follower.
   (C) a cathode follower.
   (D) all of the above.

9. The most significant aspect of a small-signal amplifier is:
   (A) its output signal is smaller than its input signal.
   (B) an input signal that is smaller than its output signal.
   (C) a power gain of less than one.
   (D) none of the above.

10. A class C amplifier is capable of producing a sine-wave output because:
   (A) it conducts during the full 360° of the input signal.
   (B) it creates its own sine wave.
   (C) the output tuned circuit has a flywheel effect.
   (D) the power supply voltage varies in accordance with the input signal.

MATCHING

Match the types of amplifiers to the devices shown in Figure 14-35.

11. Power amplifier.
12. Class B push-pull.
13. Small-signal audio amplifier.
14. Class A audio amplifier.
15. Class C RF amplifier.

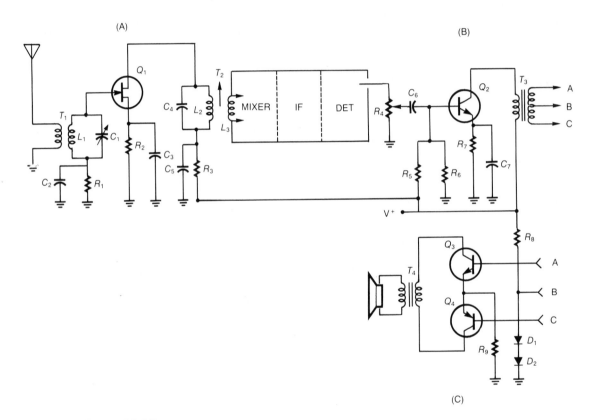

Figure 14-35

FILL IN

Fill in the blanks with the most correct answer(s).

16. An amplifier is a circuit that produces a more _____ replication of the input signal.
17. A major disadvantage of the vacuum tube is its need for a _____ to give energy to the electrons.
18. The class _____ amplifier is inefficient because it must always be _____ .
19. An advantage of a Darlington amplifier is its large _____ impedance.
20. Typically, _____ _____ amplifiers have a power rating of less than half a watt.

OPEN-ENDED

Answer the following questions as indicated.

21. In the transistor amplifier circuit in Figure 14-36, what could cause the value of V_E to decrease? to increase? What resistors affect the value of V_E?
22. For the circuit in Figure 14-36, what could cause the value of V_C to increase? to decrease? What resistors affect the value of V_C?
23. For the circuit in Figure 14-36, how does the value of R_C affect V_C? V_E?
24. What resistors in Figure 14-36 determine the amount of emitter current? the amount of collector current?
25. What would you do to determine the maximum current that could flow in the emitter–collector of the circuit in Figure 14-36? to determine the maximum voltage of V_C?

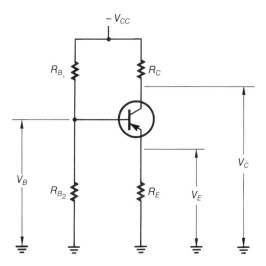

Figure 14-36

CHAPTER 15

Oscillators and Modulators

OBJECTIVES

In this chapter, you will study:

- [ ] What an oscillator is and the basic requirements of all oscillators.
- [ ] The types of oscillators used in communication circuits.
- [ ] The basic principles of modulators.
- [ ] The types of amplitude modulators.
- [ ] Modulators used in frequency and phase modulation.
- [ ] The causes of unwanted oscillations in communication circuits.

INTRODUCTION

Without oscillators, modern radio communications would not be possible. This chapter presents two very important types of circuits: the *oscillator*, used to create a carrier, and the *modulator*, used to put information on the carrier.

You will learn the principles involved in producing these two valuable circuits. In the troubleshooting and instrumentation section, you will see what causes unwanted oscillations and what can be done to prevent them.

15-1 | OSCILLATOR PRINCIPLES

Discussion

An *oscillator* is a circuit that produces its own waveform without an input signal. They can generate many different types of waveforms; see Figure 15-1. This section describes the basic principles of all oscillator circuits.

Oscillator Requirements

All oscillators must have the following:

1. *A source of energy:*
   The source could be a power supply, a battery, or a solar cell.

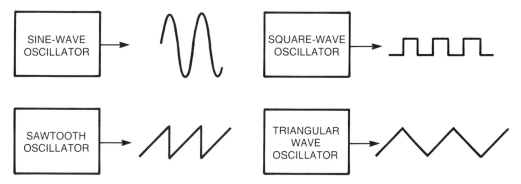

Figure 15-1 Oscillators and representative types of waveforms.

2. *A frequency-determining circuit:*
   The circuit could consist of tuned *LC* circuits, *RC* time-constant circuits, or electro-mechanical devices such as crystals.
3. *An amplifier:*
   The amplifier could be a transistor, FET, IC, or any other device from Chapter 14.
4. *A feedback circuit (positive feedback):*
   Feedback is a method of transferring energy from the output of the oscillator to the input so that the output and input signals are in phase. Wires, optical links, or electromagnetic radiation can be used.
   Figure 15-2 illustrates the basic requirements of all oscillators.

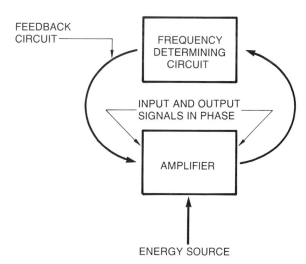

Figure 15-2 Basic oscillator requirements.

Barkhausen Criterion

The *Barkhausen criterion* states for an oscillator to oscillate (continuously produce its own signal), the product of the amplifier gain (A_V) and the amount of signal fed from the output to the input (B_V) must be equal to unity.

Mathematically it is

$$A_V B_V = 1 \qquad \textbf{(Equation 15-1)}$$

$$\begin{aligned} \textit{where} \quad A_V &= \text{Amplifier gain} \\ B_V &= \text{Feedback gain} \end{aligned}$$

The meaning of the Barkhausen criterion is shown in Figure 15-3. The overall system gain must be unity for the oscillations to be sustained.

Example 1

The amplifier of an oscillator has a gain of 10. Determine the gain of the feedback circuit to sustain circuit oscillations.

Solution

From the Barkhausen criterion:

$$A_V B_V = 1 \qquad \textbf{(Equation 15-1)}$$
$$10 B_V = 1$$
$$B_V = \frac{1}{10} = 0.1$$

Hence the gain of the feedback circuit must be 0.1.

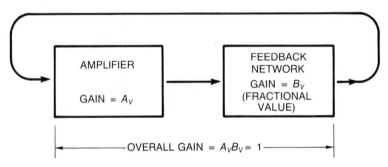

Figure 15-3 Meaning of Barkhausen criterion.

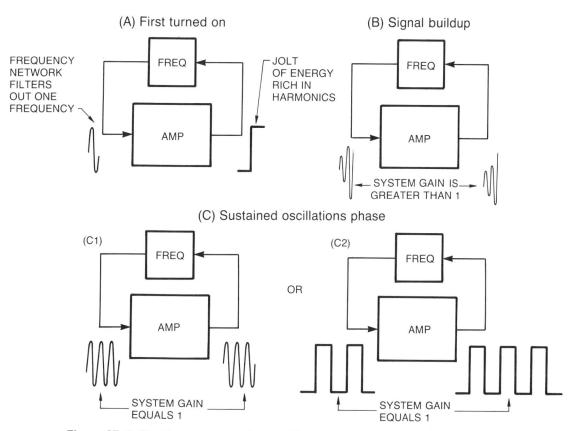

Figure 15-4 The three phases of an oscillator.

Basic Oscillator Concepts

Figure 15-4 shows the three phases of an oscillator: turn-on, signal buildup, and sustained oscillations.

When power is initially applied to any circuit, the jolt from this immediate application of external energy causes momentary *transitions* in the circuit, which are rich in *harmonics*. Since an oscillator has a frequency-determining network, one frequency is selected and intentionally fed back to the input of the oscillator. The feedback path is such that the signal fed back is in phase with the output signal and therefore reinforces it.

Oscillations are sustained when the product of the amplifier gain and the amount of feedback is equal to unity (Barkhausen criterion). You can achieve this condition by inserting a component in the circuit to cause the amplifier gain or feedback network gain to adjust automatically until the overall gain is unity. If this happens before the signal is clipped and saturated by the amplifier, then the output waveform will be a sine wave. Otherwise you achieve unity gain by forcing the circuit to produce a square wave by driving the amplifier into saturation and cutoff.

Conclusion

This section presented the fundamental requirements of all oscillators. In the next section, you will see how these principles are applied to communication circuits.

15-1 Review Questions

1. What is an oscillator?
2. State the requirements of all oscillators.
3. Explain the Barkhausen criterion.
4. Describe the three phases of an oscillator.
5. What causes the gain of an oscillator to be unity?

15-2 | TYPES OF OSCILLATORS

Discussion

Many different circuits are used for oscillators. Any device (transistor, FET, op amp, logic circuit, or other IC, etc.) can be used as the amplifier requirement for an oscillator. Vacuum tubes can also be used, but they are so outdated in this area that their circuits will not be covered here.

The electrical components that determine the resonant frequency and the method of achieving positive feedback determine the *type* of oscillator. This section presents these various oscillator types found in communication circuits.

Oscillator Using LC Networks

Table 15-1 lists the oscillators that use inductors and capacitors as their frequency-determining components.

Table 15-1	OSCILLATORS USING *LC* NETWORKS	
Oscillator Type	**Distinguishing Characteristics**	**Formula for f_r**
Armstrong (refer to Fig. 15-5)	Uses a transformer to feed back the output signal to the input. Secondary winding is sometimes called the *tickler coil*.	$\dfrac{1}{2\pi\sqrt{LC}}$
Clapp (refer to Fig. 15-6)	Uses a *capacitive* voltage divider (a refinement of the Colpitts oscillator) and an additional capacitor in series with the inductor.	$\dfrac{1}{2\pi\sqrt{LC_{equ}}}$
Colpitts (refer to Fig. 15-7)	Uses a *capacitive* voltage divider.	$\dfrac{1}{2\pi\sqrt{LC_{equ}}}$
Hartley (refer to Fig. 15-8)	Uses an *inductive* voltage divider. Usually not as stable as the Clapp or Colpitts.	$\dfrac{1}{2\pi\sqrt{L_{equ}C}}$

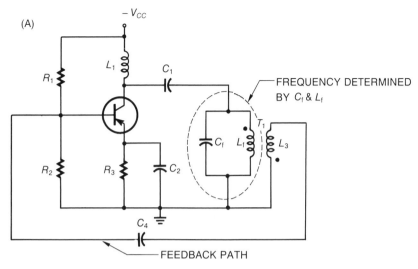

Figure 15-5 Armstrong oscillator.

Figures 15-5 through 15-8 are transistor representations of the oscillators in Table 15-1.

The frequency of the *Armstrong oscillator*, Figure 15-5, is determined by the values of C_f and L_f. The phase windings of the feedback transformer T_1 are such that the input signal to the base of the transistor is in phase with the output signal. Feedback is provided by capacitor C_4, which blocks the DC bias voltage provided by R_1 and R_2 but allows the

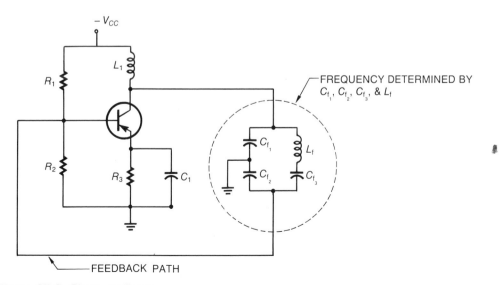

Figure 15-6 Clapp oscillator.

signal to be fed back to the input. Inductor L_1 is a *radio frequency choke* (RFC). This presents a high impedance to the resonant frequency where it is coupled to the tuned circuit through C_1.

The oscillation frequency for the Armstrong oscillator is

$$f_r = \frac{1}{2\pi\sqrt{LC}}$$ **(Equation 15-2)**

where f_r = Resonant frequency of oscillator in hertz
L = Inductance value in henrys
C = Capacitance value in farads

The frequency of the *Clapp oscillator*, Figure 15-6, is determined by C_{f_1}, C_{f_2}, C_{f_3}, and L_f. Capacitors C_{f_1} and C_{f_2} block any DC voltages from ground. The frequency of a Clapp oscillator usually does not depend on junction capacitance inside the transistor.

The oscillation frequency for a Clapp oscillator is

$$f_r = \frac{1}{2\pi\sqrt{LC_{equ}}}$$ **(Equation 15-3)**

where f_r = Resonant frequency of oscillator in hertz
L = Value of L_f in henrys

$$C_{equ} = \frac{1}{(1/C_{f_1} + 1/C_{f_2} + 1/C_{f_3})} \text{ in farads}$$

The *Colpitts oscillator*, Figure 15-7, has its frequency determined by the values of C_{f_1}, C_{f_2}, and L_f, similar to the Clapp oscillator. Its resonant frequency is more easily in-

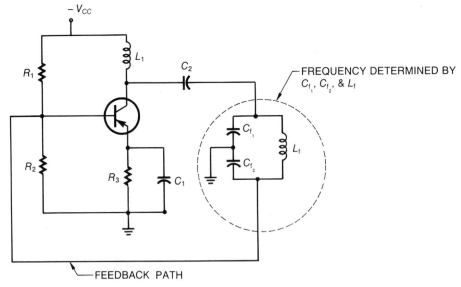

Figure 15-7 Colpitts oscillator.

fluenced by the transistor junction capacitance. Therefore, it is not used as much as the Clapp oscillator.

The oscillation frequency for the Colpitts oscillator is

$$f_r = \frac{1}{2\pi\sqrt{LC_{equ}}}$$

(**Equation 15-4**)

where f_r = Resonant frequency of oscillator in hertz

      L = Inductance of L_f in henrys

$$C_{equ} = \frac{C_{f_1}C_{f_2}}{C_{f_1} + C_{f_2}}$$

The *Hartley oscillator,* Figure 15-8, has a complex frequency-determining component that is influenced by the *mutual inductance* (L_M) of the split inductor L_{f_1} and L_{f_2}. The value of L_M is determined by the *coefficient of coupling (K).*

The resonant frequency of the Hartley oscillator is

$$f_r = \frac{1}{2\pi\sqrt{L_{equ}C}}$$

(**Equation 15-5**)

where f_r = Resonant frequency of oscillator in hertz

      $L_{equ} = L_{f_1} + L_{f_2} + 2L_M$ in henrys

      $L_M = K\sqrt{L_{f_1}L_{f_2}}$ in henrys

      C = Capacitance value in farads

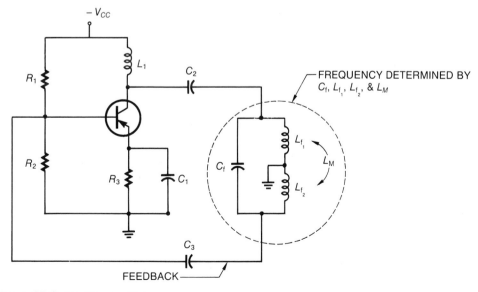

Figure 15-8 Hartley oscillator.

RC *Oscillators*

RC oscillators depend on the *RC* time constant to determine their oscillation frequency. Unlike their *LC* counterparts, these circuits are typically used for low-frequency oscillators (not much more than a few megahertz).

Table 15-2 lists the major *RC* oscillators found in communication circuits.

Table 15-2	OSCILLATORS USING *RC* NETWORKS	
Oscillator Type	**Distinguishing Characteristics**	**Formula for f_r**
Phase shift (refer to Fig. 15-9)	Uses three *RC* networks, each of which produces a phase shift of 60°, for a total phase shift of 180°.	$\approx \dfrac{1}{2\pi\sqrt{6}RC}$
Twin-T (refer to Fig. 15-10)	Uses a twin-T filter that produces a lead–lag network where the phase shift is zero at f_r.	$\dfrac{1}{2\pi RC}$
Wien bridge (refer to Fig. 15-11)	Uses a lead–lag network with an automatically variable resistor to maintain a stable output amplitude.	$\dfrac{1}{2\pi RC}$

For the *phase shift oscillator,* Figure 15-9, there will be a frequency where the phase shift through each *RC* network is exactly 60°. Since there are three of these networks, the total phase shift will be 180°. When this is combined with the 180° phase shift already provided by the amplifier, the criterion for oscillator feedback is met—but only for one frequency—and the circuit oscillates.

The formula for the resonant frequency of a phase shift oscillator is approximately

$$f_r \approx \frac{1}{2\pi\sqrt{6}RC}$$

(Equation 15-6)

where f_r = Resonant frequency of oscillator in hertz
   R = Value of one of the three equal resistors in the feedback leg in ohms
   C = Value of one of the three equal capacitors in the feedback leg in farads

The *twin-T oscillator,* Figure 15-10, consists of two *RC* filter networks, creating a *notch filter* made up of R_f and C_f. The feedback path is to the inverting input of the amplifier in the negative feedback path of the amplifier. The resistance of R_v is usually a small incandescent lamp. When the circuit is first turned on, the lamp resistance is small, thus providing the large gain required for starting oscillations. As the oscillations increase in amplitude, the resistance of the lamp increases due to the heating effect of the larger signal. The increased lamp resistance reduces the gain of the system to unity.

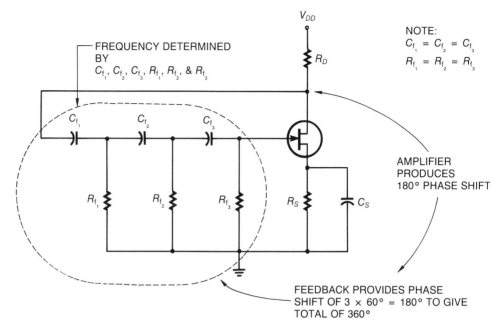

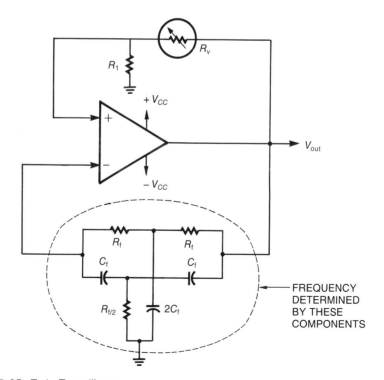

Figure 15-9 Phase shift oscillator.

Figure 15-10 Twin-T oscillator.

The formula for the resonant frequency of the twin-T oscillator is

$$f_r = \frac{1}{2\pi RC}$$
(Equation 15-7)

where f_r = Resonant frequency of oscillator in hertz
R = Value of each R_f resistor in ohms
C = Value of each C_f capacitor in farads

The *Wien bridge oscillator,* Figure 15-11, uses a lamp in its feedback path, which serves the same purpose as the lamp in the twin-T oscillator. The Wien bridge uses a *lead–lag network* consisting of resistors R_f and capacitors C_f.
The formula for the resonant frequency of the Wien bridge oscillator is

$$f_r = \frac{1}{2\pi RC}$$
(Equation 15-8)

where f_r = Resonant frequency of oscillator in hertz
R = Value of each R_f resistor in ohms
C = Value of each C_f capacitor in farads

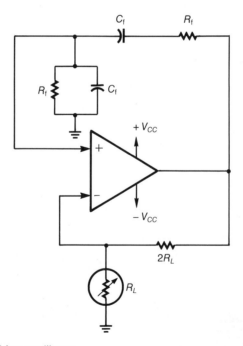

Figure 15-11 Wien bridge oscillator.

Crystals

The oscillators discussed so far have one problem in common: frequency stability. *Frequency stability* is an indication of how well the oscillator stays at the same frequency. This property is very important when the oscillator is used in communication systems where frequency accuracy is critical. Frequency instability occurs because of the changes experienced by circuit components during operation. This is where crystals come into the picture.

Crystals are physically made from substances that look like crystals. Some crystals exhibit the *piezoelectric effect:* When an AC voltage is applied across them, they mechanically vibrate at the frequency of the applied voltage. The opposite is also true: If mechanically excited, they produce an AC voltage. Figure 15-12 illustrates some of the features of quartz crystals.

The resonant frequency of a crystal is determined by its physical dimensions. This relationship is

$$f = \frac{K}{T}$$

(**Equation 15-9**)

where f = Fundamental crystal frequency
K = Constant representing specific characteristics of the crystal
T = Crystal thickness

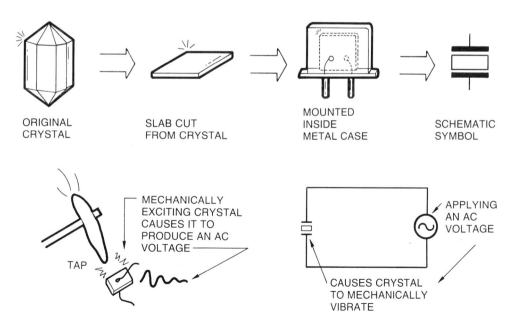

ORIGINAL
CRYSTAL

SLAB CUT
FROM CRYSTAL

MOUNTED
INSIDE
METAL CASE

SCHEMATIC
SYMBOL

MECHANICALLY
EXCITING CRYSTAL
CAUSES IT TO
PRODUCE AN AC
VOLTAGE

TAP

APPLYING
AN AC
VOLTAGE

CAUSES CRYSTAL
TO MECHANICALLY
VIBRATE

Figure 15-12 Feature of a quartz crystal.

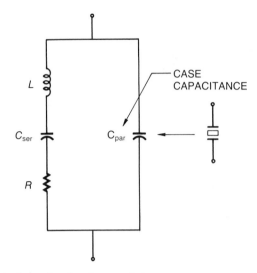

Figure 15-13 AC equivalent circuit of a crystal.

The AC equivalent circuit of a crystal is shown in Figure 15-13. The crystal can be viewed as electrical components connected in such a way that they act like series or parallel resonant circuits. The *series resonant circuit* consists of L, C_{ser}, and R. The *parallel resonant* portion is C_{par} in parallel with the rest of the circuit.

Because of this arrangement, a crystal can have *two fundamental resonant frequencies:* the parallel resonant mode and the series resonant mode. The formulas are

$$f_{par} = \frac{1}{2\pi\sqrt{LC_{equ}}}$$
(Equation 15-10)

where f_{par} = Parallel fundamental resonant frequency

         L = Equivalent inductance of the crystal

$$C_{equ} = \frac{C_{par}C_{ser}}{C_{par} + C_{ser}}$$

and for the series resonant mode

$$f_{ser} = \frac{1}{2\pi\sqrt{LC_{ser}}}$$
(Equation 15-11)

where f_{ser} = Series fundamental resonant frequency

         L = Equivalent inductance of the crystal

      C_{ser} = Equivalent series capacitance

Usually, a crystal is operated at its series or parallel fundamental frequency. However, for many applications, the crystal can be operated at a harmonic of its fundamental, called an *overtone*. In this manner, crystals can operate at frequencies up to 100 MHz.

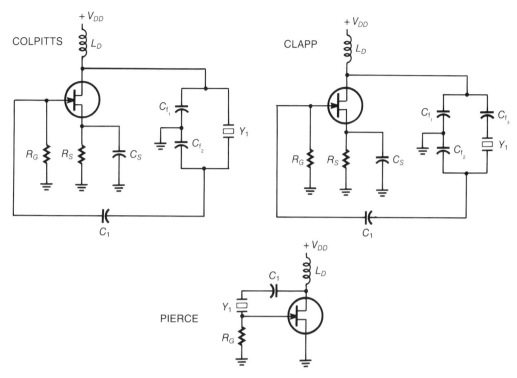

Figure 15-14 Schematic of crystal oscillators.

The equivalent values of most crystals are $L = 5\text{H}$, $C_{\text{ser}} = 0.075$ pF, $R = 1\,500\ \Omega$, and $C_{\text{par}} = 10$ pF. With such a large inductance value (in henrys!), the Q of crystals approaches 10 000 or more. Thus crystals have a very narrow bandwidth; therefore they provide good frequency stability and make excellent filters.

Crystal Oscillators

Figure 15-14 shows oscillators using crystals. The Colpitts, Clapp, and Pierce oscillators can all take advantage of crystal stability. The *Pierce* oscillator consists of only a few components while having the stability offered by the crystal.

Conclusion

This section presented the major features of oscillator circuits in communication systems. In the remaining sections, you will see some of the many applications of oscillators in communication circuits.

15-2 Review Questions

1. Name three different oscillators that use an *LC* network to determine their resonant frequencies. Describe the main differences between the oscillators.
2. Describe two different oscillators that use an *RC* time constant to determine their resonant frequency. Which one uses a phase shift of 60°?

3. State the major advantage of a crystal oscillator over that of an *LC* or *RC* oscillator.
4. What factors determine the resonant frequency of a crystal?
5. Discuss the two modes of crystal operation. What determines the resonant frequencies of these modes?

15-3 | MODULATOR PRINCIPLES

Discussion

This section presents some common elements in modulation circuits. Previous chapters have presented the reasons for modulation. In this section, you will see the underlying principles of modulation circuits.

Basic Idea

Three types of modulating circuits are presented: *amplitude, frequency,* and *phase*. You will see that there is little difference between frequency modulation and phase modulation.

Amplitude Modulating Circuits

The fundamental amplitude modulating circuit is illustrated in Figure 15-15. Since there must be distortion to produce the required harmonics necessary for amplitude modulation, the diode is included in the circuit. The RF input is coupled through C_1, where the diodes rectify the carrier waveform. Inductor L_1 presents a high-impedance path to ground for

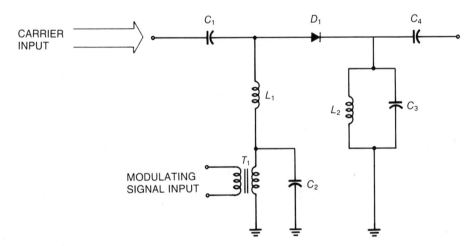

Figure 15-15 Fundamental amplitude modulating circuit.

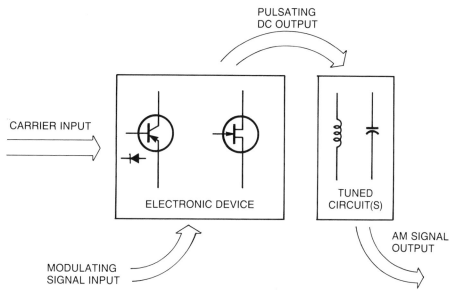

Figure 15-16 Variations of amplitude modulating circuits.

the RF waveform. Capacitor C_2 is used to bypass any RF to ground. The modulating signal is applied through a transformer. The turns ratio of the transformer is selected for optimal impedance matching between the audio circuit and the mixer. The tank circuit consisting of C_3 and L_2 is tuned to the frequency of the RF waveform, which restores the clipped sine wave through the flywheel action of the resonant circuit.

This circuit has many variations. Figure 15-16 shows block diagrams of some options. In every case in Figure 15-16, distortion is introduced because of the diode or because the amplifier is not operated as class A. Details of these circuits are given in the next section.

Balanced Modulators

The purpose of a *balanced modulator* is to amplitude modulate a carrier and transmit just the sidebands. Recall that the advantages of this are more efficient operation, because no power is used to transmit the carrier (the intelligence is contained only in the sidebands), and when no intelligence is being transmitted, no carrier is being sent so more power is saved.

Figure 15-17 shows the principle of a balanced modulator. The goal is to have a circuit that introduces distortion, has two inputs, but has an output only when both input signals are present. The important point with a balanced modulator is there is *no* carrier output at any time. In practice, there is always a very small carrier present, but the intent is to keep its output power many dB below the output power of the sidebands.

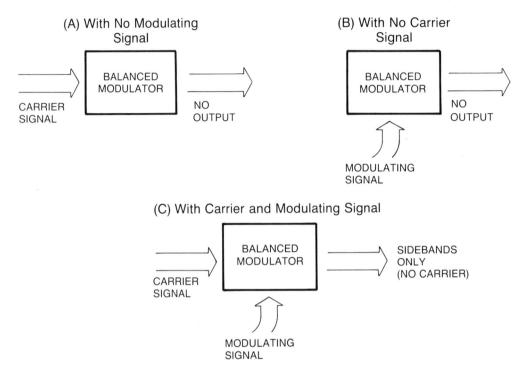

Figure 15-17 Basic idea of a balanced modulator.

FM and PM

This is a good time to examine the difference between FM and PM. The difference is slight but important. Figure 15-18 compares an FM wave to a PM wave. Observe that the FM wave has a frequency deviation proportional to the *amplitude* of the modulating signal. The frequency of the modulating signal does not affect the frequency deviation.

With the PM wave, the carrier frequency deviation is proportional to the *modulating frequency and to the amplitude* of the modulating frequency.

The waveforms of Figure 15-18 are essentially 90° out of phase. A lowpass filter is a *phase lag* network, whereas a highpass filter is a *phase lead* network. Figure 15-19 shows how a phase modulator can produce an FM waveform and how a frequency modulator can produce a PM waveform. There is very little difference between the production of FM or PM. The basic idea of producing FM or PM is illustrated in Figure 15-20. This interrelationship between FM and PM comes under the general classification of *angle modulation*.

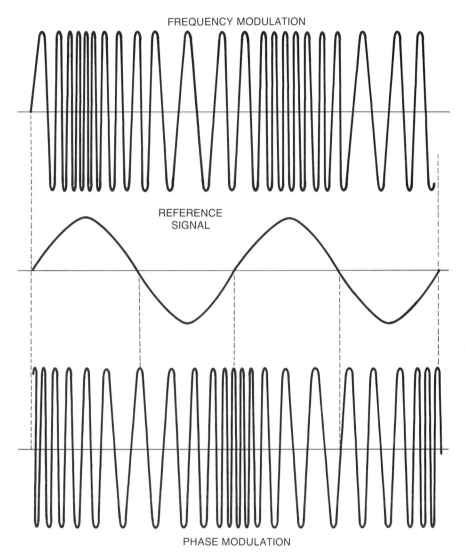

FREQUENCY MODULATION

REFERENCE
SIGNAL

PHASE MODULATION

Figure 15-18 Comparison of FM and PM waveforms.

Angle Modulation

Because of their similarity, FM and PM both come under the general classification of *angle modulation*. Angle modulation can be defined as changing the *angle* of the carrier wave from a reference value by a modulating signal.

The definition divides angle modulation into its two major components: FM, which is angle modulation where the *instantaneous frequency* of the sine-wave carrier changes

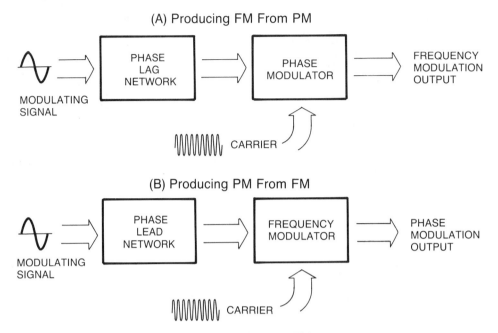

Figure 15-19 Producing FM from PM and PM from FM.

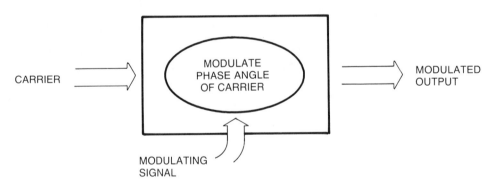

Figure 15-20 Basic idea of producing FM or PM.

from its reference frequency, and PM, which is angle modulation where the *instantaneous phase* of the sine-wave carrier changes from its reference frequency.

Conclusion

The next section describes some common methods of producing simple amplitude modulation as well as carrier-suppressed modulation (using balanced modulators).

15-3 Review Questions

1. Name three types of modulating circuits.
2. Which two circuits in question 1 are similar? Under what general heading can these two circuits be classified?
3. Describe the major requirements of a simple amplitude modulator circuit.
4. Explain the key features of a balanced modulator.
5. Discuss the main difference between FM and PM.

15-4	AMPLITUDE MODULATION CIRCUITS

Discussion

This section presents the most common forms of AM circuits used in electronic communication systems. Simple modulation circuits are given here. In the next section, balanced modulators are discussed.

Circuit Classification

Amplitude modulating circuits may be classified according to how the carrier and the modulating signal are applied to the modulating circuit. The circuits may be further classified according to the resultant output of the modulating circuit.

Table 15-3 summarizes the various types of AM circuits.

Table 15-3	TYPICAL AMPLITUDE MODULATION CIRCUITS	
Descriptive Name	**Key Features**	
Collector modulator (refer to Fig. 15-21)	Carrier signal is applied to the transistor base: modulating signal is applied to the collector.	
Series modulator (refer to Fig. 15-22)	Two transistors connected in series. One has the carrier applied to its base; the other has the modulating signal applied to its base.	
IC modulator (refer to Fig. 15-23)	Essentially an op amp with special characteristics that make it operate as an excellent but simple modulator.	

What follows is a description of the operation of each circuit in Table 15-3.

Collector Modulator

Figure 15-21 is a schematic of a *collector modulator*. The transistor is biased as a class C amplifier in order to introduce the required distortion. Recall that to modulate two frequencies, where the result is the original two frequencies and the sum and difference of

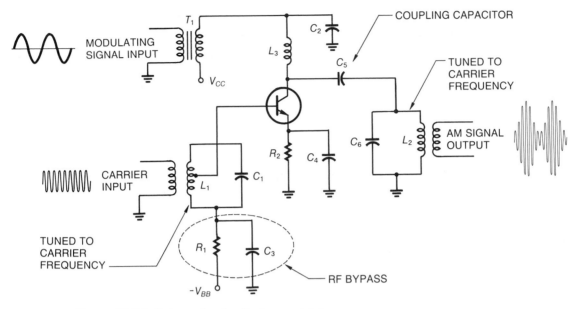

Figure 15-21 Schematic of collector modulator.

the two, you must have distortion. Capacitor C_2 must be large enough to bypass the carrier to ground but small enough to look like an open to the lower-frequency modulating signal.

This arrangement effectively changes the collector voltage of the transistor at the rate of the modulating frequency. The resulting output is an AM waveform.

Series Modulator

Figure 15-22 is a typical *series modulator*. Current flow in transistor Q_2 is controlled by the modulating input applied to the base of Q_1. Q_2 is operated as a class C amplifier, and capacitors C_1 and C_7 along with inductors L_1 and L_3 constitute a resonant circuit and an impedance-matching network. L_2 is an RF choke, and C_5 acts as an RF bypass.

IC Modulator

A widely used IC for producing AM is the CA3080A. See Figure 15-23. The amplifier/mixer contains two differential input terminals (+ and −) just like an ordinary op amp. The difference is that the output signal of this device has an output current proportional to the differential input rather than to an output voltage signal. The RF carrier is applied to the inverting input.

There is a current input to the amplifier at pin 5 that changes the gain of the amplifier/mixer. Thus, if the audio signal is applied here, the output current will represent an RF carrier whose amplitude is changing at an audio rate. The purpose of the 100-kΩ pot is to adjust the output for best signal symmetry. Since the gain of an amplifier can be calculated from its *transconductance* (g_m), it is called an *operational transconductance amplifier*.

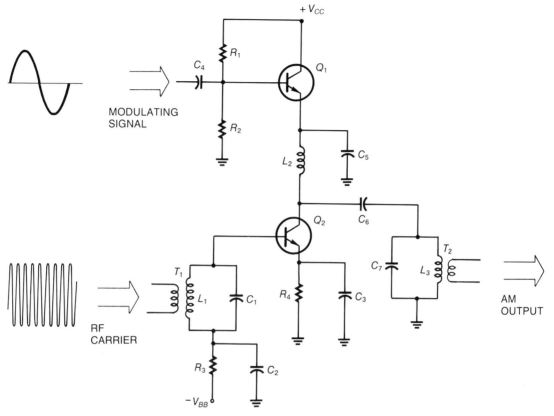

Figure 15-22 Series modulator circuit.

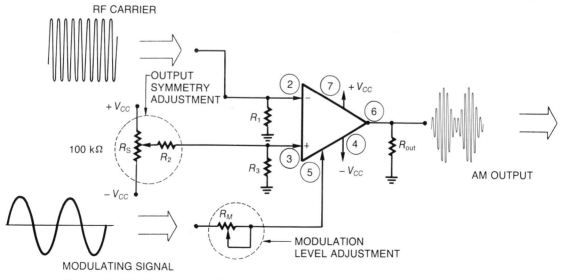

Figure 15-23 Typical CA3080A application and connections.

Conclusion

This section described some common circuits in simple AM. The next section describes some common circuits in suppressed-carrier modulation.

15-4 Review Questions

1. Name the typical AM circuits presented in this section.
2. Describe the process of collector modulation. What class of operation is used?
3. Describe the operation of a series modulator. State the class of operation for both amplifiers.
4. Discuss a popular IC used for AM.

15-5 | AM BALANCED MODULATOR CIRCUITS

Discussion

This section discusses balanced modulator circuits commonly found in communication systems. Recall that the purpose of a balanced modulator is to suppress the RF carrier and allow only the sidebands to be transmitted. This section concludes AM circuits. The next section describes FM and PM circuits.

Overview of Balanced Modulators

Several types of balanced modulator circuits are used in amplitude modulation. Table 15-4 summarizes those given here.

Table 15-4	TYPICAL BALANCED MODULATOR CIRCUITS
Descriptive Name	**Key Features**
Diode-ring balanced modulator (refer to Fig. 15-24)	Carrier and modulating signal applied to a network consisting of four diodes. The resultant output contains a suppressed carrier.
Active balanced modulator (refer to Fig. 15-26)	Consists of two active devices such as FETs or transistors. Carrier and modulating signal are applied to the gate or base. Resulting output contains a suppressed carrier.
IC balanced modulator (refer to Fig. 15-27)	A single IC chip, the LM 1596, produces a suppressed-carrier resultant signal. This offers great design and circuit simplicity.

Diode-Ring Balanced Modulator

Figure 15-24 is a schematic of a *diode-ring balanced modulator*. Note that there are two inputs to the diode-ring balanced modulator—one for the modulating frequency, and one for the carrier frequency. The tuned circuit at the output consisting of C_1 and L_1 is tuned to the frequency of the suppressed carrier. The Q of this tuned circuit must allow the

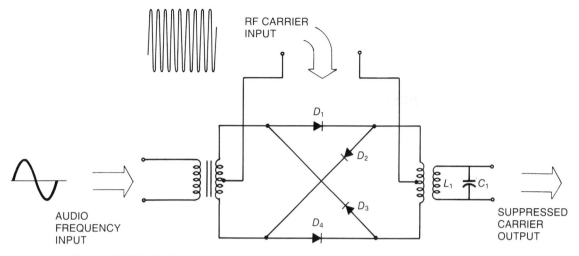

Figure 15-24 Diode-ring balanced modulator.

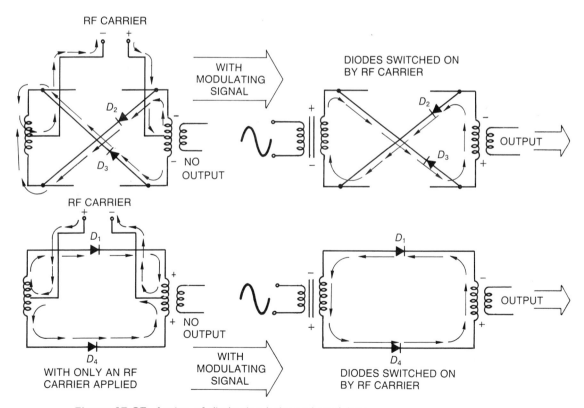

Figure 15-25 Action of diode-ring balanced modulator.

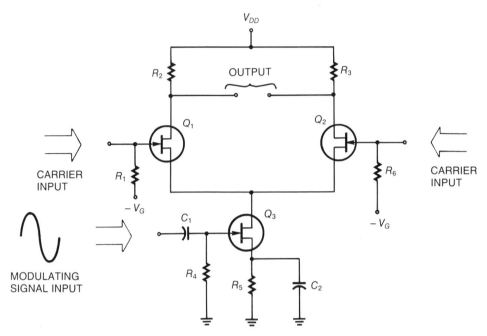

Figure 15-26 Active FET balanced modulator.

circuit bandwidth to accommodate the sideband frequencies. These frequencies are the only ones appearing at the output.

Figure 15-25 shows the action of the diode-ring balanced modulator under various circuit conditions. When only the RF carrier is present, there is no output. In actuality, the amount of carrier output depends on how well the diodes and the center taps of the transformers are matched. Since a perfect match is impossible, there is some carrier output. But, the aim is to make the carrier output power much less than the actual sideband power. In short, you want to transmit the sidebands, not the carrier.

Active Balanced Modulator

An *active balanced modulator* is shown in Figure 15-26. Q_3 acts as a current source for Q_1 and Q_2. When an in-phase carrier is applied to the gates of Q_1 and Q_2, the drain currents of both FETs are equal and the potential difference across the output is zero. This action is called *common-mode rejection,* and no carrier appears across the output.

When a modulating signal is applied to the gate of Q_3, the circuit is no longer balanced and the sidebands appear across the output terminals. Again the action is one that suppresses the carrier frequency.

IC Balanced Modulator

An IC balanced modulator consisting of the LM 1596N is illustrated in Figure 15-27. The LM 1596N is usable up to 100 MHz, and is capable of suppressing the carrier by as much as 65 dB! A unique feature of this IC is that it can also be used as a single-sideband suppressed-carrier demodulator. Thus, this versatile IC can be used in the transmitter as well as the receiver.

Schematic and Connection Diagrams

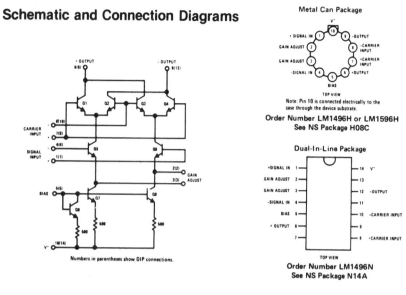

Metal Can Package

TOP VIEW
Note: Pin 10 is connected electrically to the
case through the device substrate.

Order Number LM1496H or LM1596H
See NS Package H08C

Dual-In-Line Package

+SIGNAL IN	1	14 V⁻
GAIN ADJUST	2	13
GAIN ADJUST	3	12 −OUTPUT
−SIGNAL IN	4	11
BIAS	5	10 −CARRIER INPUT
+ OUTPUT	6	9
	7	8 +CARRIER INPUT

TOP VIEW
Order Number LM1496N
See NS Package N14A

Numbers in parentheses show DIP connections.

Typical Application and Test Circuit

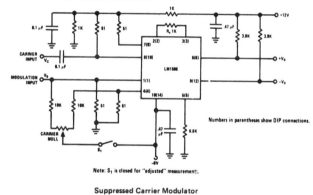

Note: S₁ is closed for "adjusted" measurement:.

Suppressed Carrier Modulator

Figure 15-27 LM1596N balanced modulator. *Courtesy* of National Semiconductor Corporation.

Conclusion

This section presented typical circuits used to produce suppressed-carrier AM signals. These modulators are called balanced modulators because they balance out the carrier. The next section describes representative circuits for producing FM and PM signals.

15-5 Review Questions

1. Explain the basic idea behind a balanced modulator.
2. Name the three kinds of balanced modulators.
3. Describe the action of a diode-ring balanced modulator.
4. Explain the operation of an active FET balanced modulator.
5. What are some features of the IC balanced modulator?

15-6 | FM AND PM MODULATORS

Discussion

As discussed in Section 15-3, there is little difference between frequency modulation and phase modulation. In essence, both systems produce a change in the phase angle of the carrier. Therefore, the circuits shown here fall under the general heading of angle modulation.

Overview of Angle Modulators

Table 15-5 summarizes the angle modulation circuits covered in this section.

Table 15-5	SUMMARY OF ANGLE MODULATION CIRCUITS
Descriptive Name	**Key Features**
Solid-state phase modulator (refer to Fig. 15-28)	Uses a single FET. The FET changes the reactance of the circuit, which in turn changes the carrier phase angle.
Varactor diode–modulator (refer to Fig. 15-29)	Uses the variable capacitance properties of a reverse-biased diode. The junction capacitance is varied by the modulated signal. This changes the resonant frequency of the RF carrier circuit.
VCO FM generator (refer to Fig. 15-30)	Complete VCO in an IC. Modulating signal changes the frequency of the VCO.

Solid-State Phase Modulator

A commonly used *solid-state phase modulator* is shown in Figure 15-28. You can analyze the circuit with phasors. Since the phase change is small, a greater deviation is obtained by feeding the output of this circuit into a frequency multiplier chain. The resultant change in amplitude is eliminated by a clipper circuit following the phase modulator. The action of the clipper has no effect on the phase of the modulated signal.

Varactor Diode Frequency Modulation

Figure 15-29 illustrates another solid-state phase modulator using an NPN transistor and a varactor diode. A *varactor diode* acts as a variable capacitor. This occurs when the diode is reverse biased. The reverse-biased junction forms a capacitor. You can change the value of this junction capacitance by changing the amount of reverse-biasing voltage. In the circuit of Figure 15-29, the modulating signal is applied across a reverse-biased varactor diode (D_1). Since this is in series with a much larger capacitor, C_3, the total circuit capacitance will be close to the value of the varactor diode capacitance. You should recognize this circuit as a *Hartley oscillator,* where the resonant frequency is determined by the component values in the tank circuit consisting of L_1, L_2, C_3, and the varactor diode D_1.

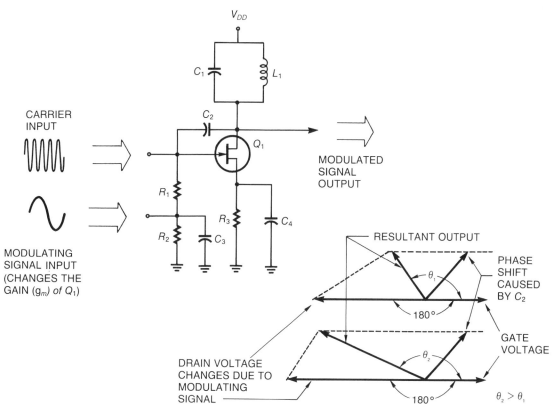

Figure 15-28 Solid-state phase modulator using an FET.

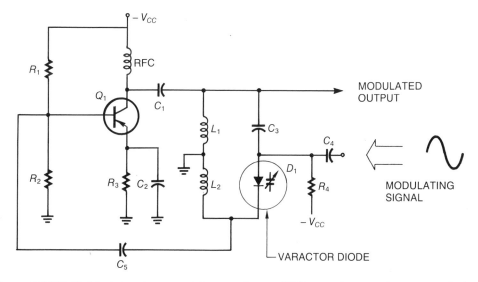

Figure 15-29 Solid-state phase modulator using an NPN transistor and varactor diode.

Schematic and Connection Diagrams

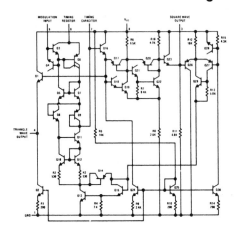

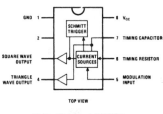

TOP VIEW

Order Number LM566CN
See NS Package N08B

Typical Application

1 kHz and 10 kHz TTL Compatible
Voltage Controlled Oscillator

Applications Information

The LM566 may be operated from either a single supply as shown in this test circuit, or from a split (±) power supply. When operating from a split supply, the square wave output (pin 4) is TTL compatible (2 mA current sink) with the addition of a 4.7 kΩ resistor from pin 3 to ground.

A .001 μF capacitor is connected between pins 5 and 6 to prevent parasitic oscillations that may occur during VCO switching.

$$f_O = \frac{2(V^+ - V_5)}{R_1 C_1 V^+}$$

where

$$2K < R_1 < 20K$$

and V_5 is voltage between pin 5 and pin 1

Figure 15-30 Complete VCO IC. *Courtesy* of National Semiconductor Corporation.

VCO FM Generator

A complete VCO IC, the LM 566, is shown in Figure 15-30. The LM 566 displays very linear modulation characteristics. That is, the carrier change is a faithful representation of the modulating signal. The simplicity of this circuit makes it easy to produce FM and PM signals.

Conclusion

This section presented the major types of circuits used in FM or PM under the general heading of angle modulation.

The next section presents the important topic of undesirable oscillations—oscillations not wanted in the communication system.

15-6 Review Questions

1. What kind of modulation describes both FM and PM?
2. Name three kinds of modulators.
3. Describe the major differences of the three kinds of modulators.
4. Explain what causes the phase shift in a solid-state FET phase modulator.
5. Describe the action of a varactor diode. Explain how it can be used to produce FM.

TROUBLESHOOTING AND INSTRUMENTATION

Undesirable Oscillations

In electronics, any circuit that meets the criteria for an oscillator will oscillate. What this means is that there are times where circuits, not intended to be oscillators, will act as oscillators. This is especially true of high-gain RF amplifiers whose purpose is to amplify a received signal. Instead of amplifying the received signal, the amplifier oscillates (produces its own signal) and ignores the intended information. This behavior can have disastrous effects on any communication system. Consider the space satellite that can no longer interpret earth signals because of undesirable oscillations in its own circuitry. This section describes the most common causes of undesirable oscillations and what can be done to prevent them.

Types of Undesirable Oscillations

Table 15-6 lists the different types of undesirable oscillations most commonly found in communications equipment.

Correcting Undesirable Oscillations

The oscillations outlined in Table 15-6 can be very difficult to analyze and cure. Some standard methods for preventing these types of oscillations follow.

Table 15-6	TYPES OF UNDESIRABLE OSCILLATIONS	
Type	**Description of Symptom**	**Most Probable Cause**
Low-frequency (refer to Fig. 15-31)	Low-frequency audio sounds like low-frequency vibrations; sometimes called *motorboating*	Poor power supply regulation
Mid-frequency (refer to Fig. 15-32)	Oscillations at the resonant frequency of the tuned circuits of the IF or RF amplifiers	Feedback from the output to the input of the amplifiers
Mid- to high-frequency (refer to Fig. 15-34)	Oscillations at the resonant frequency of the tuned circuits within the system or some other frequency	Feedback through poorly decoupled ground loops
Parasitic (refer to Fig. 15-36)	Weak oscillations that may not be consistent; usually a high-frequency that may change when test probes are brought near	Stray capacitance or circuit inductance

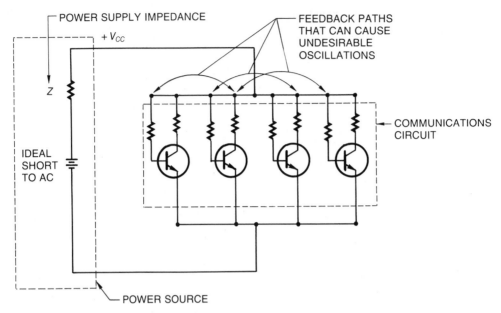

Figure 15-31 Probable cause of low-frequency oscillations.

Low-Frequency Oscillations

The cause of low-frequency oscillations is shown in Figure 15-31. The best way to prevent them is to make sure that the power supply for the communication system is well regulated. An output filter capacitor in need of replacement may also cause low-frequency oscillations (motorboating).

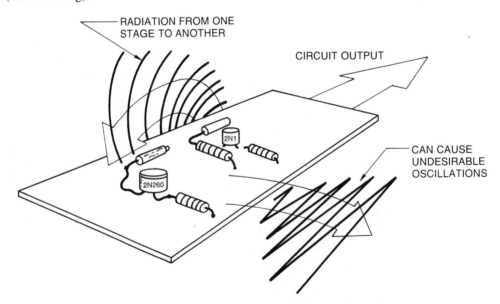

Figure 15-32 Probable cause of mid-frequency oscillations.

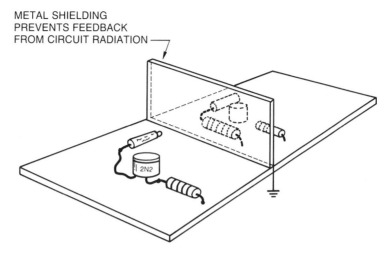

Figure 15-33 Typical method of shielding against undesirable oscillations.

Mid-Frequency Oscillations

The cause of undesirable mid-frequency oscillations is shown in Figure 15-32. These types of oscillations are best prevented by using shielding between the circuit output and input to block any feedback path from the circuit output to the circuit input for electromagnetic radiated energy. A typical shielding method is shown in Figure 15-33.

Mid- to High-Frequency Oscillations

Mid-frequency to high-frequency oscillations are usually caused by improper ground connections in the circuit. See Figure 15-34.

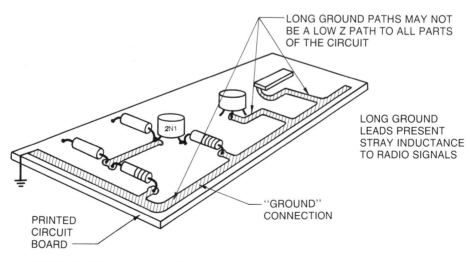

Figure 15-34 Oscillations caused by poor ground loops.

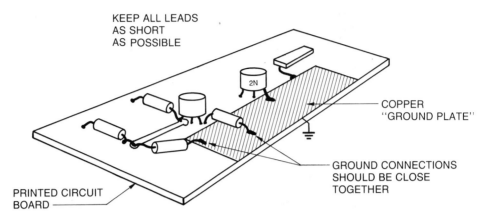

Figure 15-35 Construction techniques to prevent ground loops.

The standard acceptable cure for poor ground loops is to connect all ground connections to a single point. This single point is sometimes called a ground plate. Good circuit construction techniques are called for here. A suggested solution is shown in Figure 15-35.

Parasitic Oscillations

One reason for parasitic oscillations could be poor decoupling. See Figure 15-36. There are circuit points that do not represent a low impedance to ground for radio frequencies. For proper circuit operation, these points should be bypassed with capacitors, which creates low-impedance points to ground for radio signals. See Figure 15-37.

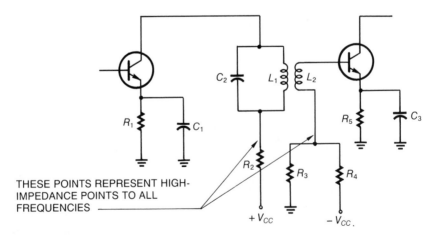

Figure 15-36 Poor decoupling causing parasitic oscillations.

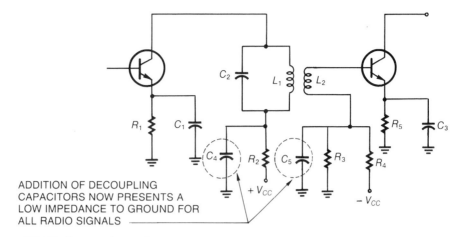

ADDITION OF DECOUPLING
CAPACITORS NOW PRESENTS A
LOW IMPEDANCE TO GROUND FOR
ALL RADIO SIGNALS

Figure 15-37 Proper decoupling.

Conclusion

Undesirable oscillations are a very real problem in the design, troubleshooting, and maintenance of communication systems. If the system is designed properly, then, when replacing components, you must use the same lead length and layout of the original component. If you don't, the system may turn into an oscillator and ignore or modify signals it is suppose to process.

15-7 Review Questions

1. What are undesirable oscillations?
2. Name three causes of undesirable oscillations.
3. What is motorboating? Describe what usually causes it.
4. Discuss what can be done to prevent undesirable oscillations caused by radiation.
5. Explain the use of a ground plate. How does it serve to prevent undesirable oscillations?

MICROCOMPUTER SIMULATION

Oscillator Design

The fifteenth troubleshooting simulation program again presents a solid state amplifier circuit; but this time, the analysis is on its AC characteristics. This simulation will help you develop troubleshooting skills in the signal analysis of a solid-state amplifier. Here you will use a simulated signal source and an RF voltmeter to analyze the AC characteristics of this amplifier. You will compare your measurements to the given specifications and determine if there is a computer generated problem, and if so, exactly what is causing the problem.

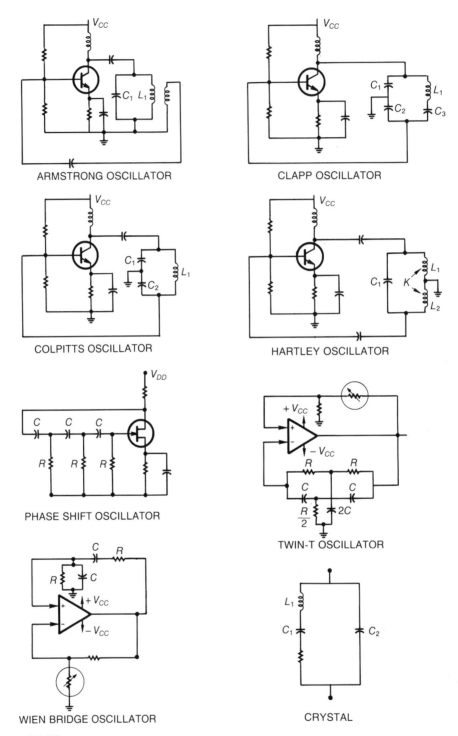

ARMSTRONG OSCILLATOR

CLAPP OSCILLATOR

COLPITTS OSCILLATOR

HARTLEY OSCILLATOR

PHASE SHIFT OSCILLATOR

TWIN-T OSCILLATOR

WIEN BRIDGE OSCILLATOR

CRYSTAL

Figure 15-38

CHAPTER PROBLEMS

(Answers to odd-numbered problems appear at the end of the text.)

1. Draw the block diagram of four different oscillators, each producing a different periodic waveform.
2. Describe the requirements all oscillators must have. Explain what each requirement means.
3. Describe three methods of achieving the required feedback circuit to produce an oscillator. What must be the phase of the feedback signal?
4. Explain the main features of the Barkhausen criterion. How is this stated mathematically?
5. For the circuit in Figure 15-39(A), what must be the gain of the feedback circuit to sustain oscillations?
6. For the circuit in Figure 15-39(B), what must be the gain of the amplifier to sustain oscillations?

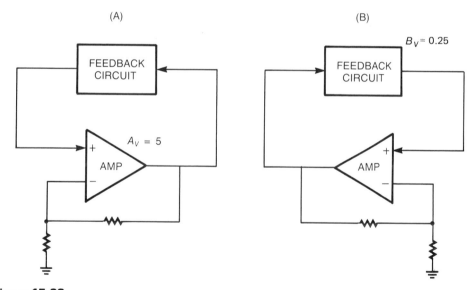

Figure 15-39

7. Describe the three phases experienced by an oscillator from the time it is first turned on.
8. Explain what would happen if the gain of an oscillator were not greater than 1 when it was first turned on.
9. What causes an oscillator to produce square waves?

10. Identify the oscillators in Figure 15-40.
11. Identify the frequency-determining components of the oscillators in Figure 15-40.
12. Determine the ideal resonant frequency of the oscillators in Figure 15-40.
13. Without changing the values of the capacitors for each oscillator in Figure 15-40(A), (C), and (D), determine the value of each inductor in order to double the ideal frequency of each oscillator.
14. Without changing the values of the inductors for each oscillator in Figure 15-40, determine the value of each capacitor in order to decrease the frequency of each oscillator by one half.

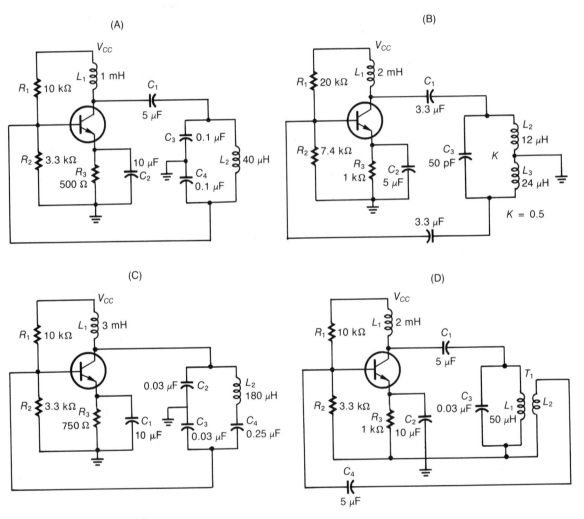

Figure 15-40

15. Identify the oscillators in Figure 15-41.
16. Identify the frequency-determining components of the oscillators in Figure 15-41.
17. Determine the ideal resonant frequency of the oscillators in Figure 15-41.
18. Without changing the values of the capacitors for each oscillator in Figure 15-41, determine the value of each resistor in order to double the ideal frequency of each oscillator.

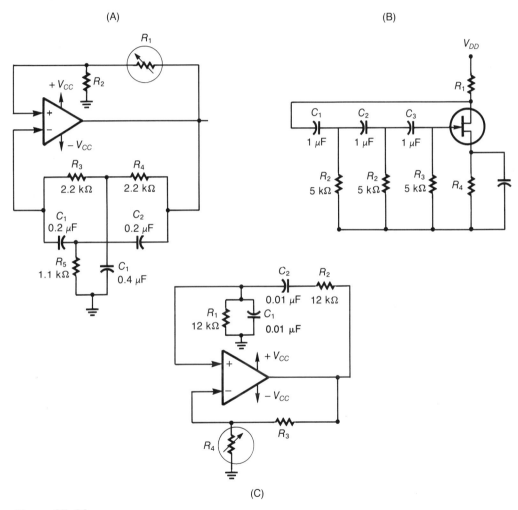

Figure 15-41

19. Without changing the values of the resistors for each oscillator in Figure 15-41, determine the value of each capacitor in order to decrease the frequency of each oscillator by one half.
20. State some advantages of using a quartz crystal as the frequency-determining component in an oscillator.
21. Determine the series and parallel resonant frequencies of the crystals in Figure 15-42(A), (B).
22. Determine the series and parallel resonant frequencies of the crystals in Figure 15-42(C), (D).
23. Find the Q of the crystals in Figure 15-42(C), (D). What is their bandwidth for the series resonant frequency mode? for the parallel resonant frequency mode?
24. Find the Q of the crystals in Figure 15-42(A), (B). What is their bandwidth for the series resonant frequency mode? for the parallel resonant frequency mode?

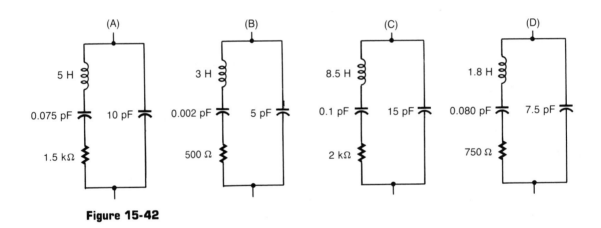

Figure 15-42

25. Identify the crystal oscillators in Figure 15-43.
26. For the crystal oscillators in Figure 15-43, identify the feedback network.

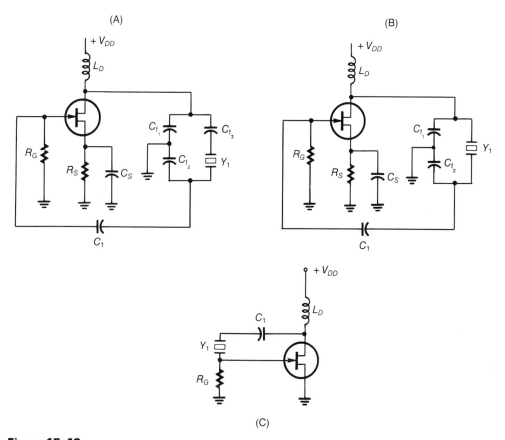

Figure 15-43

27. What is the main requirement of a circuit that must produce amplitude modulation from a carrier and a modulating signal?
28. Draw the block diagram of a balanced modulator. Illustrate the effect on the output when (A) no modulating signal is present, only a carrier, (B) both the modulating signal and carrier are present.
29. Graphically show the difference between frequency modulation and phase modulation.
30. Using block diagrams, show how you would make a phase modulator have an FM output. Show how you would convert a frequency modulator into a system with a PM output.
31. Figure 15-44 shows the schematic of a drain modulator. Show where the modulating signal and the carrier are applied.
32. For the drain modulator of Figure 15-44, show which components are tuned to the carrier frequency. What must be considered for any tuned circuits used with this type of modulator?

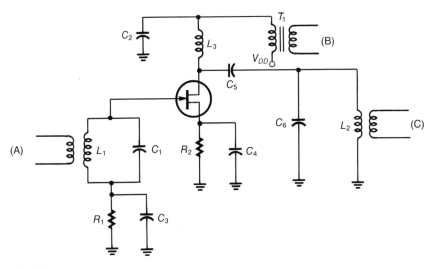

Figure 15-44

33. Identify the type of modulator in Figure 15-45.
34. For the modulator in Figure 15-45, show where the carrier is injected and where the modulating signal is injected. Which of the amplifiers is not class A?

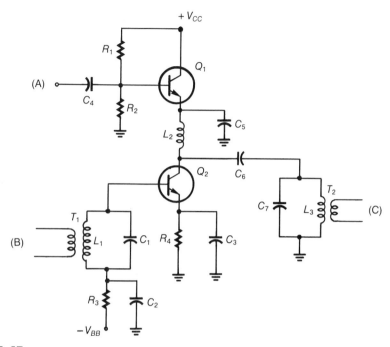

Figure 15-45

35. Identify the modulator in Figure 15-46. State the purpose of the variable resistor R_s.
36. For the modulator of Figure 15-46, show where the modulating signal and the carrier signal would be applied. What kind of signal would you expect to see on the output?

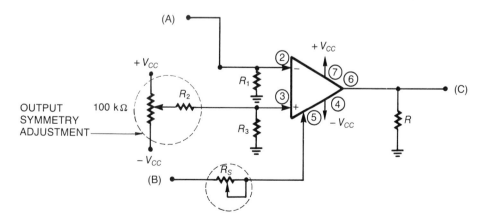

Figure 15-46

37. State the purpose of the circuit in Figure 15-47. Show where the RF carrier and the modulating signal are injected.
38. Explain by words or illustration the kind of output you would expect to observe from the circuit of Figure 15-47 with only a carrier signal present.
39. Explain what kind of output you would expect to observe from the circuit in Figure 15-47 when a carrier and a modulating signal are applied to the circuit.

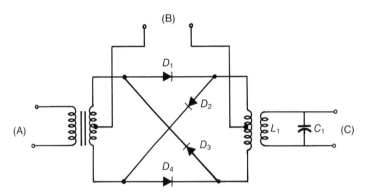

Figure 15-47

40. Identify the two modulator circuits in Figure 15-48. What kind of output would you expect to observe from each of them?
41. For the modulator circuits of Figure 15-48, show where the carrier and the modulating signal are injected.

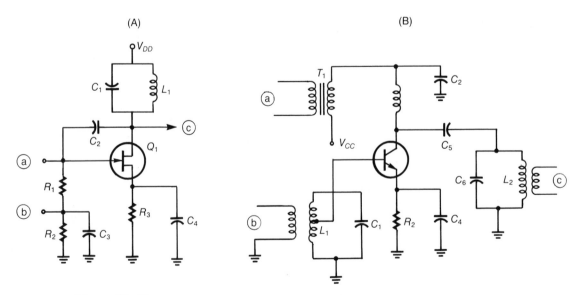

Figure 15-48

42. Identify the circuit in Figure 15-49. What components determine the resonant frequency?
43. Explain how the circuit of Figure 15-49 produces a modulating signal. Be sure to give a detailed explanation of how the diode affects the circuit operation.

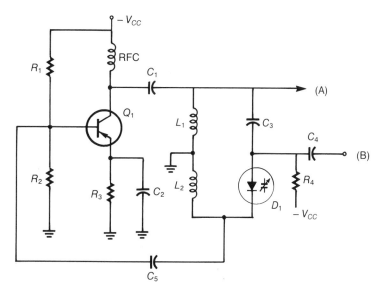

Figure 15-49

44. Explain how a high-impedance power supply can cause unwanted circuit oscillations.
45. What are unwanted circuit oscillations? What causes them?
46. What would you do to eliminate unwanted oscillations from a circuit that is getting feedback through radiation from one of its other circuits?
47. What is a poor ground loop? Show how it can be prevented.
48. What rule should be followed for the lead lengths of RF components?
49. Draw a schematic that uses capacitors for decoupling.

CHAPTER 16

Demodulators, AGC, and Filters

OBJECTIVES

In this chapter, you will study:

☐ The underlying principles of demodulators.

☐ Automatic gain control (AGC) principles and circuits.

☐ The principle AM demodulator circuits used in AM receivers.

☐ The principle FM and PM demodulator circuits used in receivers.

☐ The basic principles of filters used in communication systems.

☐ What active filters are as well as their advantages and disadvantages.

☐ The theory of operation of crystal filters.

☐ How to measure the frequency response of communication circuits.

INTRODUCTION

This chapter describes how the received signal is *demodulated* or *detected*. You will see the circuits that remove the carrier and restore the original information. You will also see how AGC works.

The subject of filters, as used in communication circuits, is also given. Filters are important circuits used in many ways in electronic communications.

16-1 DEMODULATOR PRINCIPLES AND AGC

Discussion

This section introduces the underlying principles of demodulators. Recall that a *demodulator* is a circuit that effectively restores the information contained in the modulated carrier. Another way of looking at this is to say that a demodulator extracts intelligence and, since the carrier has already done its job, gets rid of the carrier. This section also introduces the principles of automatic gain control (AGC), which is common in almost all communication receivers.

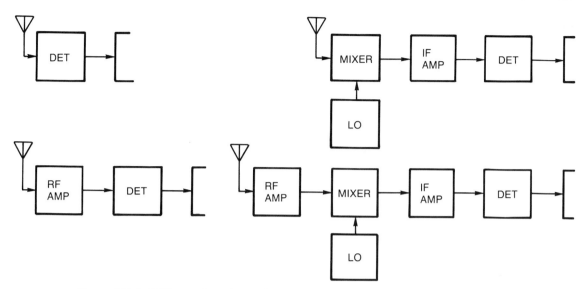

Figure 16-1 Different locations of detectors in a receiver.

Basic Idea

A *demodulator* or *detector* stage can be at several locations within a receiver, as shown in Figure 16-1.

Essentially three different classes of detectors are described in this chapter, Figure 16-2. Typical circuits for these classes are also given.

Automatic Gain Control

Automatic gain control (AGC) is used in communication receivers to keep the gain of the receiver constant. This means that within limits the volume (loudness) will not change on your radio as the signal changes from varying atmospheric conditions. It also means that

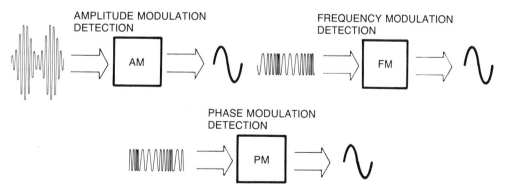

Figure 16-2 Detector classifications.

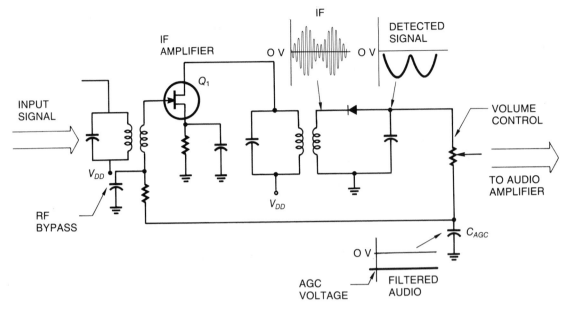

Figure 16-3 Most basic AGC circuit.

as you tune from station to station you do not have to adjust your volume control simply because one station has a weaker signal than another.

The most basic AGC circuit is illustrated in Figure 16-3. Here is how this circuit works. Part of the detected signal is filtered by the action of C_{AGC} and converted into a DC voltage level. Thus, the stronger is the signal, the greater is the amount of DC voltage. This DC voltage, called the AGC voltage, is fed back to the gate of the IF amplifier Q_1. A property of an FET is that its gain depends on the amount of gate bias voltage present. For this circuit, the greater the gate bias voltage, the smaller will be the FET gain.

Thus, for weak signals, the AGC voltage will be less, causing the FET gain to be larger. For a strong signal, more AGC voltage will be developed, causing the FET gain to decrease.

It is in this manner that the resultant output signal from the detector can be held fairly constant for a wide range of incoming signal strengths.

Sometimes this AGC voltage is amplified. Figure 16-4 shows a scheme that incorporates *amplified AGC*. A meter can be placed across the AGC line. Since the meter indicates the amount of AGC voltage present, it indicates the strength of the received signal.

Types of AGC

Two other types of AGC are found in communication systems: delayed AGC and keyed AGC.

Delayed AGC

Delayed AGC acts only on signals above a certain strength. This permits reception of weaker signals.

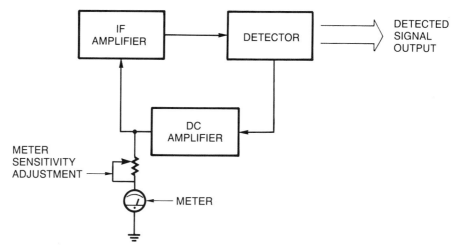

Figure 16-4 Amplified AGC.

Keyed AGC

Keyed AGC is used in TV receivers. If the AGC voltage were measured from the picture intensity level, the voltage would differ as scenes changed from dark to light. This condition is undesirable because it is not a good indication of the actual signal strength. A better method is to develop the AGC from the actual strength of the horizontal sync pulse. Thus, the AGC voltage for TV receivers is sampled or "keyed" by the horizontal sync pulse.

Conclusion

This section gave an overview of demodulation circuits. You also saw some details of AGC and what distinguishes one type from another. The next section presents various types of demodulators used to detect amplitude modulation.

16-1 Review Questions

1. Define AGC.
2. Describe the three classes of detectors.
3. Explain the action of automatic gain control. State why it is used.
4. State one use of amplified AGC.
5. Describe the action of delayed AGC and keyed AGC.

16-2 | TYPES OF AM DEMODULATORS

Discussion

Basically there are three types of AM demodulators: (1) the simple AM detector that demodulates the AM signal that has a carrier; (2) a demodulator for a suppressed-carrier AM signal; (3) a detector for a single-sideband signal. This section describes typical circuits that represent these demodulators.

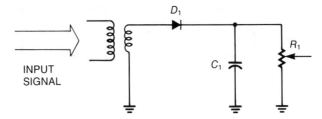

Figure 16-5 Simple AM detector.

Simple AM Detector

The simple AM detector is shown in Figure 16-5. This detector was first presented in Chapter 3. The purpose of each part is as follows:

Diode (D_1)

Converts the incoming AC into *pulsating DC*.

Capacitor (C_1)

Smooths out the pulsating DC into a waveform that is a *replication of the original modulating signal*.

Resistor (R_1)

Allows a discharge path for C_1 and also acts as a *volume control*.

Suppressed-Carrier Detector

A circuit that demodulates a suppressed-carrier signal is illustrated in Figure 16-6. This circuit needs a *reinserted carrier* that has the same frequency and phase as the original suppressed carrier. The circuit operation is such that the incoming signal is not strong

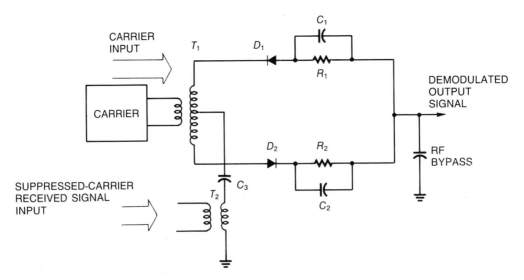

Figure 16-6 Suppressed-carrier synchronous detector.

enough to make diodes D_1 and D_2 conduct. The *RC* combinations (R_1, C_1 and R_2, C_2) allow D_1 and D_2 to conduct for only a short time. Diode conduction is caused by the reinserted carrier signal, which allows the suppressed-carrier signal to have its original modulating signal reconstructed.

The action of the suppressed-carrier synchronous detector is illustrated in Figure 16-7. During half of the cycle, the positive peaks of the received signal are sampled; during the other half, the negative peaks of the same signal are sampled. The resultant signal is a restoration of the original modulating signal.

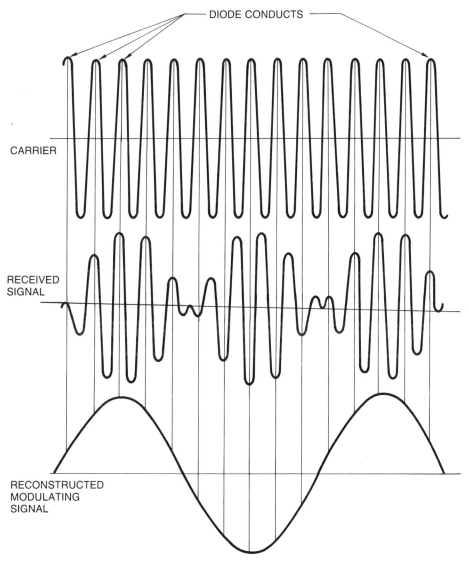

Figure 16-7 Action of suppressed-carrier synchronous detector.

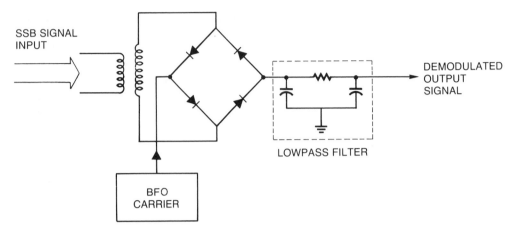

Figure 16-8 SSB detector.

Special circuits are required to make sure that the carrier is reinserted at exactly the correct frequency and phase. Recall that in TV transmission a special synchronizing signal is transmitted to ensure correct carrier reinsertion. The same thing is true of stereo FM.

An SSB detector is shown in Figure 16-8. The accuracy of the detector is not as critical as that required by the suppressed-carrier synchronous detector. The basic operation of the SSB detector requires an extra signal called the *beat frequency oscillator* (BFO). This signal is combined with the incoming SSB signal in a diode-ring mixer. If the BFO signal is within 50 kHz of the original suppressed carrier, then the resultant signal will be very close to the original modulating signal. The lowpass filter removes any remaining RF signal. This method is sometimes referred to as *product detection*. Note that a BFO signal is required.

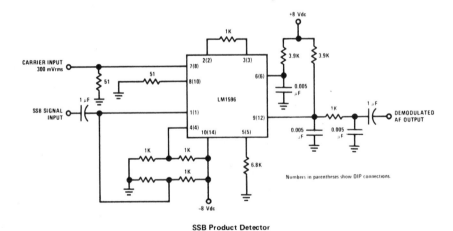

SSB Product Detector

Figure 16-9 LM 1596 SSB product detector. *Courtesy* of National Semiconductor Corporation.

IC Detectors

A popular IC AM detector is the LM 1596 balanced modulator-demodulator. This versatile IC was described in the last chapter as a balanced modulator. Figure 16-9 shows the same IC connected as an SSB product detector. The advantage of this system is its simplicity and few parts. No balancing controls are required, thus minimizing operator and servicing adjustments.

Conclusion

Two major kinds of AM demodulators were described: those with a carrier and those without one. In the next section, you'll learn about demodulation circuits for FM and PM.

16-2 Review Questions

1. Name the three different types of AM detectors.
2. Describe the action of a simple AM detector.
3. What kind of detector can be used to demodulate a suppressed-carrier transmission?
4. Describe the action of a suppressed-carrier synchronous detector.
5. Which detector requires a more critical local reference signal: the one for suppressed-carrier AM or the one for single-sideband modulation?

16-3 | FM AND PM DEMODULATORS

Discussion

This section presents typical circuits used for FM and PM detection. The main idea behind each circuit is shown in Figure 16-10. Frequency or phase changes of the incoming signal are converted to *amplitude changes,* which represent the original modulating signal.

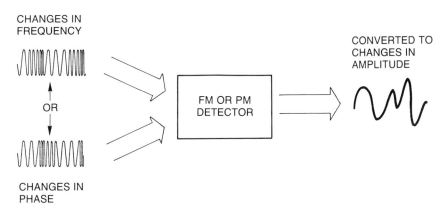

Figure 16-10 Generalized concept of FM and PM detectors.

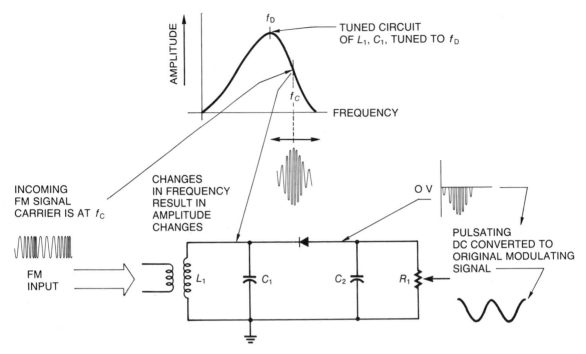

Figure 16-11 Concept of slope detector.

Slope Detector

The most straightforward means of converting frequency changes into amplitude changes is by using the response curve of a parallel tuned circuit. See Figure 16-11. The resonant circuit consisting of L_1 and C_1 resonates at a frequency that is just below the carrier frequency of the incoming signal. Since the amount of signal developed across this parallel resonant circuit is proportional to the frequency of the incoming signal, an FM signal will cause different voltage amplitudes across the circuit. For example, if the carrier frequency of the FM signal decreases, this brings it closer to the resonant frequency of the tuned circuit and there will be a larger voltage developed across it. If the carrier frequency of the FM signal increases, then this will be further away from the resonant frequency of the tuned circuit and the output amplitude will decrease. Thus, by this action, frequency changes are converted into amplitude changes.

Historically Popular Circuits

Two historically popular circuits for demodulating FM signals are the *Foster–Seeley discriminator* and the *ratio detector*. Both are shown in Figure 16-12. These are seldom used because the same function can be performed more economically and with better results with popular ICs.

Since these circuits may still be found in some older communication systems, a summary of their operation is presented here.

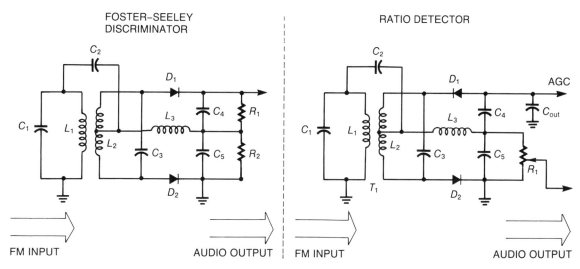

Figure 16-12 Historically popular Foster-Seeley discriminator and ratio detector.

Foster–Seeley Discriminator

The circuit is tuned to the carrier frequency. When there is no deviation of the carrier, diodes D_1 and D_2 conduct the same amount of current, thus producing zero volts across R_1 and R_2. A frequency change of the input carrier causes one diode to conduct more than the other, resulting in a corresponding change in the voltage amplitude across the output. The net result is that a frequency change is again converted into an amplitude change.

Ratio Detector

The ratio detector is similar to the Foster–Seeley discriminator. The main difference is that D_1 and D_2 are connected in series across the secondary winding of transformer T_1.

When the carrier is at its center frequency, both diodes conduct the same, causing the voltages across C_4 and C_5 to increase in series. The output capacitor C_{out} is chosen to be large. Once the circuit has been active for a while, C_{out} has then had time to charge to a voltage that is proportional to the average strength of the received signal. The time constant of this capacitive circuit is long enough to ignore any momentary amplitude changes (which in FM represent noise). The voltage across C_{out} is equal to the total series voltage of C_4 and C_5. This voltage remains relatively constant, and as the frequency of the input signal changes, one diode will conduct more than the other. This in turn will cause a changing voltage across either C_4 or C_5. Since their series *total* voltage must be constant (equal to the voltage across C_{out}), the voltage across C_5 will change, representing the change of the incoming frequency.

This presentation shows that there are two major advantages of the ratio detector over that of the Foster–Seeley discriminator: the self-limiting action that removes amplitude changes, and the availability of AGC.

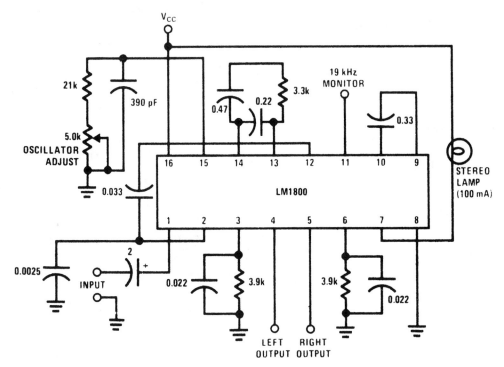

Figure 16-13 LM-1800 PLL FM stereo demodulator. *Courtesy* of National Semiconductor Corporation.

IC Demodulators

The LM 1800 is a versatile IC demodulator. It is a PLL FM stereo demodulator available in a small and inexpensive 14-pin chip. See Figure 16-13.

The LM 1800 uses PLL techniques to regenerate the 38-kHz subcarrier. Note from Figure 16-13 that pin 7 can be used as a *stereo indicator* lamp.

This chip is popular because no inductors are needed for tuning. The only required tuning is done with a variable resistor, as shown in Figure 16-13.

Conclusion

This section discussed representative circuits used for FM and PM demodulation. This function is now usually performed by IC chips. The next section begins the study of filters.

16-3 Review Questions

1. Describe the main idea behind FM and PM detectors.
2. Explain the operation of a slope detector.
3. Point out the major differences between a Foster–Seeley discriminator and a ratio detector.
4. Explain why the two circuits in question 3 are not used as often as they used to be.
5. Describe a major IC used as an FM demodulator. What kind of detection system does this IC use?

16-4	INTRODUCTION TO FILTER PRINCIPLES

Discussion

Filters play an important role in communication systems. This section reviews some material on filters, shows how filters are classified, and how they are used in communication systems.

Filter Basics

Recall from your study of electrical circuits that there are basically four types of filters: (1) lowpass, (2) highpass, (3) bandpass, and (4) band-reject. The frequency-response characteristics of these filters are shown in Figure 16-14. These filter types may be grouped

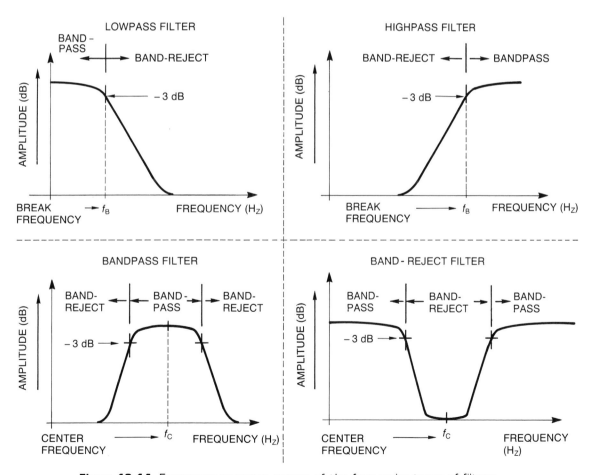

Figure 16-14 Frequency-response curves of the four major types of filters.

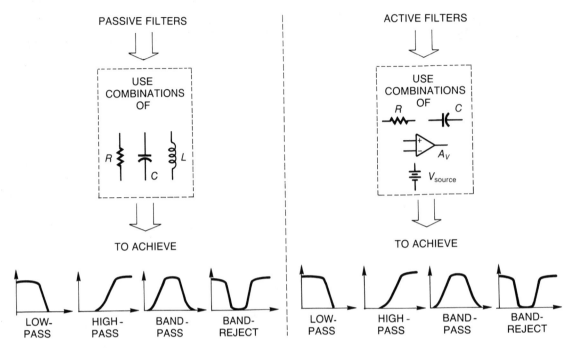

Figure 16-15 Example of filter classes.

into two *classes:* (1) passive filters and (2) active. Examples are shown in Figure 16-15. A *passive filter* consists of only passive components, such as resistors, capacitors, and inductors. They are called passive because they do not create or amplify energy. On the other hand, an *active filter* contains at least one active component, such as a transistor, FET, or op amp. They are called active components because they affect the signal by amplifying it.

The advantages of active filters are that: (1) they seldom use inductors; (2) they are easy to adjust for a wide frequency range; (3) they do not load down the previous circuit; and (4) they can deliver a zero insertion loss because they can use op amps, which have an easily adjustable gain, a high input impedance, and a low output impedance.

Today, communication systems rely on three classes of filters: (1) passive, (2) active, and (3) crystal. Passive filters are the subject of AC electrical circuits courses and will not be discussed in detail here. Active filters and crystal filters are covered in the next two sections.

Filter Terminology

Table 16-1 summarizes filter terminology.

Figure 16-16 illustrates the response of a Butterworth filter and compares this to the responses of Chebyshev and Bessel filters. Because of the popularity of the Butterworth filter, this filter will be presented in this section. Figure 16-17 shows how the *order* of a filter differs for a *decade* change.

Table 16-1	SUMMARY OF FILTER TERMINOLOGY
Term	**Definition**
Decade	A change in frequency by a factor of 10
Octave	A change in frequency by doubling or decreasing by one half
Butterworth filter (refer to Fig. 16-16)	A filter that exhibits the *flattest* response inside the passband; approaches a constant slope of 6 dB/octave or 20 dB/decade
Chebyshev filter (refer to Fig. 16-16)	A filter that exhibits a ripple in the range of frequencies it passes
Bessel filter (refer to Fig. 16-16)	Similar to the Butterworth with less decrease in amplitude beyond the break frequency
Filter order (refer to Fig. 16-17)	How a filter responds to a change in frequency; a first-order filter approaches a constant slope of 20 dB/decade; a second-order filter has 40 dB/decade

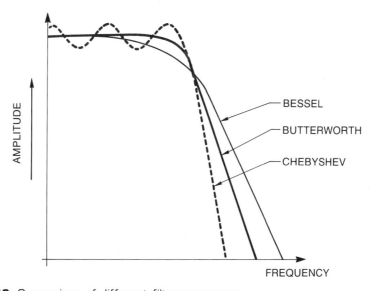

Figure 16-16 Comparison of different filter responses.

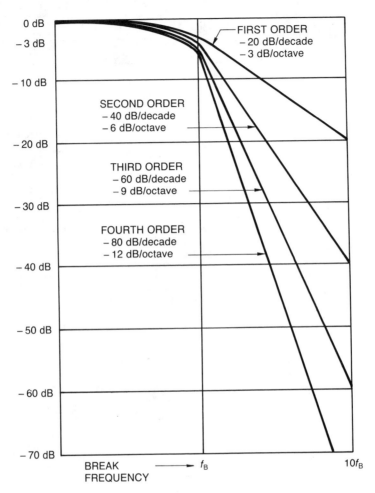

Figure 16-17 Example of filter order.

Conclusion

This section presented the fundamentals of filters and basic filter terminology.

16-4 Review Questions

1. Name the four different kinds of filters.
2. Explain the differences between passive and active filters.
3. State some of the advantages of active filters.
4. Describe the difference between a decade and an octave.
5. What is the order of a filter?

16-5 | ACTIVE FILTERS

Discussion

Active filters use active devices, such as transistors, FETs, and op amps. In this section, active filters are discussed using op amps. Op amps are the most common devices used with active filters. All four active filter types—lowpass, highpass, bandpass, and band-reject—are described.

Lowpass Active Filter

The most basic lowpass active filter is the single-pole. This circuit is shown in Figure 16-18. This is a voltage follower connection. The relationship of the output voltage to the input voltage is

$$v_{\text{out}} = \left(\frac{X_C}{\sqrt{R^2 + X_C^2}} \right) v_{\text{in}}$$

(Equation 16-1)

where v_{out} = Output voltage of op amp in volts
X_C = Reactance of capacitor in ohms
R = Resistance of resistor in ohms
v_{in} = Input voltage to op amp in volts

The filter is called single-pole because its output will decrease at the rate of 20 dB/decade.

Highpass Active Filter

The most basic highpass active filter is the single-pole. This circuit is illustrated in Figure 16-19. The main difference between this filter and the previous one is the different locations of the capacitor and resistor. The output voltage of this filter is

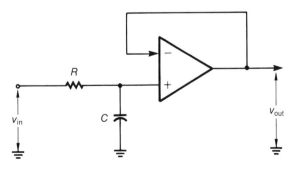

Figure 16-18 Single-pole lowpass filter.

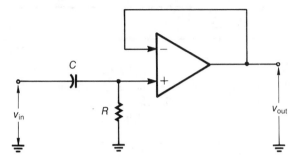

Figure 16-19 Single-pole highpass filter.

$$v_{out} = \left(\frac{R}{\sqrt{R^2 + X_C^2}} \right) v_{in}$$

(Equation 16-2)

where v_{out} = Output voltage of op amp in volts
       X_C = Reactance of capacitor in ohms
       R = Resistance of resistor in ohms
       v_{in} = Input voltage to op amp in volts

Example 1

Referring to Figure 16-20, determine the output voltage of the first-order filters for the following frequencies: (A) 1 kHz, (B) 2 kHz, and (C) 20 kHz. Construct a Bode plot response graph for each single-pole filter.

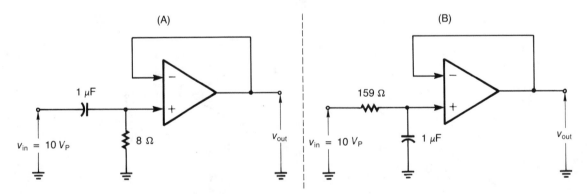

Figure 16-20 Single-pole filters.

Solution
For the highpass filter of Figure 16-20(A):

$$V_{out} = \left(\frac{R}{\sqrt{R^2 + X_C^2}} \right) V_{in}$$

(Equation 16-2)

Solve for X_C:

$$X_C = \frac{1}{2\pi f C}$$

At 1 kHz,

$$X_C = \frac{1}{6.28 \times 1 \times 10^3 \times 1 \times 10^{-6}}$$

$$X_C = \frac{1}{6.28 \times 10^{-3}} = 159 \ \Omega$$

At 2 kHz,

$$X_C = \frac{1}{6.28 \times 2 \times 10^3 \times 1 \times 10^{-6}}$$

$$X_C = \frac{1}{12.56 \times 10^{-3}} = 80 \ \Omega$$

At 20 kHz,

$$X_C = \frac{1}{6.28 \times 20 \times 10^3 \times 1 \times 10^{-6}}$$

$$X_C = \frac{1}{125.6 \times 10^{-3}} = 8 \ \Omega$$

Substituting into Equation 16-2 gives:
At 1 kHz,

$$V_{out} = \left(\frac{8 \ \Omega}{\sqrt{(8 \ \Omega)^2 + (159 \ \Omega)^2}} \right) 10 \ V$$

$$V_{out} = \left(\frac{8}{159.2} \right) 10 \ V = 0.5 \ V$$

At 2 kHz,

$$V_{out} = \left(\frac{8 \ \Omega}{\sqrt{(8 \ \Omega)^2 + (80 \ \Omega)^2}} \right) 10 \ V$$

$$V_{out} = \left(\frac{8}{80.4} \right) 10 \ V = 1.0 \ V$$

At 20 kHz,

$$V_{out} = \left(\frac{8\ \Omega}{\sqrt{8\ \Omega)^2 + (8\ \Omega)^2}} \right) 10\ V$$

$$V_{out} = \left(\frac{8}{11.3} \right) 10\ V = 7.08\ V$$

For the lowpass filter of Figure 16-20(B):

$$V_{out} = \left(\frac{X_C}{\sqrt{R^2 + X_C^2}} \right) V_{in}$$

(Equation 16-1)

At 1 kHz,

$$V_{out} = \left(\frac{159\ \Omega}{\sqrt{(159\Omega)^2 + (159\Omega)^2}} \right) 10\ V$$

$$V_{out} = \left(\frac{159}{224.8} \right) 10\ V = 7.07\ V$$

At 2 kHz,

$$V_{out} = \left(\frac{80\ \Omega}{\sqrt{(159\Omega)^2 + (80\Omega)^2}} \right) 10\ V$$

$$V_{out} = \left(\frac{80}{178} \right) 10\ V = 4.5\ V$$

At 20 kHz,

$$V_{out} = \left(\frac{8\ \Omega}{\sqrt{(159\Omega)^2 + (8\Omega)^2}} \right) 10\ V$$

$$V_{out} = \left(\frac{8}{159.2} \right) 10\ V = 0.5\ V$$

The completed graphs of the frequency responses of both single-pole filters are shown in Figure 16-21(A), (B).

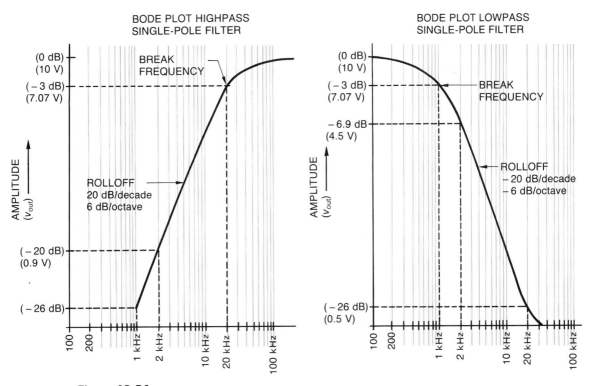

Figure 16-21

Bandpass Active Filters

Figure 16-22 shows a *bandpass active filter*. It can be thought of as a resonant circuit with a specified resonant frequency (f_r) and a specified bandwidth. This circuit can be analyzed by first selecting the desired resonant frequency and determining the Q from

$$Q = \frac{f_r}{\text{BW}}$$

(Equation 4-20)

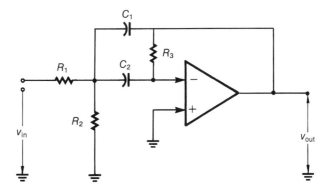

Figure 16-22 Bandpass filter.

The formulas for computing the other values are valid if you let $C_1 = C_2$. Then

$$R_1 = \frac{Q}{2\pi f_r C A_V}$$

(Equation 16-3)

$$R_2 = \frac{Q}{2\pi f_r C(2Q^2 - A_V)}$$

(Equation 16-4)

$$R_3 = 2R_1 A_V$$

(Equation 16-5)

where R_1, R_2, R_3 = Values of circuit resistors in ohms

Q = Circuit Q

A_V = Voltage gain of op amp

C = Values of circuit capacitors in farads

The voltage gain of the op amp must be less than $2Q^2$ to ensure that the denominator of Equation 16-4 is not zero or negative.

Example 2

Design a bandpass unity-gain active filter with a center frequency of 1 kHz and a bandwidth of 50 Hz. Use a 1.0-μF capacitor value for the feedback capacitors.

Solution

$$Q = \frac{f_r}{\text{BW}}$$

$$Q = \frac{1 \text{ kHz}}{50 \text{ Hz}} = 20$$

Solve for R_1:

$$R_1 = \frac{Q}{2\pi f_r C A_V}$$

(Equation 16-3)

$$R_1 = \frac{20}{2 \times 3.141\ 59 \times 1 \times 10^3 \times 1 \times 10^{-6} \times 1}$$

$$R_1 = \frac{20}{6.283\ 18 \times 10^{-3}}$$

$$R_1 = 3.18 \times 10^3 = 3.2 \text{ k}\Omega$$

Solve for R_2:

$$R_2 = \frac{Q}{2\pi f_r C(2Q^2 - A_V)}$$ **(Equation 16-4)**

$$R_2 = \frac{20}{2 \times 3.141\ 59 \times 1 \times 10^3 \times 1 \times 10^{-6}(2 \times 20^2 - 1)}$$

$$R_2 = \frac{20}{6.283\ 18 \times 10^{-3}(800 - 1)}$$

$$R_2 = \frac{20}{5027 \times 10^{-3}} = \frac{20}{5.027}$$

$$R_2 = 3.98 = 4\ \Omega$$

Solve for R_3:

$$R_3 = 2R_1 A_V$$ **(Equation 16-5)**
$$R_3 = 2 \times 3.2\ \text{k}\Omega \times 1 = 2 \times 3.2 \times 10^3 \times 1$$
$$R_3 = 6.4 \times 10^3 = 6.4\ \text{k}\Omega$$

The resulting band-pass filter is shown in Figure 16-23.

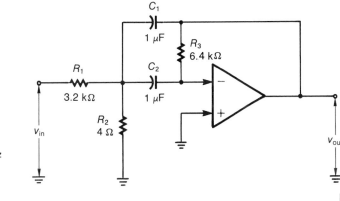

CENTER FREQUENCY = 1 kHz
BW = 50 Hz
UNITY-GAIN BANDPASS

Figure 16-23

Band-Reject Active Filter

Figure 16-24 shows a *band-reject active filter*. This circuit can also be thought of as a resonant circuit. It will reject a band of frequencies and pass all frequencies above and below that band. This type of filter is also referred to as a bandstop filter.

The mathematical relationships for this filter are

$$R_1 = \frac{R_4}{4Q^2}$$ **(Equation 16-6)**

$$R_2 = R_1$$ **(Equation 16-7)**

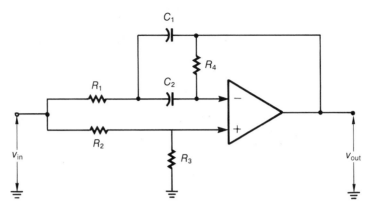

Figure 16-24 Band-reject active filter.

$$R_3 = 2Q^2R_1 \qquad \text{(Equation 16-8)}$$

$$R_4 = \frac{Q}{\pi f_r C} \qquad \text{(Equation 16-9)}$$

where $\qquad Q = \dfrac{f_r}{\text{BW}}$

f_r = Resonant frequency of filter in hertz
R_1, R_2, R_3, R_4 = Resistor values of filter in ohms
$C_1 = C_2$ = Capacitance values of filter in farads

Example 3

Design a bandstop filter that has a center frequency of 25 kHz and a bandwidth of 5 kHz. Use a 0.01 μF capacitor in the filter.

Solution
Solve for the circuit Q:

$$Q = \frac{f_r}{\text{BW}}$$

$$Q = \frac{25 \text{ kHz}}{5 \text{ kHz}} = 5$$

First, solve for R_4:

$$R_4 = \frac{Q}{\pi f_r C}$$ **(Equation 16-9)**

$$R_4 = \frac{5}{3.141\,59 \times 25 \times 10^3 \times 1 \times 10^{-8}}$$

$$R_4 = \frac{5}{78.5 \times 10^{-5}} = 0.064 \times 10^5 = 6.4 \text{ k}\Omega$$

Next, solve for R_1:

$$R_1 = \frac{R_4}{4Q^2}$$ **(Equation 16-6)**

$$R_1 = \frac{6.4 \text{ k}\Omega}{4 \times 5^2}$$

$$R_1 = \frac{6.4 \times 10^3}{100} = 64 \ \Omega$$

Now

$$R_2 = R_1 = 64 \ \Omega$$

Solve for R_3:

$$R_3 = 2Q^2 R_1$$ **(Equation 16-8)**
$$R_3 = 2 \times 5^2 \times 64$$
$$R_3 = 50 \times 64 = 3200$$
$$R_3 = 3.2 \text{ k}\Omega$$

The resulting circuit is shown in Figure 16-25.

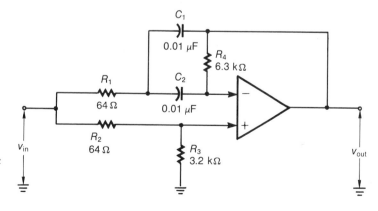

CENTER FREQUENCY = 25 kHz
BW = 5 kHz
UNITY-GAIN BAND-REJECT

Figure 16-25

Conclusion

This section presented examples of the four different kinds of active filters: highpass, lowpass, bandpass, and band-reject. In the next section, you will see the fundamental properties of *crystal filters*.

16-5 Review Questions

1. Describe the difference in connections between an active lowpass filter and an active highpass filter.
2. Define Q as it applies to active filters.
3. What is the main difference in the mathematical relationships of highpass and lowpass active filters?
4. Explain what restrictions apply to the gain of a bandpass filter.

16-6 | CRYSTAL FILTERS

Discussion

Crystal filters are used when more selectivity is needed than can be achieved with active filters. Another limitation of active filters is the frequency response of the op amp. In general, crystal filters can operate at frequencies up to hundreds of megahertz. This section presents the fundamental principles of crystal filters and their application in communication circuits.

Crystal Response

Recall from Chapter 15 that the equivalent circuit of a crystal is as shown in Figure 16-26. Each crystal presents different impedance characteristics for various frequencies. For example, when the crystal is at its series resonant mode, it appears as a minimum impedance (equal to the crystal R). When the crystal is at its parallel resonant mode, it exhibits a very large impedance. The change in impedance of a crystal with frequency is shown in Figure 16-27. At very low frequencies, the impedance of the crystal is very large

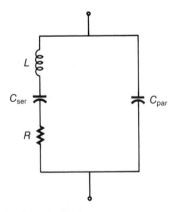

Figure 16-26 Equivalent circuit of a crystal.

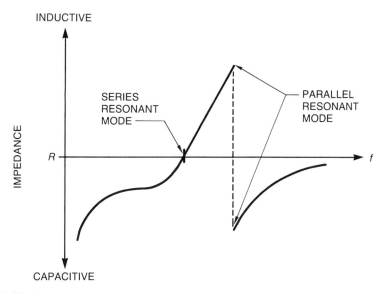

Figure 16-27 Impedance characteristics of a typical crystal.

and is *capacitive*. As the frequency increases, the impedance decreases until the crystal reaches series resonance. Beyond series resonance, the impedance of the crystal again increases and becomes *inductive*. Once parallel resonance is reached, the impedance of the crystal becomes very large. Just above parallel resonance, the crystal again appears *capacitive*.

These properties of the crystal can be used to form a very selective filter with a Q of 1 000 or more. With such a large Q, the crystal filter becomes very selective and produces extremely *narrow bandwidths*.

Crystal Filter

Figure 16-28 shows the schematic of a *crystal filter*. The filter has two pairs of crystals that have matched frequency characteristics. This means that Y_{1A} and Y_{1B} are matched crystals and that Y_{2A} and Y_{2B} are matched crystals.

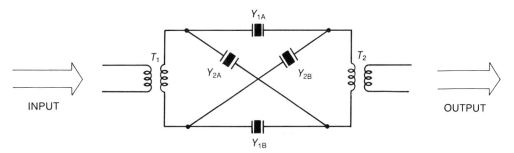

Figure 16-28 Schematic of crystal filter.

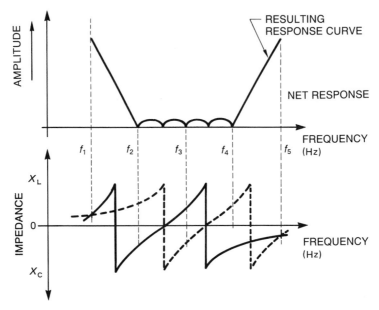

Figure 16-29 Response of crystals and crystal filter.

Table 16-2	SUMMARY OF CRYSTAL FILTER OPERATION
Frequency	**Reason for Resultant Response**
f_1	Both crystal pairs present equal impedances to the input signal. This results in equal signals to each side of the output transformer and cancels any output signal.
f_2	Both crystal pairs again present equal impedances. As shown in Figure 16-29, one crystal pair is inductive while the other pair is capacitive. This brings the opposite phases of the signal from T_1 in phase. One pair causes a 90° lead, the other a 90° lag, producing a large signal at T_2.
f_2 to f_3	Within this frequency range, the crystal pair formed by Y_{1A} and Y_{1B} is in series resonance. The other pair is in parallel resonance. The signal passes to the output of T_1 through the series resonant pair.
f_3	Same conditions as f_2, but with opposite crystal pairs.
f_3 to f_4	Same conditions as with f_2 to f_3, but with opposite crystal pairs at series resonance.
f_4	Same conditions as with f_2.
f_5	Both crystal pairs present equal impedances, and resulting signal cancels at the output.

Figure 16-29 shows the response curve of each crystal and the total response of the crystal filter. Each crystal pair has *two parallel* resonant frequency points due to the inductance of the input and output transformers.

Table 16-2 summarizes the operation of the crystal filter.

If greater selectivity is required, the crystal filter can be cascaded with other crystal filters.

Conclusion

This section presented the properties of crystal impedance for different frequencies. You saw how these impedance characteristics could be used to construct a crystal filter. In the next section, you will see how to measure the frequency response of an amplifier.

16-6 Review Questions

1. State the main advantage of crystal filters over other types of filters.
2. Describe the impedance of a crystal in its series resonant mode and in its parallel resonant mode.
3. When does a crystal appear inductive? resistive? capacitive?
4. Describe the construction of a crystal filter.
5. Briefly describe the operation of a crystal filter.

TROUBLESHOOTING AND INSTRUMENTATION

Measuring Frequency Response

Often you want to know the frequency response of a communication circuit when a sweep generator is not available. The operation and application of sweep generators were given in the troubleshooting and instrumentation section of Chapter 6. This section presents information for measuring the frequency response with a sine-wave generator and a voltmeter or oscilloscope.

General Idea

Figure 16-30 shows the general setup to test the frequency response of an amplifier. What you want to do is measure the amplitude of the amplifier output as you change the frequency of the input. Thus, a voltage reading will be made for a given frequency input. The resultant readings will then be displayed in graphical form as shown in Figure 16-31.

At the breakpoints, several frequency readings are made to determine the shape of the curve. At the points where the response is flat, taking readings every octave is satisfactory. Also note from Figure 16-31 that the vertical axis is linear and measured in dB, and the horizontal axis is logarithmic and measured in hertz. This, you should recall, is called a *Bode plot*.

You must observe the following points to obtain accurate measurements:

1. The output of the signal generator must always have exactly the same amplitude as its frequency is changed.

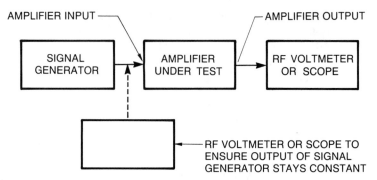

Figure 16-30 General setup for measuring frequency response.

2. The input signal must be made small enough so that the amplifier under test is not saturated. This can easily be checked with a scope connected to the output of the amplifier while a sine wave is inserted at the input.
3. If the circuit contains any AGC, you must break the AGC loop. If you don't, you will read the action of the AGC along with the frequency response of the amplifier.
4. The frequency response of the voltmeter used to measure the output voltage of the amplifier must be flat over the range of frequency measurements to be made.

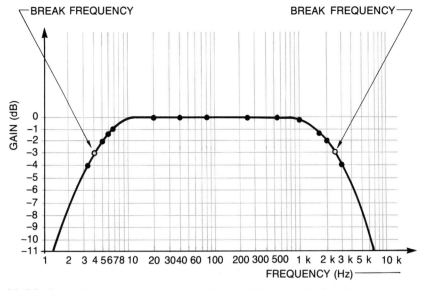

Figure 16-31 Recording measurements and resulting graph for frequency-response measurements.

Performing the Measurements

Table 16-3 lists the steps for making a frequency-response measurement of a typical amplifier.

Table 16-3	STEPS FOR MAKING FREQUENCY-RESPONSE MEASUREMENT	
Step	**Action**	**Comments**
1	Make sure frequency range of meter or scope is within the frequency range to be measured.	Read owner's manual to see if your voltmeter has the required frequency range. If not, use a scope.
2	Connect the signal generator to equipment under test. Change its frequency from below the amplifier frequency response to above it. Adjust the amplitude as small as possible to prevent saturation of the amplifier.	Make sure that the output amplitude of the generator stays constant. Check this with a scope or voltmeter. Make sure you break the AGC connection.
3	Without adjusting the amount of output signal from the generator, start at the lowest measurable output signal and gradually increase the frequency. Note the change in output amplitude of the amplifier under test.	If the amplitude of the generator does change during these readings, it will be necessary to adjust generator signal amplitude to keep it constant.
4	As you increase the input frequency, the output signal strength of the amplifier should increase.	You are now measuring the low end of the amplifier response.
5	You will arrive at a point where the output signal strength of the amplifier will stay constant as you increase frequency.	You are now measuring the midband of the amplifier response.
6	Keep increasing the frequency until the amplifier output signal starts to decrease.	You are now measuring the high end of the amplifier response.
7	Change the frequency of the generator to find at what frequencies the output voltage is 0.707 of the midband frequencies. Do this for the low end and the high end.	These are the breakpoints or 3-dB points used to measure bandwidth of the amplifier.
8	Transfer your readings to a semilog graph. Be sure to indicate the 3-dB points on the graph.	You will now have a Bode plot of the amplifier.

Conclusion

This section explained how to make a frequency-response reading of a typical communication amplifier with a signal generator and a voltmeter or scope.

16-7 Review Questions

1. Describe what equipment you would use to measure the frequency response of an amplifier if a sweep generator were not available.
2. For the measurement in question 1, what precautions should be taken if any AGC is present in the amplifier?
3. What must you do as the frequency of the signal generator is changed?
4. Describe what results you may get in doing a frequency response while the amplifier under test is saturated. How do you relieve this condition?
5. What is a Bode plot?

MICROCOMPUTER SIMULATION

Troubleshooting simulation sixteen again presents the solid state amplifier and allows you to use your DC and AC troubleshooting analysis skills to determine if there is a circuit problem. This simulation brings together the skills you acquired in the previous two troubleshooting simulation programs. This program is an important summary for utilizing all of the troubleshooting skills you have acquired up to this point. As before, be sure to first view the **instructions** and then the **demonstration** before proceeding on with the test portion of this troubleshooting simulation.

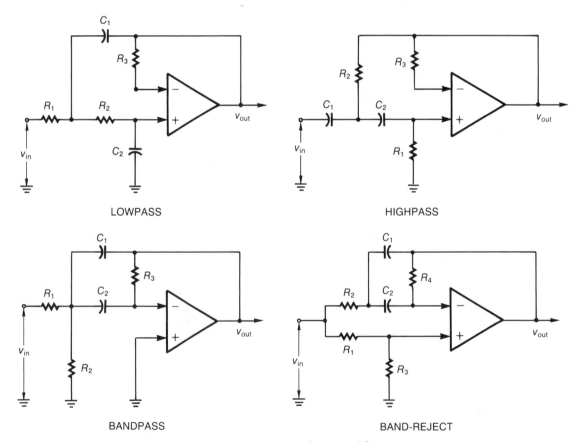

Figure 16-32 Active unity-gain filters for Chapter 16 program.

CHAPTER PROBLEMS

(Answers to the odd-numbered problems appear at the end of the text.)

TRUE/FALSE

1. In a radio receiver, a detector must be located immediately after the IF amplifier.
2. Detectors are used only in AM receivers.
3. For a detector to function properly, it must have an AGC circuit.
4. AGC means automatic gain control.
5. The purpose of AGC is to try and maintain a constant volume for different received signals.

MULTIPLE CHOICE

Answer the following questions by selecting the most correct answer.

6. Delayed AGC is AGC that:
   (A) takes place only after a specified amount of time.
   (B) has no effect on weak signals.
   (C) gets delayed on its way from the detector.
   (D) None of the above are correct.
7. Keyed AGC is AGC that:
   (A) works only with Morse code.
   (B) is activated when keyed by the transmitting signal.
   (C) is used in TV receivers so that transmitted picture brightness does not effect the AGC.
   (D) is used in color TV receivers so that the transmitted color has no effect on the AGC.
8. The three kinds of demodulators for AM are:
   (A) simple, medium, and complex.
   (B) simple, suppressed-carrier, and single sideband.
   (C) frequency, phase, and amplitude.
   (D) (A) and (C) but not (B).
9. A suppressed-carrier synchronous detector requires:
   (A) the presence of a reinserted carrier of exactly the same frequency and phase as the original.
   (B) the presence of a reinserted carrier that has a frequency close to the original carrier.
   (C) a BFO that is at the exact frequency difference between the original carrier and modulating frequency.
   (D) a BFO that is near the frequency difference between the original carrier and modulating frequency.
10. A singe-sideband detector requires:
   (A) the presence of a reinserted carrier of exactly the same frequency and phase as the original.
   (B) the presence of a reinserted carrier that has a frequency close to the original carrier.
   (C) a BFO that is at the exact frequency difference between the original carrier and modulating frequency.
   (D) a BFO that is near the frequency difference between the original carrier and modulating frequency.

MATCHING

Match the filter responses in Figure 16-33 to the corresponding filter in questions 11–15.

11. Band-reject filter
12. Lowpass filter
13. Bandpass filter
14. Highpass filter
15. Lowpass, highpass filter

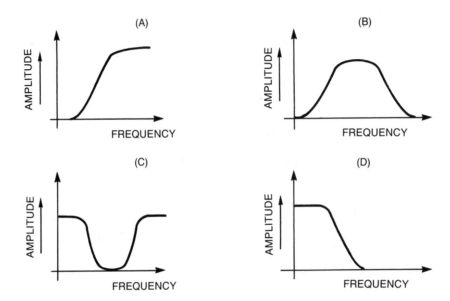

Figure 16-33

FILL-IN

Fill in the blanks with the most correct answer(s).

16. A/An _____ represents a doubling of frequency.

17. A Butterworth filter has the _____ response curve when compared to other types of filters.

18. A/An _____ exhibits a ripple in its passband.

19. A/An _____ represents an increase in frequency by a factor of 100.

20. How a filter responds to a change in frequency is the _____ of the filter.

OPEN-ENDED

Answer the following questions as indicated.

21. For the filters in Figure 16-34, construct a Bode plot for frequencies from 100 Hz to 10 kHz.

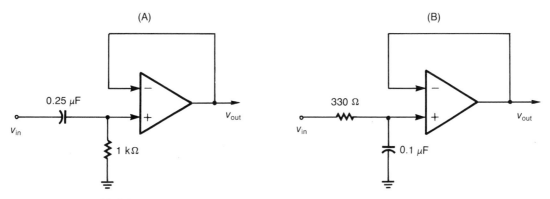

Figure 16-34

22. For the filters in Figure 16-35, construct a Bode plot for frequencies from 100 kHz to 10 MHz.

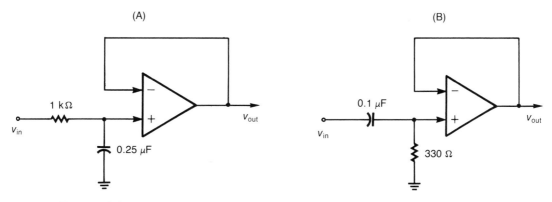

Figure 16-35

23. Design a bandpass filter that has a center frequency of 25 kHz, a bandwidth of 500 Hz, and uses 0.01-μF capacitors.
24. Design a bandpass filter that has a center frequency of 100 kHz, a bandwidth of 1 kHz, and uses 0.05-μF capacitors.
25. Design a band-reject filter that has a center frequency of 25 kHz, a bandwidth of 100 Hz, and uses 0.15-μF capacitors.
26. Design a band-reject filter that has a center frequency of 250 kHz, a bandwidth of 1 kHz, and uses 0.033-μF capacitors.
27. A certain crystal has the equivalent circuit shown in Figure 16-36. Sketch its frequency-response curve.

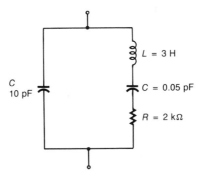

Figure 16-36

28. For the crystal lattice filter of Figure 16-37, one set of crystals is series resonant at 10 MHz, and parallel resonant at 9 MHz and 11 MHz. The other set of crystals is series and parallel resonant 500 kHz above the first two. Sketch the frequency response of the crystal pairs for the filter.

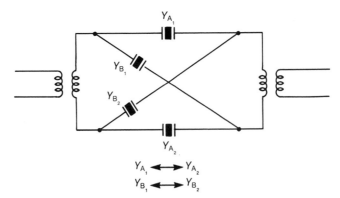

Figure 16-37

29. Sketch the connections of the equipment you would use to plot the frequency response of an amplifier if a sweep generator were unavailable.
30. Explain the steps and precautions you would take to measure the frequency response of an amplifier without using a sweep generator.

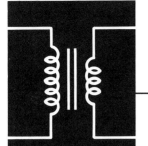

CHAPTER 17

Power Supply Regulation

OBJECTIVES

In this chapter, you will study:

- ☐ The fundamentals of power supplies used in communication systems.
- ☐ The basic concepts of series regulators.
- ☐ The basic concepts of parallel (shunt) regulators.
- ☐ The IC devices used in power supplies.
- ☐ The fundamentals of switching regulators.
- ☐ Applications of switching regulators.
- ☐ How to troubleshoot and analyze power supply problems.

INTRODUCTION

The numerous advances in power supply design in the past few years make a separate chapter necessary in a communications text. What was once accomplished by discrete components is now easily done with ICs. These circuits greatly simplify the design and troubleshooting of communication power supplies. This chapter presents developments in the rectification and regulation of power supplies. The importance of switching regulators is introduced along with their applications in power sources for communication equipment.

17-1 | POWER SUPPLY REQUIREMENTS

Discussion

This section presents the basic requirements of power supplies used in communication systems. A good power supply must provide

■ A steady source of voltage, that *does not change with changes in current demands from the circuit.*

■ An output impedance that looks like *zero ohms to any changing voltage or current.*

■ A safety mechanism that will shut off the power source due to *any circuit overload or other condition that may be hazardous to equipment or personnel.*

Ideal Power Source

An *ideal power source* is shown in Figure 17-1. The output impedance of an ideal power source is zero ohms. This means that the power supply output voltage will always be the same value no matter what the current requirement is from the rest of the circuit. To achieve this is impossible. Consider what happens when a short is placed across a power supply. If the resistance of the short is zero ohms, then the power supply must be able to deliver infinite current, which would require all the energy in the universe!

Hence, all practical power supplies have *some* internal impedance that is inherent in the components used to construct it. The idea is to keep this internal impedance as small as possible so that good regulation can be achieved. This is illustrated in the following example.

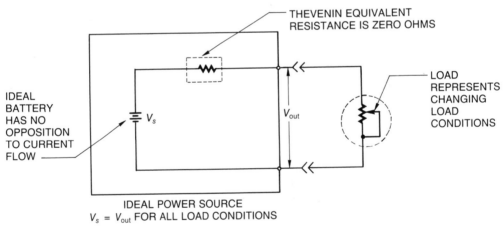

Figure 17-1 Thevenin equivalent of ideal power source.

Example 1

Determine the no-load and full-load output voltages for the two power supplies (A and B) in Figure 17-2. Which one has better regulation? Why? What can you conclude about the relationship of the internal impedance of a power supply and its voltage regulation?

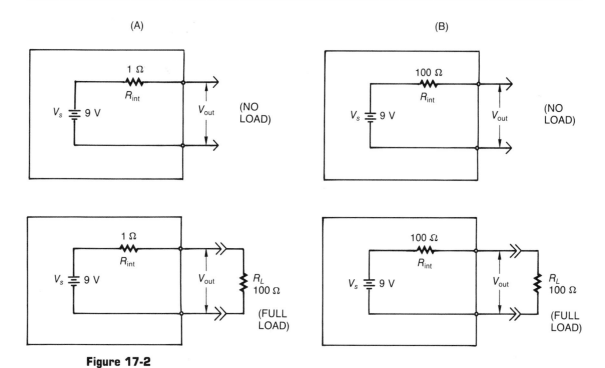

Figure 17-2

Solution

Power supply A
No-load conditions

$$V_{out} = V_s = 9 \text{ V}$$

Full-load conditions

$$V_{out} = \frac{R_L}{R_{int} + R_L} \times V_s \quad \text{(voltage divider formula)}$$

$$V_{out} = \frac{100 \ \Omega}{1 \ \Omega + 100 \ \Omega} \times 9 \text{ V}$$

$$V_{out} = \frac{100}{101} \times 9 = 0.99 \times 9 = 8.91 \text{ V}$$

Power supply B
No-load conditions

$$V_{out} = V_s = 9 \text{ V}$$

Full-load conditions

$$V_{out} = \frac{R_L}{R_{int} + R_L} \times V_s \quad \text{(voltage divider formula)}$$

$$V_{out} = \frac{100 \ \Omega}{100 \ \Omega + 100 \ \Omega} \times 9 \ V$$

$$V_{out} = \frac{100}{200} \times 9 = 4.5 \ V$$

Power supply A has better regulation because the change in its output voltage from no load to full load is much less than that of power supply B.

The reason is that the internal impedance of power supply A is much less than that of power supply B.

You can conclude that the smaller the internal impedance of a power supply, the better is the voltage regulation.

As the example shows, the power supply that maintains an almost constant output voltage is the one with the lowest internal impedance.

Voltage Regulation

Voltage regulation can be a measure of how well a power supply will maintain the same output voltage for different amounts of current demands. This is expressed mathematically as

$$\%V_{reg} = \frac{V_{NL} - V_{FL}}{V_{NL}} \times 100\% \qquad \textbf{(Equation 17-1)}$$

where $\%V_{reg}$ = Percent voltage regulation

V_{NL} = Output voltage with no load (open-circuit voltage) in volts

V_{FL} = Output voltage with full load connected in volts

Example 2

Calculate the percent voltage regulation for both power sources of Example 1.

Solution

Power source A:

$$\%V_{reg} = \frac{V_{NL} - V_{FL}}{V_{NL}} \times 100\% \qquad \textbf{(Equation 17-1)}$$

$$\%V_{reg} = \frac{9 \ V - 8.91 \ V}{9 \ V} \times 100\%$$

$$\%V_{reg} = \frac{0.09}{9} \times 100\% = 1\%$$

Power source B:

$$\%V_{\text{reg}} = \frac{V_{\text{NL}} - V_{\text{FL}}}{V_{\text{NL}}} \times 100\%$$

(Equation 17-1)

$$\%V_{\text{reg}} = \frac{9 \text{ V} - 4.5 \text{ V}}{9 \text{ V}} \times 100\%$$

$$\%V_{\text{reg}} = \frac{4.5}{9} \times 100\% = 50\%$$

As illustrated in Example 2, the power source with the best regulation also has the lowest internal impedance.

Conclusion

This section introduced the basic idea of voltage regulation. The rest of this chapter shows you how voltage regulation is achieved in modern communications equipment. The next section introduces series regulators. They are an important addition to power supply regulation. By using a device, such as a transistor, the power source can appear to have a lower internal impedance because it will give better regulation.

17-1 Review Questions

1. State the three main criteria for a good power supply.
2. Explain the concept of an ideal power source. Why is this impossible to achieve?
3. Describe the difference in the output voltage of a practical power supply between no-load and full-load conditions.
4. What factor determines the regulation of a power supply?

17-2 SERIES REGULATION

Discussion

There are two basic ways of regulating the output voltage of a power supply: (1) connecting a regulating device in series with the power source—called *series regulation;* (2) placing a regulating device in parallel with the power source—called *parallel* or *shunt* regulation. This section shows how series regulated power supplies function. Parallel regulation is presented in the next section.

Basic Idea

The basic idea behind a series regulated power supply is shown in Figure 17-3. The circuit is basically a simple series circuit with R_{int} and R_L. Therefore, the voltage across R_{int} and R_L must *always* equal the source voltage V_s. Thus, if the voltage across one resistor increases, then the voltage across the other must decrease. This *must* happen because the *sum of the two voltages is always equal to V_s.*

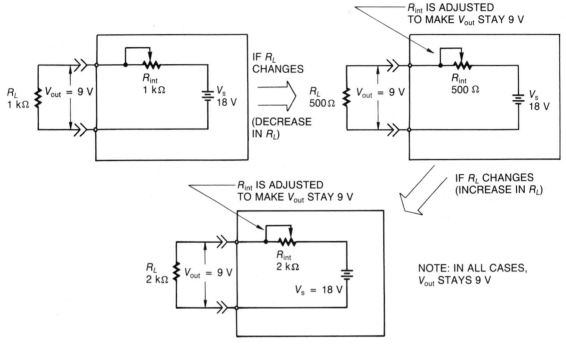

Figure 17-3 Basic idea of series regulator.

If the internal resistance of the power source (R_{int}) were changed each time to accommodate different current demands, then the output voltage would appear constant. This would occur because the change in internal resistance would bring the output voltage back to its original value. This process is illustrated in Example 1.

Example 1

For the hypothetical power source in Figure 17-4, what must be done to the variable resistor to maintain a constant output voltage if the load (A) increases or (B) decreases?

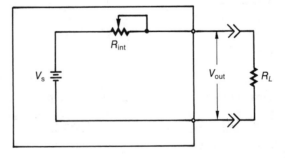

Figure 17-4

Solution

(A) When the load increases, there will be less current from the power source and hence less voltage drop across R_{int}. Therefore, V_{out} increases. Hence, R_{int} must be made *larger*. Increasing R_{int} produces the same amount of output voltage for less current.

(B) When the load decreases, there will be more current from the power source and therefore more voltage drop across R_{int}. Hence, V_{out} will decrease. Therefore, R_{int} must be made *smaller*. Decreasing R_{int} produces the same amount of output voltage for more current.

Example 1 illustrates the fundamental idea behind all series regulators. Instead of a variable resistor to regulate the amount of output voltage for different load conditions, a variable-resistance device, such as a transistor, can be used.

The concept of a transistor acting as a variable resistor is shown in Figure 17-5. If the base of the NPN transistor is made *more* positive, more current flows (internal resistance is *decreasing*). If the base is made *less* positive, less current flows (internal resistance is *increasing*). Just the opposite is true for a PNP transistor.

Therefore, you only need a transistor instead of the series variable resistor in Example 1. But now you need a way to control the amount of transistor internal resistance so that the output voltage will remain nearly constant. The method uses a zener diode and a differential amplifier. A *differential amplifier* produces an output voltage proportional to

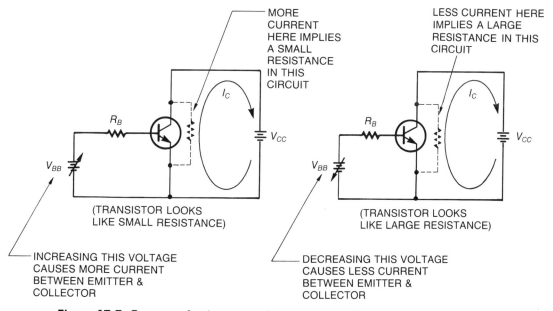

Figure 17-5 Concept of using a transistor as a variable resistor.

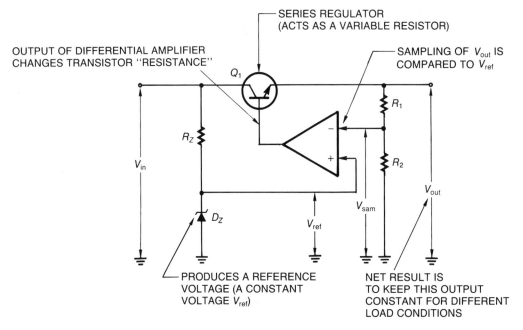

Figure 17-6 Using a zener diode and a comparator.

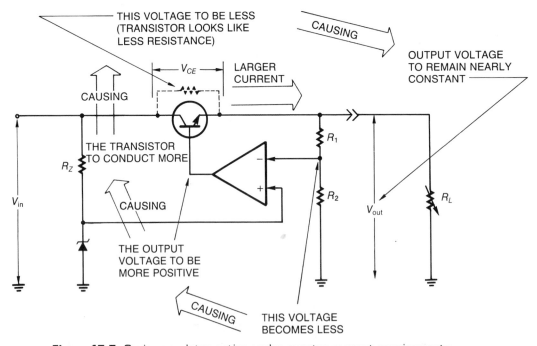

Figure 17-7 Series regulator action under greater current requirements.

the difference between an input voltage and a reference voltage. This arrangement is shown in Figure 17-6. The *zener diode* maintains a constant reference voltage to the (+) input of the differential amplifier. The (−) input of the differential amplifier is connected to a voltage divider consisting of R_1 and R_2. Ideally, the ratio of R_1 and R_2 is such that the voltage to the (−) input of the differential amplifier is slightly more than the zener reference voltage. This gives the proper amount of output voltage from the differential amplifier to supply the transistor with just the right amount of internal resistance.

Figure 17-7 shows what happens when the output load requires more current. The *pass transistor* (the transistor that allows the supply current to pass to the load) and the load resistor R_L make up a series circuit. Thus the sum of the voltage drop across the transistor (Q_1) and the load resistor (R_L) must equal the input voltage (V_{in}). Thus, if the voltage across the load tends to increase, then the voltage across the pass transistor will decrease. This is the same idea used for the first example of a series regulator using a variable resistor in place of the pass transistor.

Figure 17-8 shows what happens when the output load requires less current.

Computing the Output Voltage

The output voltage of a series regulator can be computed if you consider that the differential amplifier is acting as a *noninverting op amp*. This means that the gain can be expressed as

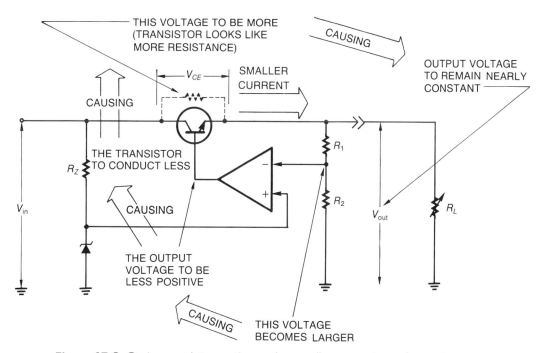

Figure 17-8 Series regulator action under smaller current requirements.

$$A_V = 1 + \frac{R_1}{R_2}$$

(Equation 17-2)

where A_V = Voltage gain of the amplifier

$\quad\quad R_1, R_2$ = Resistor values of voltage divider

If the small voltage drop from the transistor's base to emitter is neglected, the output voltage of the op amp becomes

$$V_{out} \approx \left(1 + \frac{R_1}{R_2}\right)V_{ref}$$

where V_{out} = Output voltage of the op amp

$\quad\quad R_1, R_2$ = Resistor values of voltage divider

$\quad\quad V_{ref}$ = Reference voltage of the zener diode

Example 2

Calculate the output voltage of the series regulator in Figure 17-9.

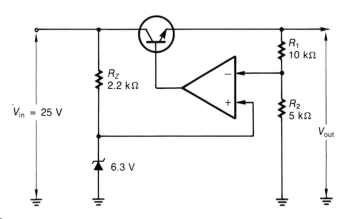

Figure 17-9

Solution

$$V_{out} \approx \left(1 + \frac{R_1}{R_2}\right)V_{ref}$$

$$V_{out} \approx \left(1 + \frac{10 \text{ k}\Omega}{5 \text{ k}\Omega}\right)6.3 \text{ V}$$

$$V_{out} \approx (1 + 2)6.3 \text{ V} = (3)6.3$$

$$V_{out} \approx 18.9 \text{ V}$$

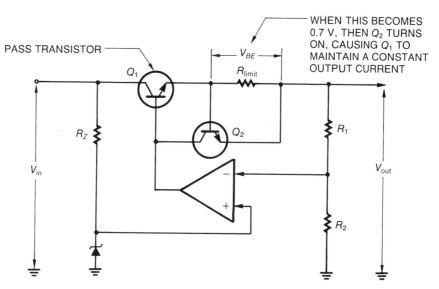

Figure 17-10 Addition of current-limiting transistor.

Current Limiting

There is always the possibility that the current limitations of the series pass transistor can be exceeded. When this happens, the transistor is damaged. To avoid this, you need to add a *current-limiting transistor* to the series regulator circuit, as shown in Figure 17-10. The voltage across R_{limit} cannot be more than 0.7 volt. When there is enough current in R_{limit} to cause a voltage drop of 0.7 volt, transistor Q_2 will then be forward biased and start to conduct. This causes the internal resistance of Q_2 to decrease. When this happens, the forward biasing of Q_1 becomes fixed resulting in a constant current output. The net effect of this is to maintain a maximum current value on the output and protect the pass transistor Q_1.

Conclusion

This section presented the fundamental operation of a series regulated power supply. In the next section, you will see how the shunt regulated power supply functions.

17-2 Review Questions

1. Explain the basic idea behind a series regulator. Use a variable resistor as the series regulator.
2. Describe the operation of a transistor in terms of the resistance it presents to the power source.
3. Describe the action of a zener diode. What role does it play in a series regulator?
4. Explain the action of a series regulator that uses a pass transistor, op amp, and zener diode.

17-3 | SHUNT REGULATORS

Discussion

The last section introduced a method of regulating a power source using a series variable-resistance device. In this section, you will see how to use a variable-resistance device, such as a transistor, in parallel or shunt with the power supply load.

Basic Idea

The basic idea behind a *shunt regulated* power supply is shown in Figure 17-11. A variable resistor R_{var} is added inside the power supply in parallel or in shunt with the load. It will affect the voltage drop across R_{int} to maintain a constant output voltage from the power supply. This action is illustrated in Example 1.

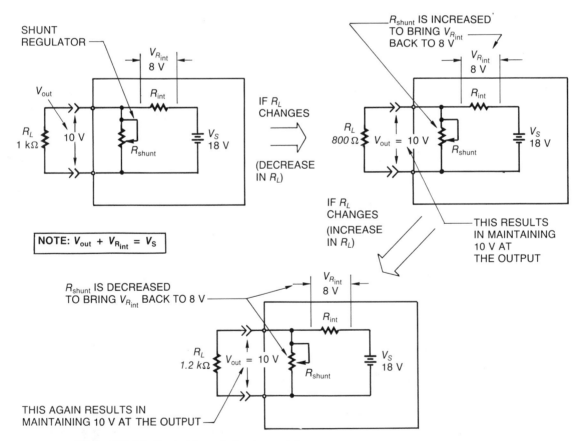

Figure 17-11 Basic idea of shunt regulator.

Example 1

For the hypothetical power source in Figure 17-12, what must be done to the variable resistor to maintain a constant output voltage if the load (A) increases and (B) decreases?

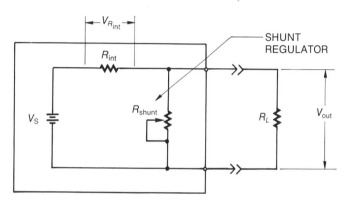

Figure 17-12

Solution

(A) Increase in the load. If the value of the load increases, there will be less current on the output, causing a smaller voltage drop across R_{int}. Thus, V_{out} will tend to increase, so the shunt regulating resistor R_{shunt} must decrease so that more current flows in R_{int}. This in turn causes more voltage across R_{int}, resulting in a decrease of V_{out} thus tending to bring the output voltage back to its original value.

(B) Decrease in the load. If the value of the load decreases, there will be more current on the output, causing a larger voltage drop across R_{int}. Thus, V_{out} will tend to decrease, so the shunt regulating resistor R_{var} must increase so that less current flows in R_{shunt}. This decreases the voltage drop across it and causes V_{out} to remain constant.

Example 1 illustrates the fundamental idea behind all shunt regulators. Although not as efficient as the series regulator, the shunt regulator does have the advantage of giving overload protection, since if the output is shorted, the current is limited by the internal voltage and the value of the internal series resistor (R_{int}).

Shunt Regulator Circuit

A *shunt regulator* circuit is shown in Figure 17-13. The principle of operation presented in the previous discussion is involved in the differential amplifier, zener diode, and shunt transistor. If the output voltage tends to decrease, the voltage fed to the base of the tran-

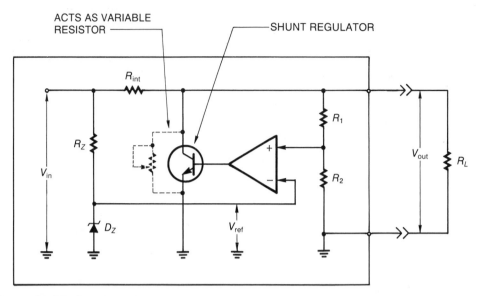

Figure 17-13 Complete shunt regulator circuit.

sistor will be less, causing the internal resistance of the transistor (R_{int}) to increase. This causes less current flow in R_{int} and less voltage drop. Since the voltage across R_{int} will now be less, then the output voltage (V_{out}) will be brought back to its normal value. Just the opposite occurs if the output voltage tends to increase. In this case, the voltage fed to the base of the shunt transistor will cause it to conduct more, causing more current in R_{int}. This will cause a greater voltage drop across R_{int} with correspondingly less output voltage (V_{out}). Again, the net result is to tend to keep the V_{out} constant with varying load conditions.

Conclusion

This section described shunt regulators. You should now be familiar with the basic concepts of series and shunt regulators.

17-3 **Review Questions**

1. Explain the basic idea behind a shunt regulator. Use a variable resistor as the shunt regulator.
2. What must happen to the value of the shunt regulating resistor if the output voltage tends to increase?
3. Describe the action of a shunt regulator using a transistor in place of the variable shunt resistor.
4. Compare the operation of a shunt regulator to that of a series regulator.

| 17-4 | IC REGULATORS |

Discussion

A variety of excellent IC regulators in four general classes is now available:

1. Fixed, positive voltage
2. Fixed, negative voltage
3. Fixed, positive and negative voltages (dual-tracking)
4. Variable output voltage

The most common voltages for these regulators are $+5$ volts, $+12$ volts, and -12 volts. These values are so common because they are frequently used by ICs. The $+5$ volts is used primarily for digital circuits, the ± 12 volts for op amps. This section presents a representative cross section of some of the most popular IC voltage regulators.

Packaging

The kind of packaging used for the IC regulators depends on the power dissipation requirements. Table 17-1 summarizes the different package types and the nominal power dissipation.

Table 17-1	IC PACKAGE TYPES AND POWER DISSIPATION	
Package Type	**Power Dissipation (W)**	**Maximum Current* (A)**
TO-3	20 and above	Up to 50
TO-220	15	1.5
TO-202	7.5	0.5
TO-39	2	0.1
TO-92	1.2	0.1

*This value may vary according to the actual regulator used.

The different package types are shown in Figure 17-14. *Heat sinking* of these regulators is critical for their proper operation. Heat must be transferred from the actual IC to the surrounding media. Failure to do this will cause heat buildup inside the IC chip, which usually shuts off the regulator causing no output voltage from the device.

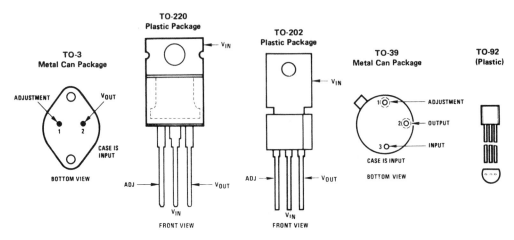

Figure 17-14 IC package for voltage regulation. *Courtesy* of National Semiconductor Corporation

Fixed-Voltage Regulators

The LM 140A/LM 140/LM 340A/LM 340 is a popular series of three-terminal fixed-positive-voltage regulators. A typical application is illustrated in Figure 17-15. Even though designed as fixed regulators, these devices can be used with external components to form a variable regulated output voltage.

The LM 79XX series three-terminal negative voltage regulators provide a fixed negative output voltage. Typical connections are shown in Figure 17-16.

The series includes the LM 7905 −5-volt regulator, the LM 7912 −12-volt regulator, and the LM 7915 −15-volt regulator.

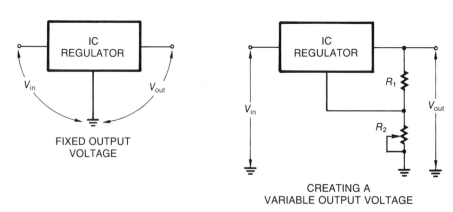

Figure 17-15 Typical application of three-terminal fixed positive voltage regulator.

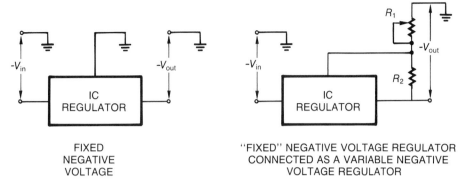

FIXED
NEGATIVE
VOLTAGE

"FIXED" NEGATIVE VOLTAGE REGULATOR
CONNECTED AS A VARIABLE NEGATIVE
VOLTAGE REGULATOR

Figure 17-16 Typical applications of three-terminal fixed negative voltage regulator.

Fixed Positive and Negative Regulators

The LM 125/LM 325/LM 325A/LM 326 series provides a dual-polarity tracking output. This means that the positive voltage is designed to be the same value above ground as the negative voltage is below ground. These devices have the capabilities of *tracking,* which means that if one output voltage changes due to load conditions, so will the other.

Figure 17-17 shows a typical connection of a dual-tracking voltage regulator.

Variable-Output-Voltage Regulators

Variable-output-voltage regulators are available in a wide range of voltage outputs, both positive and negative.

As an example, the LM 196/LM 136 series is a 10-amp adjustable positive voltage regulator. Its output voltage may be adjusted from 1.25 to 15 volts. It has a maximum power rating of 70 watts. A typical circuit connection is shown in Figure 17-18.

Circuit Protection

These IC voltage regulators are virtually immune to *blowout* caused by circuit overload. This means that even if the output is shorted, they will be able to recover. These circuits are protected by one or more of the following methods:

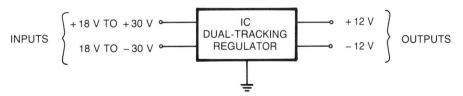

Figure 17-17 Typical connection of dual-tracking voltage regulator.

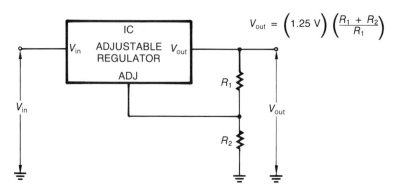

Figure 17-18 Typical circuit connection of variable output positive voltage regulator.

Thermal Shutdown

The regulator actually *shuts down* (turns itself off) if the case temperature reaches a certain point. Normal output operation is not resumed again until the regulator has cooled to a specified temperature.

Current Limiting

Regardless of the amount of overload, the regulator current output will not exceed a certain amount. This applies even if the output of the regulator is short circuited.

Current Foldback

Current foldback is usually used in high-current regulators. In this protection scheme, an overload or short causes the output current of the regulator to *decrease* to a small value. This process prevents the load or regulator from overheating.

Remote Shutdown

IC regulators contain an extra control lead that will cause them to turn off when a specified voltage is applied to the control lead. They are used where power can quickly be removed during a system emergency, such as fire or water immersion.

Conclusion

This section gave an overview of the various types of IC regulators available for communications systems. The next section gives the theory of operation of *switching* regulators. These regulators are being used more and more in all areas of electronics primarily because of their excellent regulation, efficiency, and economy, provided by ICs.

17-4 Review Questions

1. Name the different classes of IC regulators that are commercially available.
2. Explain the relation between the package type used for IC voltage regulators and the power dissipation of the regulator.
3. Describe how a fixed-voltage regulator can be used as a variable-voltage regulator.
4. What is a dual-tracking regulator?
5. Name the methods of IC regulator protection.

17-5 | SWITCHING FUNDAMENTALS

Discussion

As you discovered in the preceding two sections, a series voltage regulator is better than a parallel voltage regulator. However, because all of the current used by the load must pass through the series regulator, it dissipates a lot of power. See Figure 17-19. Therefore the series voltage regulator is not the most efficient approach. For large-current applications, a *switching regulator* is better.

This section presents the basic idea behind switching regulators. The next section presents applications.

The Pulse Train

Switching regulators are biased upon class S operation. These regulators cost more than the series voltage regulator, but this is compensated by the increase in system efficiency and reduced heat dissipation requirements of the regulating components.

At this time, it's best to review some of the details concerning a train of pulses. Figure 17-20 shows such a pulse train. Note that all of the pulses are above 0 volts. The *average*

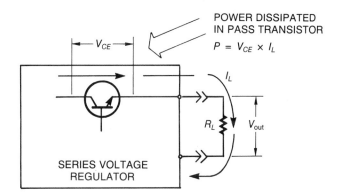

Figure 17-19 Power dissipation in series regulator circuit.

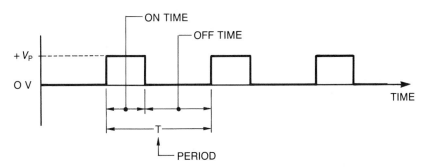

Figure 17-20 Characteristics of a typical pulse train.

value of the pulse is

$$P_{avg} = \frac{A}{T}$$

(Equation 17-3)

where P_{avg} = Average value of the pulse in volts
  A = Area under the pulse in volts per second
  T = Period of the pulse in seconds

The *area under the pulse* is

$$A = (PW)V_P$$

(Equation 17-4)

where A = Area under the pulse in volts per second
  PW = Pulse width in seconds
  V_P = Peak voltage of pulse in volts

Example 1

Determine the average value of the pulse trains in Figure 17-21.

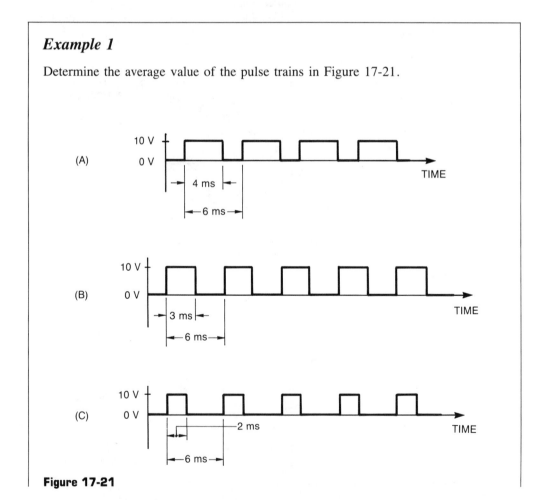

Figure 17-21

Solution

$$P_{avg} = \frac{A}{T}$$
(Equation 17-3)

$$A = (PW)V_P$$
(Equation 17-4)

Substituting (17-4) into (17-3) yields

$$P_{avg} = \frac{(PW)V_P}{T}$$

For pulse train A

$$P_{avg} = \frac{(PW)V_P}{T}$$

$$P_{avg} = \frac{4 \text{ ms} \times 10 \text{ V}}{6 \text{ ms}} = \frac{40 \text{ V}}{6}$$

$$P_{avg} = 6.67 \text{ V}$$

For pulse train B

$$P_{avg} = \frac{3 \text{ ms} \times 10 \text{ V}}{6 \text{ ms}} = \frac{30 \text{ V}}{6}$$

$$P_{avg} = 5 \text{ V}$$

For pulse train C

$$P_{avg} = \frac{2 \text{ ms} \times 10 \text{ V}}{6 \text{ ms}} = \frac{20 \text{ V}}{6}$$

$$P_{avg} = 3.33 \text{ V}$$

There is an important relationship to see from Example 1: the wider the pulses, the greater the amount of *average output voltage*. This relationship is useful in the operation of switching regulators.

Recall from basic electrical circuits that the *duty cycle* of a pulse train is

$$D = \frac{PW}{T}$$
(Equation 17-5)

> *where* D = Duty cycle (no units)
> PW = Pulse width in seconds
> T = Period of the pulse train in seconds

Therefore, the greater the duty cycle, the greater will be the average voltage.

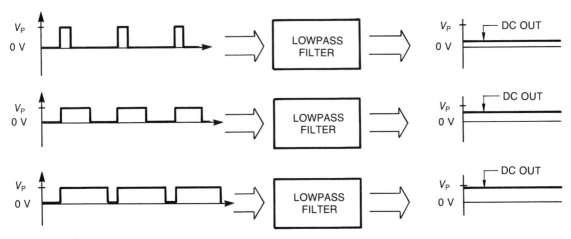

Figure 17-22 Result of filtering pulse trains.

What all of this is leading up to is that if you can control the duty cycle of a pulse train, you can also control the average DC voltage of that train. Furthermore, if you send this pulse train through a lowpass filter, you can filter out the changes and pass on the resulting DC voltage. This is shown in Figure 17-22.

Advantages of Class S Operation

Figure 17-23 contains a transistor series voltage regulator operated at class S. A pulse train is fed into the base of the transistor. This causes the transistor to operate between saturation and cutoff. The resulting small power dissipation of the transistor operating

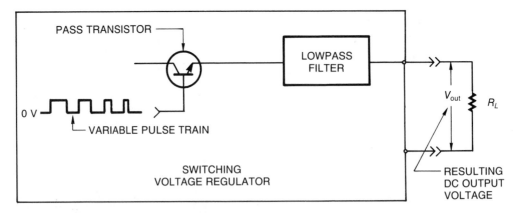

Figure 17-23 Series class S regulator.

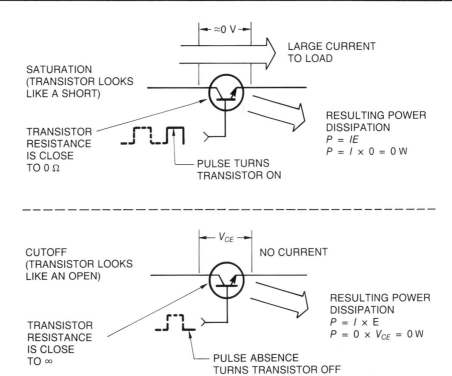

Figure 17-24 Resulting power dissipation of class S operation.

between saturation and cutoff is shown in Figure 17-24. Passing the output of the circuit in Figure 17-24 through a lowpass filter gives a DC voltage output. The amount of DC voltage will be directly proportional to the duty cycle of the pulse train.

The relationship of the duty cycle and the DC output voltage is

$$V_{\text{out}} = DV_{\text{in}}$$ **(Equation 17-6)**

where V_{out} = DC voltage output in volts
D = Duty cycle of pulse train (no units)
V_{in} = DC input voltage to the regulator in volts

Example 2

Determine the amount of DC output voltage for the switching regulators in Figure 17-25.

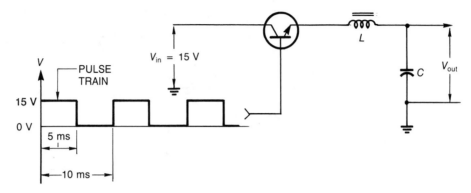

Figure 17-25

Solution

Solve for the duty cycle:

$$D = \frac{PW}{T} \qquad \text{(Equation 17-5)}$$

$$D = \frac{5 \text{ ms}}{10 \text{ ms}} = 0.5$$

Then we have

$$V_{out} = DV_{in} \qquad \text{(Equation 17-6)}$$

$$V_{out} = 0.5 \times 15 \text{ V}$$

$$V_{out} = 7.5 \text{ V}$$

To improve the efficiency of the inductor and capacitor and to allow smaller values of these components to be used, most switching regulators operate from 20 kHz to 50 kHz. Thus, the values of the filter capacitor and inductor can be much smaller (which makes them less expensive and more reliable), a significant improvement over _LC_ filters for 60-Hz supplies.

Conclusion

This section presented the basic concepts behind switching regulators. The next section presents actual switching regulator circuits.

17-5 Review Questions

1. What causes the power dissipation in the pass transistor of a series regulator?
2. Discuss the characteristics of a typical pulse train. What determines the average value of a pulse train?
3. Describe the operation of a class S amplifier. Why does this device have a low-power dissipation?
4. Why is a lowpass filter used in a switching regulator?
5. What frequencies are usually used for switching regulators? Why are they better than 60 Hz?

17-6 | SWITCHING REGULATORS

Discussion

This section introduces some practical switching regulators. The last section presented the basic theory of operation of switching regulators. This section discusses applications of that theory.

Generalized Circuit

Figure 17-26 shows a generalized circuit for a switching regulator. Effectively the circuit is a *pulse-width modulator*. The width of the pulse is determined by the voltage level of the comparator. The lower the output voltage, the longer the comparator will be on and the wider the pulse width will be. Recall from the last section that a wider pulse width produces a higher output voltage. If the output voltage is too large, the differential amplifier will make the output pulses narrower. This action reduces the output voltage.

A more detailed schematic of a switching regulator is shown in Figure 17-27. The *triangular-to-pulse-width converter* is controlled by the output of a *differential amplifier*.

Figure 17-28 illustrates the LH 1605/LH 1605C series of switching regulators. The switching regulator contains all of the control circuits.

Circuit Application

Figure 17-29 illustrates a circuit application of a switching regulator. It uses a standard full-wave rectifier power supply with an *LC* filtered output.

Figure 17-30 illustrates a switching regulator used with a *push-pull* transformer coupled circuit. The transformer can provide either a stepup or stepdown output voltage. Class B push-pull transistors increase the operating efficiency of each transistor.

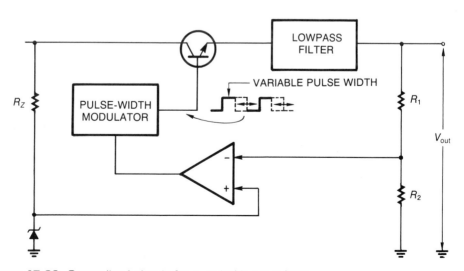

Figure 17-26 Generalized circuit for a switching regulator.

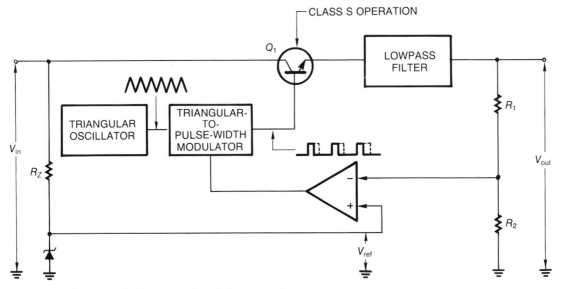

Figure 17-27 Details of switching regulators.

Block Diagram and Connection Diagram

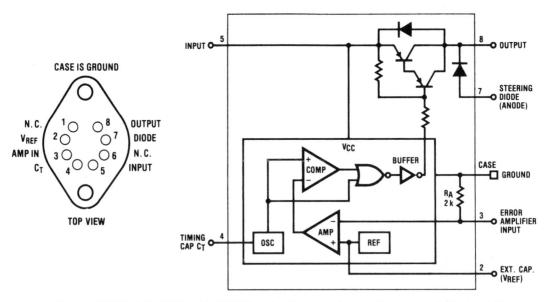

Figure 17-28 LH 1605/LH 1605C switching regulator. *Courtesy* of National Semiconductor Corporation

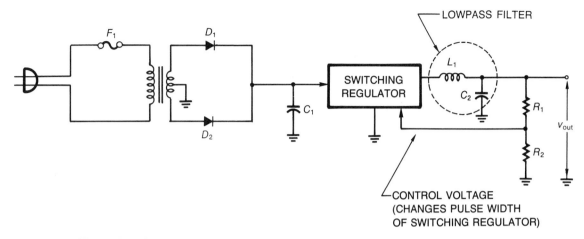

Figure 17-29 Circuit application of switching regulator.

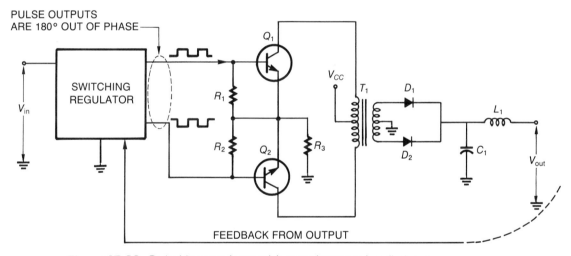

Figure 17-30 Switching regulator with transistor push-pull circuit.

DC-to-DC Converters

A DC-to-DC converter converts a DC voltage to another DC voltage. Remember that a transformer requires a *changing* current in order to couple a voltage from its primary to its secondary. Thus a DC-to-DC converter first converts the original DC voltage to a changing voltage (AC or pulsating DC), uses a transformer to step this voltage up or down, and converts the changing voltage back to DC. This process is illustrated in Figure 17-31.

These converters are used in many ways in mobile communications equipment where a 12-volt supply is available but different voltages are required for equipment operation.

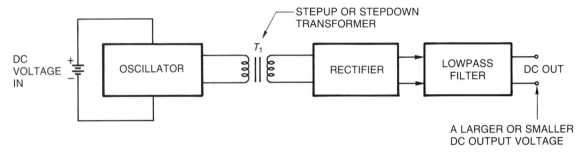

Figure 17-31 Basic process of DC-to-DC converter.

Conclusion

This section presented some of the practical applications of switching regulators. The next section gives some tips for troubleshooting power supplies for communication systems.

17-6 Review Questions

1. What role does a pulse-width modulator play in a switching regulator?
2. Discuss how a pulse-width modulator can be constructed from a triangular waveform.
3. Why are push-pull amplifiers used with a switching regulator?
4. Describe the general construction of a DC-to-DC converter.

TROUBLESHOOTING AND INSTRUMENTATION

Power Supply Troubleshooting

Power supply problems can cause many different symptoms in a communication system. Thus, always check the power supply first when troubleshooting any electronic system.

This section suggests methods of troubleshooting five major types of power supplies:

1. Unregulated
2. Series regulated
3. Shunt regulated
4. Switching regulator
5. Battery supplied

The first part of this section introduces troubleshooting methods that are common to two or more types of power supplies.

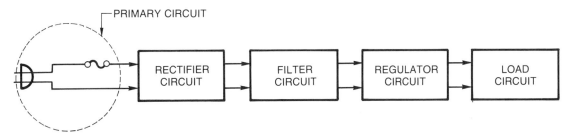

Figure 17-32 Block diagram of typical power supply.

General Procedures

Figure 17-32 shows the block diagram of a typical power supply. There are five major sections that can cause a problem in a power supply system. They are listed in Table 17-2.

Table 17-2	MAJOR CAUSES OF POWER SUPPLY FAILURE	
Circuit	**Reason for Failure**	**Symptom**
Primary	Blown fuse or open line cord	No output voltage
Rectifier	Faulty rectifier or open or shorted secondary	Low output or no output voltage
Filter	Open or shorted filter capacitor	Audible hum or no output voltage
Regulator	Open, shorted, or faulty regulator	Poor regulation or no output voltage
Load	Short or open in load	No output for short. Voltage larger than normal output voltage for open.

Figure 17-33 shows a method of isolating a short in a power supply. If the power supply consistently blows a fuse, it usually means a short or overload is in the circuit. The best way to deal with this is to break open the power chain at the indicated points, starting with point A. Then, with the power off, replace the fuse and apply power. Keep repeating this procedure, going next to point B, until the fuse no longer blows. The problem is then in the circuit to the right of the last break you made.

Series and Shunt Regulated Power Supplies

Be cautious when troubleshooting voltage regulators. Often the problem is not in the regulator but in the load itself. If the load is taking too much current, the voltage regulator will undergo *thermal shutdown* or *current foldback*. In the first case, the output voltage of the regulator will be close to zero. In the second case, the output voltage will be very

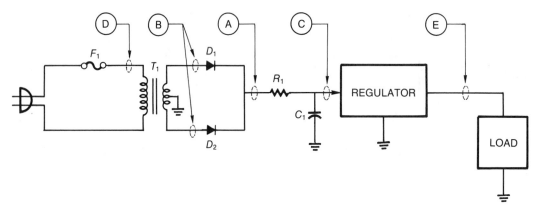

Figure 17-33 Method of isolating a short in typical power supply.

low. The best way to check for these symptoms is to remove the load from the power source and check the output voltage of the regulator *after it has cooled down*.

IC voltage regulators, because of the built-in protective circuits, are very difficult to destroy. You should suspect these circuits last as a probable cause of power supply problems.

Switching Regulators

A switching regulator can be analyzed with an oscilloscope. One method of troubleshooting is to remove the load from the regulator and replace it with a rheostat in series with a fixed resistor. Be sure that the *power rating of the rheostat is large enough to accommodate the power output of the power supply*. This method is illustrated in Figure 17-34.

Batteries

For the purpose of this section, there are three major types of batteries used as a power source in communication equipment:

1. Dry cell
2. Nickel–cadmium (nicad) battery
3. Lead–acid cell

The problems encountered with each type are summarized as follows:

Dry Cells

Dry cells usually develop a high internal resistance. The symptom here is that the cell will measure an open-circuit voltage that appears to be right, but its output voltage will then decrease the moment it is connected to the load. The rule here is to measure the voltage of these cells when they are connected in the circuit.

Nickel–Cadmium Battery

Nickel–cadmium batteries are rechargeable over a long useful life. These cells produce hydrogen and oxygen gases when being recharged. These two gases are *potentially highly*

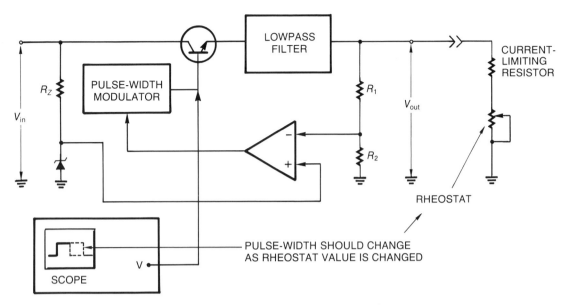

Figure 17-34 Method of troubleshooting switching regulator.

explosive near an open flame. Charging cells must therefore be well vented and kept away from sparks or open flames.

These cells will usually experience an internal short, indicated by excessive charging current, at which time they should be replaced.

Lead–Acid

Lead–acid batteries are low-voltage high-current types used to start internal combustion engines in vehicles. They are extremely dangerous because of their potentially high current output and corrosive liquid contents.

Conclusion

This section gave some tips for troubleshooting power sources for electronic communication systems. The microcomputer simulation contains a structured BASIC program that helps you analyze a power source.

17-7 Review Questions

1. Name the five major types of power supplies.
2. Describe the five major sections of a power supply. What is the most likely problem and the cause of the problem for each section?
3. How would you isolate a short in a power supply?
4. Discuss the precautions you should exercise when troubleshooting a series or shunt regulated power supply.
5. Describe the major types of batteries in communications equipment.

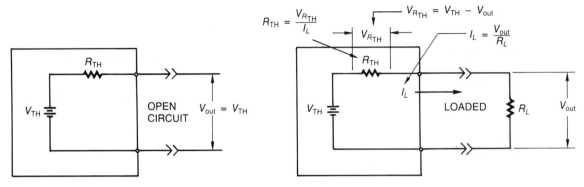

Figure 17-35 Power supply analysis.

MICROCOMPUTER SIMULATION

Power Supply Analysis

The final troubleshooting simulation program on your disk helps you develop troubleshooting skills in the analysis of DC power supplies. Here you will have the opportunity to determine if the computer generated power supply is working according to a given set of specifications. The ability to analyze power supply operation is an important skill that will round out the troubleshooting skills you have already developed up to this point. Again it is important to first do the **instructions** part of this simulation as well as the **demonstration** before starting the **test** portion. After successfully completing this simulation, you will have acquired an amount of experience in the fundamentals of troubleshooting that would be difficult to achieve solely in a laboratory setting. Congratulations, you have come a long way in the development of your skill as well as knowledge in the exciting field of electronic communications.

CHAPTER PROBLEMS

(Answers to odd-numbered problems appear at the end of the text.)

1. What conditions determine the quality of a power supply?
2. Explain why the output impedance of a power supply is an important factor. Can the output impedance ever be zero? Explain.
3. Draw the schematic of an ideal power source. Show its Thevenin equivalent.
4. Determine the no-load and full-load output voltages for the power supplies in Figure 17-36(A), (B).
5. For the power supplies in Figure 17-36(C), (D), determine the no-load and full-load output voltages.

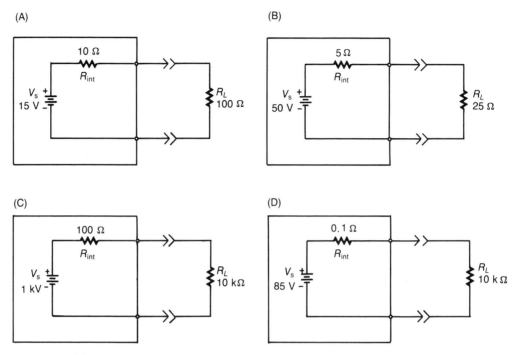

Figure 17-36

6. Which of the power sources in Figure 17-36 has the best regulation? Why?
7. Which of the power sources in Figure 17-36 has the poorest regulation? Why?
8. Determine the percentage regulation for each power source in Figure 17-36(A), (B).
9. For each power source in Figure 17-36(C), (D), determine the percentage of voltage regulation.
10. What factors of a power supply influence its voltage regulation capabilities?
11. What kind of regulator is shown in Figure 17-37?
12. What is the purpose of the variable resistor in the regulator circuit of Figure 17-37?
13. To maintain a constant output voltage for the regulator circuit of Figure 17-37, what must be done to the variable resistor if the load resistor decreases? if the load resistor increases?

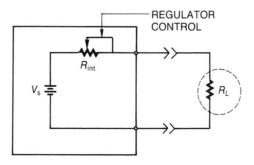

Figure 17-37

14. Explain how a transistor (NPN and PNP) can act as a variable resistor.
15. Explain the operation of a zener diode. How is this device used in a voltage regulator?
16. Draw the schematic of a series voltage regulator using a pass transistor. Illustrate the action of the pass transistor for an increase in the load value and a decrease in the load value.
17. Using diagrams, explain how the power dissipation of a pass transistor varies under different load conditions.
18. Calculate the output voltage of the series regulator in Figure 17-38(A). Ignore the transistor voltage drop.
19. For the series regulator in Figure 17-38(B), determine the amount of output voltage (ignore the voltage drop across the transistor).

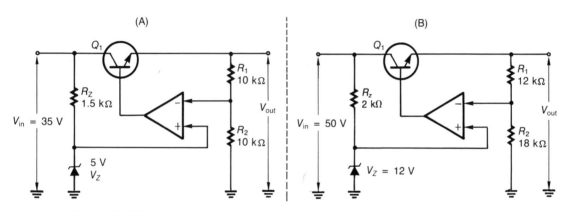

Figure 17-38

20. Draw the schematic of a series voltage regulator that uses current limiting. Explain the action of the current-limiting transistor.
21. State some advantages of current-limiting in a voltage regulator.

22. What kind of regulator is shown in Figure 17-39?
23. What is the purpose of the variable resistor in the regulator circuit of Figure 17-39?
24. To maintain a constant output voltage for the regulator circuit of Figure 17-39, what must be done to the value of the variable resistor if the load resistor decreases? if the load resistor increases?

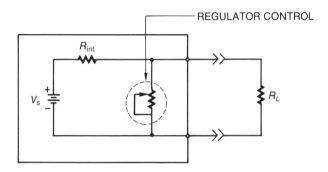

Figure 17-39

25. Sketch the schematic of a shunt regulator using a transistor. Explain the circuit operation.
26. Show by illustration what determines the power dissipation of a shunt regulator transistor. Do this for varying load conditions.
27. Describe the four types of IC regulators that are commonly used.
28. Using an available technical manual, list a specific regulator type (such as LM 140A) for each package type in Table 17-1.
29. From the list in problem 28, state which regulators dissipate the most power and the least power.
30. Describe the action of a typical fixed positive IC voltage regulator. Give a typical type and sketch how it should be connected.
31. Referring to problem 30, do the same thing for a fixed negative voltage regulator.
32. Sketch the diagram of a dual-tracking voltage regulator. Using an available technical manual, list a specific dual-tracking regulator type. Make sure it's a different type from the one given in the chapter.
33. Describe the action of a dual-tracking voltage regulator.
34. Sketch the schematic of the proper connection for a variable-voltage IC regulator. Using an available technical manual, list a specific type. Make it different from the one given in the chapter.
35. List the IC regulator protection schemes. State the operation of each one.

36. For the pulse trains in Figure 17-40(A), (B), determine the average values.
37. Determine the average values for the pulse trains in Figure 17-40(C), (D).
38. For the pulses in Figure 17-40(C), (D), determine the duty cycles.
39. Determine the duty cycles for the pulse trains of Figure 17-40(A), (B).

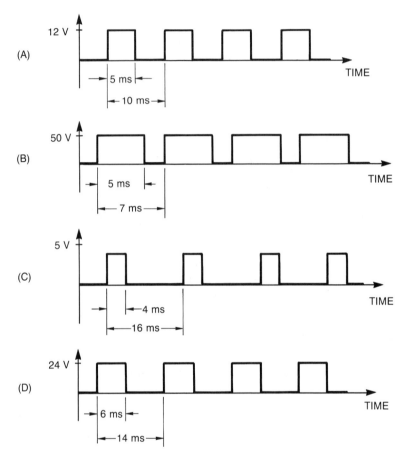

Figure 17-40

40. Sketch the schematic of a class S amplifier. Explain its basic operation. What advantage does this class of operation have?
41. Sketch the schematic of a class S switching regulator. Using diagrams, explain how this helps maintain a steady output voltage under changing load conditions.

42. Determine the DC output voltage for the switching regulator in Figure 17-41(A).
43. For the switching regulator in Figure 17-41(B), determine the DC output voltage.

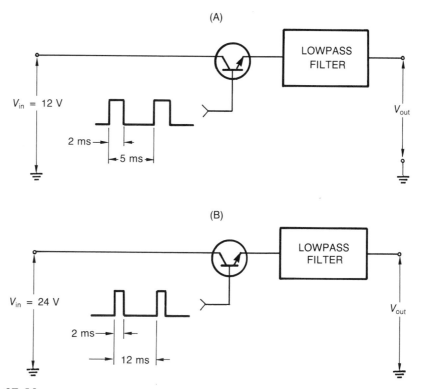

Figure 17-41

44. Sketch the schematic of a switching regulator. Explain the theory of operation for each section.
45. Using a technical manual, list an available IC switching regulator. Use a type other than the one given in the chapter.
46. Using a technical manual, show two applications of switching regulators other than the ones given in the chapter.
47. Using diagrams, explain the basic operation of a DC-to-DC converter. Why is it necessary to use an oscillator?
48. What precautions are necessary when troubleshooting power supplies containing IC voltage regulators?
49. List the most common types of batteries used in communication equipment. What precautions must be taken with each kind?

APPENDIX A
Logic Circuits for Communications

OBJECTIVES

In this appendix, you will study:

- [ ] How to convert from binary to decimal and decimal to binary.
- [ ] The identification and meaning of standard logic circuit symbols.
- [ ] The identification of standard flip-flop symbols and what they mean.
- [ ] How to read the schematics representing binary registers and counters.
- [ ] The functional block diagram of a frequency counter.

INTRODUCTION

Logic circuits have found their way into communications through many different applications. The programming of your VCR, digital systems that find and display your selected radio station, and circuits that convert voice into digital sound are all examples of systems that use logic circuits. This appendix offers a brief but important introduction to this useful topic.

| A-1 | BINARY NUMBERS |

Discussion

Binary numbers are used to represent numbers in electronic circuits that have two states: on or off. In this section, you will learn how to represent binary numbers and how to use them.

Our Number System

Before starting with binary numbers, you should review some facts about our decimal numbering system. The system is called *decimal* because 10 symbols are used:

0, 1, 2, 3, 4, 5, 6, 7, 8, 9

The notation N_{10} indicates that a number is decimal. For example, 12_{10} means 12 is a decimal number.

Binary Number

A binary number has only two symbols:

   0, 1

The notation N_2 means that a number is binary. For example, 11_2 means 11 is a binary number. With these two symbols, you can represent any number.

   Here's how. Look at Figure A-1. It has four empty cups. Each cup has a different "value." The values are first cup = 1, second cup = 2, third cup = 4, fourth cup = 8. The condition for representing any number is that each cup is either full or empty. This condition is indicated by a stirring stick. When a stirring stick is in the cup, it is full; when a stirring stick is not in the cup, it is empty. Only full cups are counted. Using this method and only four cups, you can represent any whole number from 0 to 15, as shown in Figure A-2.

Figure A-1 Cups used to represent binary numbers.

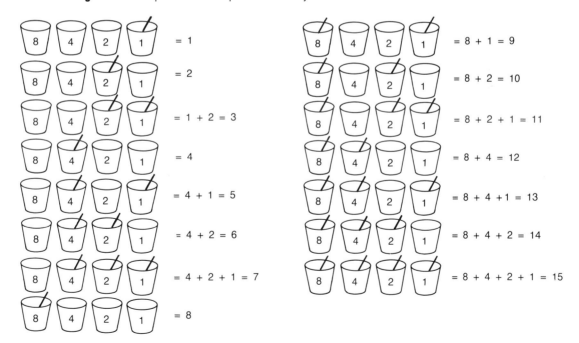

Figure A-2 Using cups to represent binary numbers.

Example 1

What decimal numbers are represented by the binary numbers (A) 1001_2, (B) 0110_2, (C) 1101_2, (D) 1100_2?

Solution

Use the place value of each "cup" and add the values of all the "full" cups to get the answer.

(A) Place value: 8 4 2 1
    Given number: 1 0 0 1
    Adding the full cups: $8 + 0 + 0 + 1 = 9_{10}$

(B) Place value: 8 4 2 1
    Given number: 0 1 1 0
    Adding the full cups: $0 + 4 + 2 + 0 = 6_{10}$

(C) Place value: 8 4 2 1
    Given number: 1 1 0 1
    Adding the full cups: $8 + 4 + 0 + 1 = 13_{10}$

(D) Place value: 8 4 2 1
    Given number: 1 1 0 0
    Adding the full cups: $8 + 4 + 0 + 0 = 12_{10}$

More cups could be added to represent larger numbers. For each new cup added, its place value is double that of the preceding cup. The numbering sequence of cups would be 1, 2, 4, 8, 16, 32, 64, etc. Mathematically this may be expressed as 2^N.

Example 2

Convert the following decimal numbers to their binary equivalents: (A) 8_{10}, (B) 5_{10}, (C) 10_{10}.

Solution

Imagine using the four cups (or more if you need them); let the leftmost cup represent a number that is as large as possible without being larger than the given decimal number. Subtract what's left and repeat this process until your answer is zero.

(A) Original number: 8_{10}
    Largest valued cup: 8
    Difference: 0
    Answer: $8_{10} = 1000_2$

(B) Original number: 5_{10}
Largest valued cup: 4
Difference: 1
Largest valued cup: 1
Difference: 0
Answer: $5_{10} = 0101_2$

(C) Original number: 10_{10}
Largest valued cup: 8
Difference: 2
Largest valued cup: 2
Difference: 0
Answer: $10_{10} = 1010_2$

Conclusion

Binary numbers are a way of having on–off circuits represent numbers. This section used the concept of cups for these circuits. In actual practice, these circuits consist of many transistors and other electrical components housed in a small integrated circuit.

A-1 Review Questions

1. Why are binary numbers used in electronics?
2. State the major difference between the decimal number system and the binary number system.
3. What notation is used to distinguish a binary number from a decimal number?
4. State the place value of each cup when four cups are used to represent binary numbers. What is the largest number that can be represented?
5. What binary numbers represent 2_{10}, 6_{10}, 12_{10}?
6. What decimal numbers represent 11_2, 101_2, 1011_2?
7. If eight cups were used to represent binary numbers, what is the largest number that could be represented?

A-2 | COMMUNICATION LOGIC CIRCUITS

Discussion

Logic circuits are used in communications to control signal processing and time events. Many schematics of communication circuits use the standard logic symbols presented in this section. A working knowledge of these logic symbols is essential for the communication technician.

Logic Notation

Logic circuits can be thought of as switches that are either on or off. The nomenclature used to designate these circuit conditions is given in Table A-1.

Table A-1	LOGIC CIRCUIT NOTATION	
On		**Off**
Closed		Open
1		0
True		False
High		Low
+5 V		0 V

The AND Function

Observe the circuit in Figure A-3. Note that the lamp will light only when switches A and B are closed.

A *truth table* for a logic circuit indicates all possible conditions of the inputs and what effect the input has on the output. The following truth table is for the AND circuit in Figure A-3.

TRUTH TABLE FOR AND FUNCTION		
Switch A	**Switch B**	**Lamp**
Off	Off	Off
Off	On	Off
On	Off	Off
On	On	On

The truth table shows that the lamp is on (true) only when both input switches are on (true).

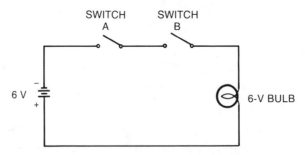

Figure A-3 The AND function.

Figure A-4 is the symbol of an *AND gate*. The corresponding truth table for the AND gate is

A	B	X
0	0	0
0	1	0
1	0	0
1	1	1

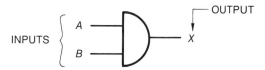

Figure A-4 AND gate symbol.

OR Gate

The operation of an *OR gate* can also be represented by switches, as shown in Figure A-5. Note that the output lamp is true (on) when either one or both of the switches are closed. The logic symbol for an OR gate is shown in Figure A-6. The truth table for an OR gate is

OR GATE TRUTH TABLE		
A	B	X
0	0	0
0	1	1
1	0	1
1	1	1

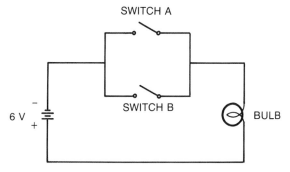

Figure A-5 Switch representation of OR function.

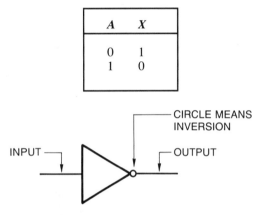

Figure A-6 OR gate logic symbol.

Inverter

The logic symbol for an *inverter* is shown in Figure A-7. The truth table for an inverter is

A	X
0	1
1	0

Figure A-7 Inverter logic symbol.

The inverter seems to have a very trivial function. But it really serves a very important purpose in practical logic circuits.

The NAND Gate

The logic symbol of a *NAND gate* is shown in Figure A-8. Note that a NAND gate is nothing more than an AND gate followed by an inverter. The truth table for a NAND gate is

NAND GATE TRUTH TABLE		
A	B	X
0	0	1
0	1	1
1	0	1
1	1	0

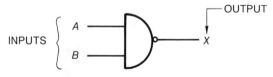

Figure A-8 Logic symbol for NAND gate.

NOR Gate

The logic symbol of a *NOR gate* is shown in Figure A-9. Note that a NOR gate is nothing more than an OR gate followed by an inverter. The truth table for a NOR gate is

NOR GATE TRUTH TABLE		
A	*B*	*X*
0	0	1
0	1	0
1	0	0
1	1	0

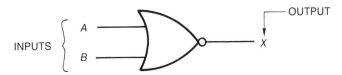

Figure A-9 Logic symbol for NOR gate.

Exclusive OR (XOR) Gate

The *exclusive OR (XOR)* gate finds many applications in logic circuits. The difference between the OR gate and the exclusive OR gate is that the output is true on the two-input XOR gate only when the inputs are different. The XOR gate symbol is shown in Figure A-10.

The truth table for an XOR gate is

XOR GATE TRUTH TABLE		
A	*B*	*X*
0	0	0
0	1	1
1	0	1
1	1	0

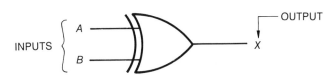

Figure A-10 Logic symbol for exclusive OR (XOR) gate.

Exclusive NOR (XNOR) Gate

The *exclusive NOR (XNOR)* gate can be thought of as an XOR gate followed by an inverter. The logic symbol of an XNOR gate is shown in Figure A-11.

The truth table for an XNOR gate is

XNOR GATE TRUTH TABLE		
A	*B*	*X*
0	0	1
0	1	0
1	0	0
1	1	1

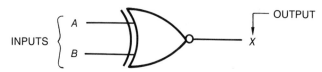

Figure A-11 Logic symbol for exclusive NOR (XNOR) gate.

Note from the truth table that output of a two-input XNOR gate is true only when the inputs are the same.

Conclusion

This section introduced you to the basic logic gates commonly found in communication circuits. The logic circuits presented here can have many more inputs and be combined in various ways. The combination of these logic circuits is beyond the scope of this book, but you should be able to recognize these symbols in communication circuits and know what they mean.

A-2 Review Questions

1. State some uses of logic circuits in communication circuits.
2. Describe some of the nomenclature used to designate an on condition and an off condition.
3. Give an example of a circuit that simulates the AND function. Do the same for the OR function.
4. Explain the difference between an AND gate and an OR gate.
5. What does an inverter do?
6. Describe the function of a NAND gate. Do the same for a NOR gate.
7. Explain how an exclusive OR gate functions. Do the same for an exclusive NOR gate.

A-3 | FLIP-FLOPS

Discussion

A *flip-flop* is a circuit used to store a single binary number. Recall that a single binary number can be represented by a cup that is either empty or full. Hence, you'd expect the flip-flop to be either empty or full.

In electronics, when a flip-flop is full, it is said to be a 1 or true; when it is empty it is a 0 or false. For convenience, flip-flops have two outputs that are used to measure its condition (either 1 or 0). These two outputs are always opposites; when one is true, the other is false.

Most Basic Flip-Flop

The equivalent circuit of a most basic flip-flop is shown in Figure A-12. Note that the output of a flip-flop is considered a 1 (or true) when it is +5 V, and a 0 (or false) when its output is 0 V. These values are industry standards for manufacturing flip-flops.

Flip-flops are not actually constructed with switches; instead, circuit elements such as transistors are used. The important thing about flip-flops is that they have two outputs that are always different. Figure A-13 shows an IC with four different flip-flops. Each flip-flop has two outputs.

Clear-Set Flip-Flop

The *Clear-Set* flip-flop (*C-S* F/F) is the most basic flip-flop available. Its logic representation is shown in Figure A-14. The outputs of a flip-flop are designated as Q and $\bar{Q}$. The bar ($^-$) means NOT Q. This is just another way of emphasizing that the two flip-flop outputs are always *different*. Thus if Q is true, then $\bar{Q}$ (NOT Q) is false.

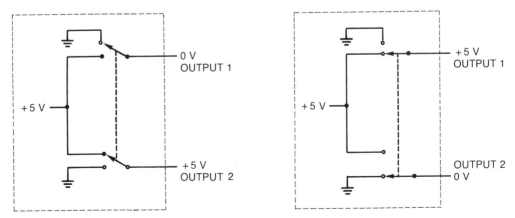

Figure A-12 Equivalent circuit of most basic flip-flop.

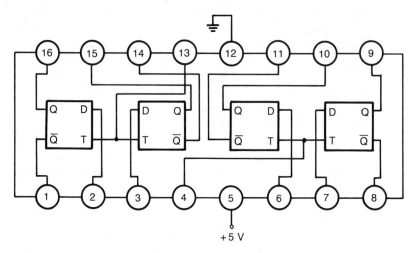

Figure A-13 IC circuit with four flip-flops.

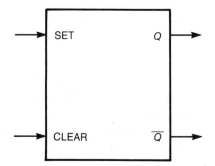

Figure A-14 Logic symbol of Clear-Set flip-flop.

The conditions of the C-S F/F are shown in the following truth table:

C	S	Q	$\bar{Q}$
0	0	No change	
0	1	1	0
1	0	0	1
1	1	Not allowed	

The meaning of the table is as follows:

■ *Clear and Set both false:*
In this state the flip-flop is storing its condition. If Q is true, it will stay true; if Q is false, it will stay false with $\bar{Q}$ (NOT Q) maintaining the opposite condition.

■ *Clear false and Set true:*
Here, the Q output of the flip-flop goes true while the $\bar{Q}$ output goes false. If the Q output was already true, it will stay that way. This input condition simply ensures that the Q output is true. This procedure is called *setting* the flip-flop.

■ *Clear true and Set false:*
For this condition, the Q output of the flip-flop goes false while the $\bar{Q}$ output goes true. If the flip-flop was already in this condition, it will stay that way. This input ensures that the Q output is false. This procedure is called *clearing* the flip-flop.

■ *Clear true and Set true:*
This condition is not allowed. The reason is that the output state of the flip-flop may not be predictable.

Clock Pulses

Flip-flops are normally used together rather than in isolation. Therefore, they are usually timed to change their inputs at specific intervals. This timing is controlled from an oscillator circuit called a *clock*. Recall that an oscillator is a circuit that creates its own signal. When an oscillator is used as a clock (to generate timing pulses), its output is a square wave. See Figure A-15.

Clocked Clear-Set Flip-Flop

A *clocked C-S* F/F is shown in Figure A-16. Note that its output will not pay any attention to the inputs until the clock pulse is true (+5 V).

With a clock, flip-flops can now be made to control their outputs in unison. In the next section, on registers, you will see why this is done.

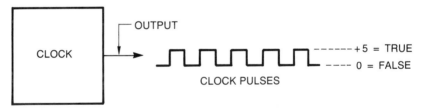

Figure A-15 Block diagram of a clock.

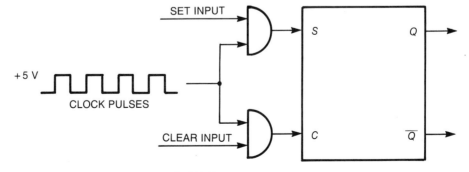

Figure A-16 Action of a clocked *C-S* F/F.

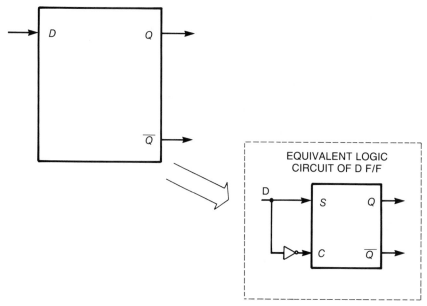

Figure A-17 Diagram of a *D* flip-flop.

Data Flip-Flop

The *data* flip-flop, usually called the *D* flip-flop, is shown in Figure A-17. The equivalent circuit of a *D* flip-flop is a *C-S* flip-flop with an inverter at the Clear input.

The operation of a *D* flip-flop is shown in the following table:

D F/F		
D	*Q*	*Q̄*
0	0	1
1	1	0

The essence of a *D*-type flip-flop is that the *Q* output stores the condition of the *D* input.

Leading and Trailing Clock Edges

When logic circuits are used, they usually require more precise timing than can be acquired from the *level* of the clock pulse (being +5 V or 0 V). For a more exact timing, most flip-flops will respond to either the *leading edge* or the *trailing edge* of the clock pulse. The leading and trailing edges of a clock pulse are shown in Figure A-18.

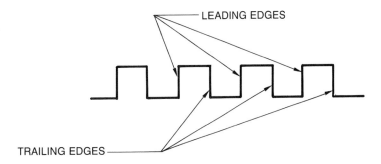

Figure A-18 Leading and trailing edges of clock pulses.

The symbol used in this text for indicating that a flip-flop is activated by a leading or trailing clock pulse is shown in Figure A-19. The inputs of an edge-triggered flip-flop cannot affect the output until the correct *edge* of the clock appears. Thus, the timing of flip-flops is very precise.

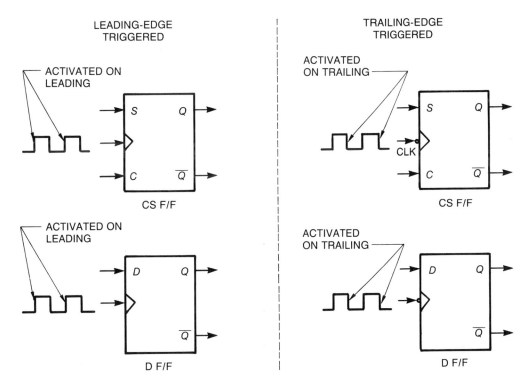

Figure A-19 Symbols for edge-triggered flip-flops.

J-K *Flip-Flop*

The *J-K* flip-flop is one of the most versatile flip-flops used in communications. Its logic symbol is shown in Figure A-20. The truth table of a *J-K* flip-flop is

CP	J	K	Q	Q̄
	0	0	No change	
↓	0	1	0	1
↓	1	0	1	0
↓	1	1	Toggles	

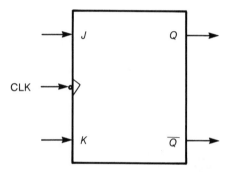

Figure A-20 Logic symbol for *J-K* flip-flop.

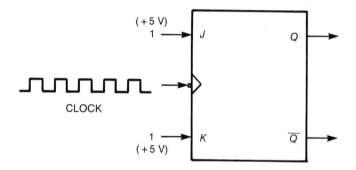

Figure A-21 *J-K* flip-flop in the toggle mode.

The down arrow (↓) in the CP (clock pulse) column means that the output will change state only on the *trailing* edge of the clock pulse. The output *toggles* when the J and K inputs are both true. This condition is shown in Figure A-21. A J-K flip-flop in the toggle mode acts just like a frequency divider. If the input frequency of the clock is 2 kHz, then the output frequency (Q or $\bar{Q}$ output) will be 1 kHz.

Clear and Set Inputs

Most *J-K* flip-flops come with two other inputs, called Clear and Set. These inputs are different from the *J-K* inputs in that they ignore the clock pulse. As such, these inputs are called *asynchronous,* which means they are *not* synchronized. The logic symbol of a *Clear-Set (C-S) J-K* flip-flop is shown in Figure A-22. The truth table for a *C-S J-K* flip-flop is

CP	CLEAR	SET	J	K	Q	$\bar{Q}$
	0	0	0	0	No change	
↓	0	0	0	1	0	1
↓	0	0	1	0	1	0
↓	0	0	1	1	Toggles	
Ignored	0	1	X	X	1	0
Ignored	1	0	X	X	0	1
	1	1	Not allowed			

The X means *don't care*. In other words, no matter what the condition of J or K is, the condition of the output will be as indicated. For example, if Clear is false and Set is

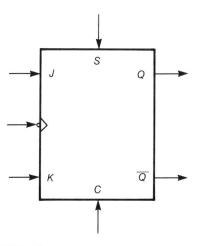

Figure A-22 A Clear-Set *J-K* flip-flop.

true, then the Q output will become true no matter what the conditions of J and K happen to be. Only when the Clear and Set inputs are 0 will the J and K inputs have control of the flip-flop outputs.

Conclusion

This section is very important in the study of communications. Flip-flops are used in communications equipment and in the instruments used to measure and repair this equipment. You will see a very useful application of flip-flops in the instrumentation and troubleshooting section of this chapter.

A-3 Review Questions

1. Describe how a flip-flop is used.
2. Give an example of a circuit that describes the operation of a flip-flop output.
3. Explain the operation of a *C-S* F/F.
4. What is a clock? What is a clock pulse?
5. Explain the operation of a clocked *C-S* F/F.
6. Explain the operation of a *D* F/F. In what way is a *D* F/F similar to a *C-S* F/F?
7. Explain the difference between the leading and trailing edges of a clock pulse.
8. Describe the operation of an edge-triggered F/F. Explain why this is an advantage.
9. Explain the operation of a *J-K* F/F.
10. Describe the effects of the Clear and Set inputs on a *J-K* F/F.
11. What does the "don't care" notation mean with an F/F?

| A-4 | REGISTERS |

Discussion

A *register* is nothing more than a group of flip-flops. Recall that a flip-flop held a single binary digit. A group of flip-flops can hold several binary digits and can therefore represent a binary code, such as a number.

Figure A-23 shows a register consisting of four flip-flops. Note that the present contents of the register can be viewed as the number 9.

Register Words

A *register word* is the contents of any group of flip-flops treated as a unit. Figure A-24 illustrates several different registers and their corresponding *words*.

The size of a register word is given various names, as shown in Table A-2.

Table A-2	WORD NAMES	
	NUMBER OF F/F	**WORD SIZE NAME**
	1	Bit
	4	Nibble
	8	Byte

Hence, a *byte* requires eight flip-flops, but a *nibble* requires only four.

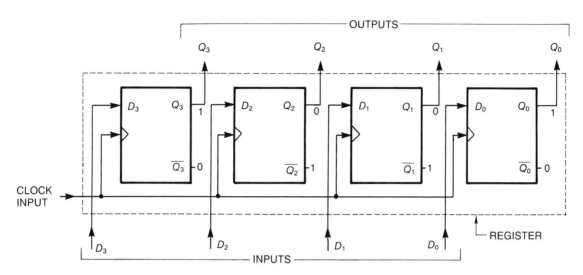

Figure A-23 A register.

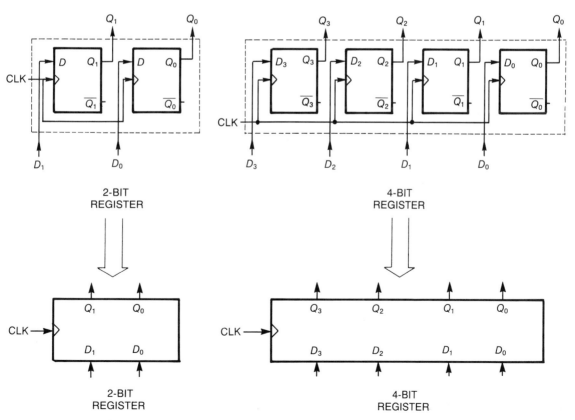

Figure A-24 Registers with different words.

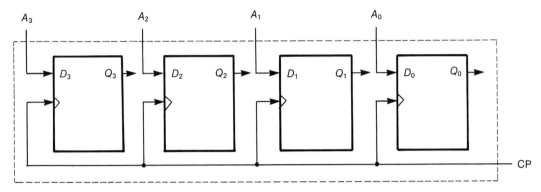

Figure A-25 Broadside loading a word into a register.

Broadside Loading

A *4-bit* register is shown in Figure A-25 (its word size is a nibble). The method of loading the register in Figure A-25 is sometimes referred to as *parallel* or broadside loading. Note that it takes only one clock pulse to store the word into the register. The binary number present on the A_0 through A_3 lines will be stored in the register at the leading edge of the clock pulse.

Serial Loading

Another 4-bit register is shown in Figure A-26. This register is *serially* loaded. Since the register in Figure A-26 holds four bits, it will take four clock pulses to completely store the word. If the Clear input is first activated, the contents of the register will be 0000. During the next four clock pulses, the condition of the D input will be transferred to the register. If the D input stays true (as in Figure A-26), all flip-flops in the register become true within four clock pulses.

The differences between broadside and serial loading are that it requires more clock pulses but less wires for a serial load, and only one clock pulse but more wires for a broadside load.

Parallel-in–Serial-out

Figure A-27 shows a parallel-in–serial-out shift register. When the *Read* input is true, the contents of the parallel input lines are stored in the register. When the *shift* line is true, the contents of the register are shifted out through the D output. One bit is shifted out at the leading edge of each clock pulse.

One use of a parallel-in–serial-out shift register in digital communications is when information must be stored quickly but transmitted out over telephone lines. The telephone line requires the serial output.

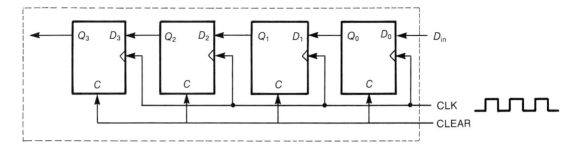

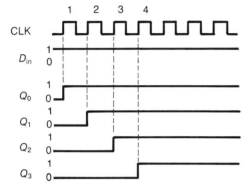

Figure A-26 Serially loading a word into a register.

Serial-in–Parallel-Out

A *serial-in–parallel-out* shift register is shown in Figure A-28. This arrangement could be used for receiving data from telephone lines and then transferring the complete *word* into a device such as a computer or a printer.

When the *out* line is true, the contents of the register appear on the output lines. When the *shift* line is true, the condition of the *D* input is shifted into the register at the leading edge of each clock pulse.

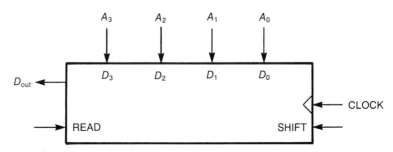

Figure A-27 Parallel-in–serial-out shift register.

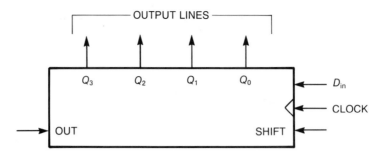

Figure A-28 Serial-in—parallel-out shift register.

Binary Counter

A *binary counter* is a group of flip-flops arranged to present a binary count for each clock pulse. See Figure A-29. This binary counter is a 4-bit binary counter. Note that when the count reaches 1111_2 (15_{10}), it "resets" itself to 0000_2. Binary counters serve useful purposes in instrumentation used for communications, such as frequency counters, which are presented in the next section.

Conclusion

This section presented the marriage of logic gates and flip-flops to form some very useful circuits called registers. In your work in communications, you will see many applications for registers.

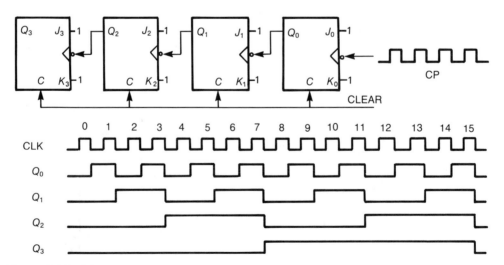

Figure A-29 A binary counter.

A-4 Review Questions

1. Define a register.
2. Define a register word.
3. Define bit, nibble, byte.
4. Explain broadside loading. What is another name for it?
5. Explain serial loading.
6. What are the advantages and disadvantages of broadside loading? serial loading?
7. Give an example of a parallel-in–serial-out register.
8. Give an example of a serial-in–parallel-out register.
9. What is an application of the registers used in problems 7 and 8?
10. Describe the operation of a binary counter.

TROUBLESHOOTING AND INSTRUMENTATION

Frequency Measurement

Frequency measurement is necessary in electronic communications. Due to frequency-division multiplexing and the necessary rigorous FCC requirements that transmitters stay on their assigned frequency allocations, frequency measurement becomes very important.

Many methods are used to measure radio frequencies. The method shown here uses the information presented in this chapter. Frequency counters are commonplace in the communications industry.

Frequency Counter

The block diagram of a frequency counter is shown in Figure A-30. The more precise the one-shot is, the more accurate the count is. A *one-shot* is a flip-flop that stays true for a given amount of time and then returns to false. Since the one-shot in Figure A-30 stays true for 1 s, the binary counter will count for the number of cycles of the unknown frequency that occur in 1 s. For the frequency counter shown in the figure, this is only good

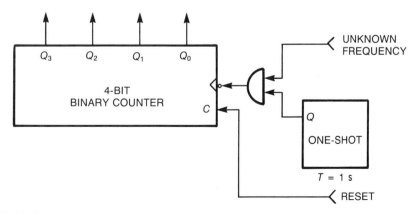

Figure A-30 Block diagram of a frequency counter.

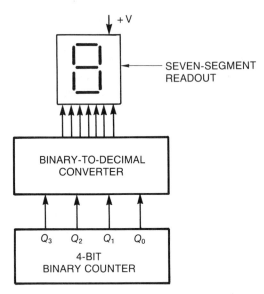

Figure A-31 Using binary-to-decimal converters.

for frequencies less than 15 Hz. The one-shot could be designed to remain true for 1 ms, and thus the number appearing in the binary counter would be the number of cycles of the unknown waveform that occur in 1 ms. In this case, the highest frequency measured would be 15 kHz.

Practical frequency counters use a *binary-to-decimal* converter to display the output count in decimal form. This arrangement is shown in Figure A-31. Observe that the output is now displayed using *seven-segment* readouts. These readouts are commonly found in communication equipment. Many radios use seven-segment readouts to indicate the selected station.

Seven-Segment Readout

Figure A-32 shows the details of a seven-segment readout. The readout is usually connected in series with a resistor value recommended by the manufacturer. This resistor limits the current in each segment to a safe value. Each segment is nothing more than a *light-emitting diode* (LED). When current flows in the LED, it emits energy in the form of light.

Conclusion

In this section, you saw a practical application of a binary counter. You will find that many other forms of digital electronics are now being used in electronic communications, from satellite communications to digital audio recordings. The information in this chapter

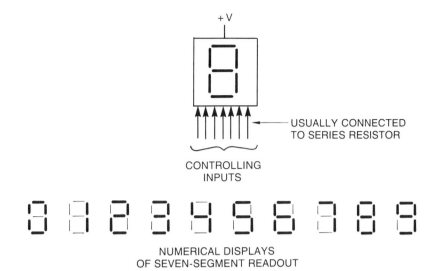

Figure A-32 Typical seven-segment readout.

has wider applications to electronic communications than ever before. Your knowledge will pay great dividends in the future.

A-5 Review Questions

1. Why is frequency measurement important?
2. Describe the main sections of a frequency counter.
3. Name a main factor that determines the accuracy of a frequency counter.
4. Explain the purpose of a binary-to-decimal converter.
5. Describe the details of a seven-segment readout.
6. What is an LED?

APPENDIX PROBLEMS

(Answers to the odd-numbered problems appear at the end of the text.)

TRUE/FALSE

Answer the following questions true or false.

1. Binary numbers are used to represent numbers in electronic circuits that have two states.
2. The decimal number system uses 10 different symbols to represent numbers.
3. The binary number system has two symbols: 1 and 2.
4. Using cups to represent binary numbers requires at least four cups to represent the decimal number 15.
5. The largest binary number that can be represented with three cups is 8.

MULTIPLE CHOICE

Answer the following questions by selecting the most correct answer.

6. The decimal number that represents the binary number 1101 is:
   (A) 5. (B) 10. (C) 12. (D) none of these.
7. The binary number that represents the decimal number 14 is:
   (A) 1011. (B) 1110. (C) 1010. (D) none of these.
8. The numbers 0011_2, 0100_2, and 0101_2 are represented by:
   (A) 5_{10}, 4_{10}, and 3_{10}.
   (B) 1_{10}, 2_{10}, and 3_{10}.
   (C) 3_{10}, 4_{10}, and 5_{10}.
   (D) none of these.
9. The numbers 10_{10}, 11_{10}, and 12_{10} are represented by:
   (A) 10_2, 11_2, and 12_2.
   (B) 10_2, 11_2, and 100_2.
   (C) 1010_2, 1011_2, and 1100_2.
   (D) none of these.
10. The largest decimal number that can be represented by five cups is:
   (A) 11111_2. (B) 31_{10}. (C) 16_{10}. (D) none of these.

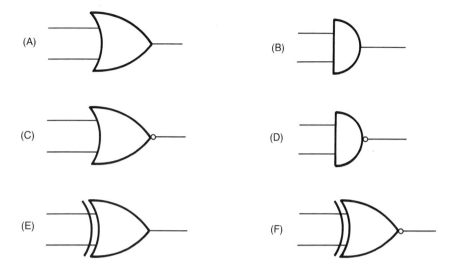

Figure A-33

MATCHING
Match each logic circuit in Figure A-33 to its correct name.
11. AND gate
12. NAND gate
13. XOR gate
14. OR gate
15. NOR gate

FILL IN
Fill in the blanks with the most correct answer.
16. All flip-flops have two outputs called _____ and _____ .
17. When one output of a flip-flop is true, the other output is _____ .
18. When the Clear and Set inputs of a *C-S* F/F are both false, the output of the flip-flop will _____ change.
19. The square wave produced by an oscillator used to control the action of flip-flops is called a _____ pulse.
20. If both inputs to a _____ flip-flop are true, then the flip-flop will toggle on each clock pulse.

OPEN ENDED

Answer the following questions as indicated.

21. What will be the contents of the register in Figure A-34 at the trailing edge of the next clock pulse?

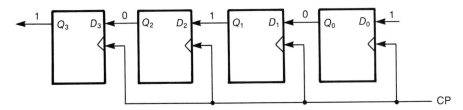

Figure A-34

22. What will be the contents of the register in Figure A-35 after the second clock pulse?

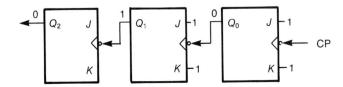

Figure A-35

23. Figure A-36 shows two registers. The contents of register A are to be transferred into register B. How many clock pulses will this take?

24. If the registers in Figure A-36 start in the conditions shown, what will be the contents of each register after three clock pulses?

25. Draw the logic diagram of a circuit that will give the period of an unknown waveform.

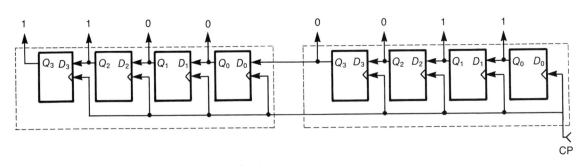

Figure A-36

APPENDIX B
Formulas

EQ.	TO DETERMINE	PG.
	Capacitor Current:	
(1-1)	$i_C = C \dfrac{\Delta v}{\Delta t}$	6
	Capacitive Reactance:	
(1-2)	$X_C = \dfrac{1}{2\pi f C}$	6
	Inductor Voltage:	
(1-3)	$v_L = L \dfrac{\Delta i}{\Delta t}$	7
	Inductive Reactance:	
(1-4)	$X_L = 2\pi f L$	7
	Impedance (Series Circuit):	
(1-5)	$Z = \sqrt{R^2 + (X_L - X_C)^2}$	9
	Phase Angle (Series Circuit):	
(1-6)	$\theta = \arctan \dfrac{X_T}{R}$	10
	Total Current (Parallel Circuit):	
(1-7)	$I_T = \sqrt{I_R^2 + (I_C - I_L)^2}$	11
	Resonant Frequency:	
(1-8)	$f_r = \dfrac{1}{2\pi\sqrt{LC}}$	13
	Harmonic Voltage Content of a Square Wave:	
(3-1)	$V_N = \dfrac{2A}{n\pi}$ (n is odd)	60

EQ.	TO DETERMINE	PG.
	Fourier Series:	
(3-2)	$v = V_0 + V_1 \sin(\omega t + \phi_1)$ $+ V_2 \sin(2\omega t + \phi_2)$ $+ V_3 \sin(3\,\omega t + \phi_3) + \ldots$ $+ V_N \sin(N\omega t + \phi_N)$	61
	Modulation Factor:	
(4-1)	$m = \dfrac{B}{A}$	74
	Percent Modulation:	
(4-2)	$m_p = \dfrac{B}{A} \times 100\%$	76
(4-3)	$m_p = \dfrac{V_{\max} - V_{\min}}{V_{\max} + V_{\min}} \times 100\%$	78
	Upper sideband:	
(4-4)	$f_{USB} = f_C + f_M$	80
	Lower sideband:	
(4-5)	$f_{LSB} = f_C - f_M$	80
	Bandwidth:	
(4-6)	$BW = f_{USB} - f_{LSB}$	81
(4-7)	$BW = 2f_M$	81
	Voltage Gain:	
(4-8)	$A_V = \dfrac{V_{out}}{V_{in}}$	83

EQ.	TO DETERMINE	PG.
	Current Gain:	
(4-9)	$A_I = \dfrac{i_{out}}{i_{in}}$	83
	Power Gain:	
(4-10)	$A_P = \dfrac{P_{out}}{P_{in}}$	83
	Signal-to-Noise Ratio:	
(4-11)	$SNR = \dfrac{S}{N}$	84
	Noise Figure:	
(4-12)	$NF = \dfrac{S_{in}/N_{in}}{S_{out}/N_{out}}$	85
(4-13)	$NF(dB) = 10 \log (NF)$	85
	Total Gain:	
(4-14)	$A_T = A_1 A_2 A_3 \cdots A_N$	87
	Decibel Power Gain:	
(4-15)	$A_P(dB) = 10 \log(A_P)$	87
	Total Gain in dB:	
(4-16)	$A_T(dB) = A_1(dB) + A_2(dB) + A_3(dB) + \cdots + A_N(dB)$	87
	dBm:	
(4-17)	$P(dBm) = 10 \log \left(\dfrac{P}{1 \text{ mW}} \right)$	89
	Decibel Voltage Gain:	
(4-18)	$A_{V(dB)} = 20 \log(A_V)$	90
	Bandwidth (Resonant Circuit):	
(4-19)	$BW = f_U - f_L$	91

EQ.	TO DETERMINE	PG.
	Q:	
(4-20)	$Q = \dfrac{f_r}{BW}$	92
	Circuit Q:	
(4-21)	$Q = \dfrac{X_L}{R_L}$	93
	Bandwidth:	
(4-22)	$BW = \dfrac{R_L f_r}{X_L}$	93
	Local Oscillator Frequency:	
(4-23)	$f_{LO} = f_R + f_{IF}$	97
	Image Frequency:	
(4-24)	$f_I = 2f_{IF} + f_R$	97
	Power Content in Each Sideband:	
(5-1)	$P_{SB} = \dfrac{m^2 P_C}{4}$	112
	Total Power of Transmitted AM Wave:	
(5-2)	$P_T = \dfrac{m^2 P_C}{4} + \dfrac{m^2 P_C}{4} + P_C$	112
(5-3)	$P_T = P_C \left(1 + \dfrac{m^2}{2} \right)$	113
	Number of Stations (Standard AM):	
(5-4)	$N_S = \dfrac{BW_S}{2f_M}$	115
	Number of Stations (SSB):	
(5-5)	$N_S = \dfrac{BW_S}{f_M}$	115

EQ.	TO DETERMINE	PG.
	Frequency of Reconstructed Audio from SSB with no Carrier:	
(5-6)	$f_R = f_{USB} - f_{LSB}$	122
	Carrier Suppression:	
(5-7)	$S_C dB = 20 \log\left(\dfrac{V_{p-p}}{V_R}\right)$	135
	Carrier Swing of FM Signal:	
(6-1)	$f_{CS} = 2f_D$	145
	Upper and Lower Frequency Reached by FM Carrier:	
(6-2)	$f_H = f_C + f_D$	149
(6-3)	$f_L = f_C - f_D$	149
	FM Modulation Index:	
(6-4)	$m_I = \dfrac{f_D}{f_M}$	151
	Deviation Ratio:	
(6-5)	$R_D = \dfrac{f_{D(max)}}{f_{M(max)}}$	155
	Percent Modulation:	
(6-6)	$M_{FM} = \dfrac{f_{D(actual)}}{f_{D(max)}} \times 100\%$	155
	Bandwidth of Narrow Band FM:	
(6-7)	$BW_{FM(narrow)} = 2f_M$	156
	Frequency Deviation from Frequency Multiplication:	
(6-8)	$f_{D(out)} = N_f f_{D(MA)}$	160

EQ.	TO DETERMINE	PG.
	Scanning Frequency:	
(7-1)	$f_S = L_T F_T$	184
	Minimum Sampling Frequency:	
(8-1)	$f_s = 2f_{N(max)}$	218
	Bits per Second:	
(9-1)	$BPS = N(\text{baud rate})$	280
	Transmission Time:	
(10-1)	$t_s = \dfrac{l}{vel_s}$	293
	Characteristic Impedance of Transmission Line:	
(10-2)	$Z_0 = 276 \log_{10}\left(\dfrac{d}{r}\right)$	296
	Deriving Characteristic Impedance:	
(10-3)	$Z_0 = \sqrt{\dfrac{L}{C}}$	298
	Voltage Standing-Wave Ratio:	
(10-4)	$VSWR = V_{rms\ max} : V_{rms\ min}$	302
	Current Standing-Wave Ratio:	
(10-5)	$ISWR = I_{rms\ max} : I_{rms\ min}$	302
	Standing-Wave Ratio:	
(10-6A)	$SWR = Z_L : Z_0$	303
(10-6B)	$SWR = Z_0 : Z_L$	303
	Reflection Coefficient:	
(10-7)	$K_{ref} = \dfrac{V_{ref}}{V_{inc}}$	304

EQ.	TO DETERMINE	PG.
	Relationship between SWR and K_{ref}:	
(10-8)	$\text{SWR} = \dfrac{K_{ref} + 1}{1 - K_{ref}}$	304
	Reflection Coefficient and Characteristic Impedance:	
(10-9)	$K_{ref} = \left\lvert \dfrac{Z_L - Z_0}{Z_L + Z_0} \right\rvert$	304
	Reflection Coefficient and Power Ratio:	
(10-10)	$K_{ref}^2 = \dfrac{P_{ref}}{P_{inc}}$	306
	Percent of Reflected Power:	
(10-11)	$\%P_{ref} = K_{ref}^2 \times 100\%$	306
	Percent of Absorbed Power:	
(10-12)	$\%P_{abs} = 100 - \%P_{ref}$	306
	Wavelength:	
(10-13)	$\lambda = \dfrac{\text{Vel}}{f}$	309
	Frequency of Given Wavelength:	
(10-14)	$f = \dfrac{\text{Vel}}{\lambda}$	309
	Antenna Radiation Resistance:	
(10-15)	$R_{rad} = \dfrac{P}{I^2}$	313
	Antenna Q:	
(10-16)	$Q = \dfrac{f_0}{\text{BW}}$	317
	Antenna Gain:	
(10-17)	$A(\text{dB}) = 10 \log_{10}\left(\dfrac{P_2}{P_1}\right)$	323

EQ.	TO DETERMINE	PG.
	Receiving Antenna Gain:	
(10-18)	$A(\text{dB}) = 20 \log_{10}\left(\dfrac{V_2}{V_1}\right)$	324
	Maximum Distance Between Antennas:	
(10-19)	$d = \sqrt{2h_{trans}} + \sqrt{2h_{rec}}$	328
	Design Equations for a Resistive Pad:	
(10-20)	$R_1 = Z_1 \sqrt{1 - \dfrac{Z_2}{Z_1}}$	333
(10-21)	$R_2 = \dfrac{Z_2}{\sqrt{1 - Z_2/Z_1}}$	333
	Stub Matching Transformer:	
(10-22)	$Z_T = \dfrac{Z_0^2}{Z_L}$	337
	Cutoff Wavelength:	
(11-1)	$\lambda_0 = \dfrac{2a}{m}$	349
	Guide Wavelength:	
(11-2)	$\lambda_G = \dfrac{\lambda}{\sqrt{1 - \left(\dfrac{\lambda}{\lambda_0}\right)^2}}$	350
	Angle of Incidence and Angle of Reflection:	
(12-1)	$\alpha_{inc} = \alpha_{ref}$	381
	Snell's Law:	
(12-2)	$\dfrac{\sin \alpha_{inc}}{\sin \alpha_{refr}} = \dfrac{\text{Vel}_1}{\text{Vel}_2}$	382

EQ.	TO DETERMINE	PG.	EQ.	TO DETERMINE	PG.
	Index of Refraction:			*555 Output Frequency:*	
(12-3)	$n = \dfrac{\sin \alpha_{inc}}{\sin \alpha_{refr}}$	384	(13-2)	$f = \dfrac{1.44}{(R_1 + 2R_2)C}$	433
	Critical Angle:			*555 Duty Cycle:*	
(12-4)	$\sin \alpha_c = \dfrac{n_2}{n_1}$	386	(13-3)	$D = \dfrac{R_1 + R_2}{R_1 + 2R_2} \times 100\%$	434
	Numerical Aperture:			*Gain of Inverting Op Amp:*	
(12-5)	$NA = \sin \phi$	397	(13-4)	$A_V = -\dfrac{R_f}{R_{in}}$	436
	Numerical Aperture in Terms of Index of Refraction:			*Gain of Noninverting Op Amp:*	
(12-6)	$\sin \phi = \dfrac{\sqrt{n_1^2 - n_2^2}}{n_0}$	398	(13-5)	$A_V = \dfrac{R_f + R_{in}}{R_{in}} = \dfrac{R_f}{R_{in}} + 1$	437
	Power:			*Phase Detector Gain:*	
(12-7)	$P = \dfrac{E}{t}$	402	(13-6)	$A_\phi = \dfrac{V_\phi}{\theta_{rad}}$	439
	Irradiance:			*Phase Detector Voltage Output:*	
(12-8)	$IRD = \dfrac{P}{A}$	402	(13-7)	$V_\phi = A_\phi \theta_{rad}$	440
	Radiance:			*Voltage to Frequency Conversion of VCO:*	
(12-9)	$RA = \dfrac{P}{\omega A_{ap}}$	403	(13-8)	$V_f = \dfrac{f_2 - f_1}{V_2 - V_1}$	441
	Solid Angle:			*PLL Cutoff Frequency (NE565 PLL):*	
(12-10)	$\omega = \dfrac{A}{r^2}$	403	(13-9)	$f_c = \dfrac{1}{2\pi R_{int}C_2}$	442
	Energy Emitted:			*Barkhausen Criterion:*	
(12-11)	$E = hf$	408	(15-1)	$A_V B_V = 1$	498
	Degrees to Radians:			*Armstrong Oscillator Frequency:*	
(13-1)	$\theta_{rad} = \left(\dfrac{\pi}{180°}\right)\theta$	428	(15-2)	$f_r = \dfrac{1}{2\pi\sqrt{LC}}$	502

EQ.	TO DETERMINE	PG.	EQ.	TO DETERMINE	PG.
	Series Regulator Output Voltage (Noninverting Op Amp):			*Area under Pulse:*	
			(17-4)	$A = (PW)V_P$	594
(17-2)	$A_V = 1 + \dfrac{R_1}{R_2}$	584		*Duty Cycle:*	
			(17-5)	$D = \dfrac{PW}{T}$	595
	Average Value of Pulse Train:			*DC Output Voltage of Pulse Train:*	
(17-3)	$P_{\text{avg}} = \dfrac{A}{T}$	594	(17-6)	$V_{\text{out}} = DV_{\text{in}}$	597

APPENDIX C
Answers

Review Questions Chapter 1

1-2:

1] Voltage = Resistance × Current. 2] Current decreases. Current increases. Total resistance increases. Total resistance decreases. 3] Current decreases. Current increases. Resistance increases. Resistance decreases. 4] 2.4 mA, 28.8 mW, 28.8 mW. 5] 1.26 mA, 1.26 V, 4.17 V, 6.57 V, 15.16 mW. 6] 668 Ω, 12 mA, 3.6 mA, 2.3 mA, 216 mW. 7] 16.97 V, 83 μs, 1.9 W. 8] 1.8 kV. 9] 18 A, 270 A, 32.4 kW, 32.4 kW.

1-3:

1] Frequency-dependent opposition to current flow. $X_L = 2\pi fL$, $X_C = 1/2\pi fC$. 2] (A) Current increases, current decreases. (B) Current increases, current decreases. 3] (A) Voltage increases, voltage decreases. (B) Voltage increases, voltage decreases. 4] (A) 1 MΩ (B) 1 kΩ. 5] (A) 33.9 μA (B) 33.9 mA. 6] (A) 0.75 Ω (B) 7.5 kΩ. 7] (A) 22.6 A (B) 2.26 mA. 8] 1.8 pF. 9] 1.7 H.

1-4:

1] Total opposition to current flow. 2] Impedance includes resistance. 3] (A) V leads I (B) I leads V. 4] (A), (B), (C) all in phase. 5] (A) I lags V (B) V lags I (C) In phase. 6] 1.77 kΩ, 47.3°. 7] 1 kΩ, 56°.

1-5:

1] That frequency where the impedance is minimum for a series RLC circuit or maximum for a parallel RLC circuit. 2] Minimum impedance for series, maximum impedance for parallel. 3] At resonance. 4] No. Current in series circuit is everywhere the same. 5] At resonance. 6] No. Voltage in parallel circuit is everywhere the same. 7] 650 kHz. 8] About 0.022 μH, in parallel. 9] About 0.02 pF, in series.

1-6:

1] Act of locating and repairing a fault. The use of equipment to measure electrical quantities. 2] Complete failures, poor system performance, tampered equipment, intermittent, massive traumas. 3] Complete failure. Intermittent. 4] An inconsistent electrical problem. A problem that changes with temperature.

Chapter Problems (odd-numbered)

1] (A) 0.5 A (B) 6 mA (C) 0.5 mA. 3] (A) 6 Ω (B) 5 kΩ (C) 250 MΩ. 5] (A) 20 V (B) 59.4 V (C) 9.6 V. 7] (A) 250 W (B) 62 mW (C) 36 μW. 9] (A) 106 Ω (B) 21.2 kΩ. 11] (A) 340 mA (B) 5.2 mA. 13] (A) V_{R1} = 7.48 V, V_{R2} = 11.22 V, V_{R3} = 17.34 V (B) V_{R1} = 27 V, V_{R2} = 44.6 V, V_{R3} = 38.4 V. 15] (A) 12 W (B) 571 mW. 17] (A) 5 Ω (B) 660 Ω. 19] (A) I_{R1} = 1.85 mA; I_{R2} = 1.39 mA; I_{R3} = 426 μA (B) I_{R1} = 12 mA; I_{R2} = 5 mA; I_{R3} = 7 mA. 21] 1.43 kΩ. 23] (A) V_{TH} = 12 V, R_{TH} = 660 Ω. (B) V_{TH} = −15 V, R_{TH} = 3.3 kΩ. 25] (A) I_N = 2 A, R_N = 52 Ω, (B) I_N = 5 mA, R_N = 1 kΩ. 27] (A) V_{TH} = 7.2 V, R_{TH} = 2 kΩ, P_{max} = 6.5 mW (B) P_{max} = 13 W. 29] Figure A: (A) 3 V (B) 6 V (C) 2.1 V (D) 125 Hz (E) 8 ms. Figure B: (A) 0.4 V (B) 0.8 V (C) 0.29 V (D) 1.25 kHz (E) 0.8 ms. 31] (A) 2.37 W (B) 20 mW. 33] (A) 100 V, 0.9 A (B) 4.16 V, 0.416 mA. 35] (A) 5.5 Ω (B) 120 kΩ. 37] (A) 30.9 Ω (B) 64.6 Ω. 39] (A) 18.5 Ω (B) 138.9 Ω. 41] (A) 14.3° (B) −33.7°. 43] (A) 2.59 W (B) 430 mW. 45] (A) 1.2 kΩ (B) 44.4 kΩ. 47] (A) −65.3° (B) 74°. 49] (A) 795.7 Hz (B) 205 kHz. 51] (A) 0.56 A (B) 80 mA. 53] (A) 57 mA (B) 2.08 mA. 55] (A) ideally 0 A (B) ideally 0 A.

Review Questions Chapter 2

2-1:

1] The human voice. 2] Advantage: Ease of use. Disadvantages: Short distances, moves slowly, not very private at long distances. 3] Smoke signals, tribal drums, shiny object. 4] Radio, television, recordings, and long-distance two-way communications.

2-2:

1] Advantage: It could cover very long distances quickly. Major disadvantages: Needed a code and required connecting wires. 2] To energize the receiver. 3] A series circuit consisting of a voltage source, switch (transmitter), and relay (receiver).

2-3:

1] Advantage: Could cover very long distances quickly and did not need a special code. Disadvantage: Transmitter and receiver had to be connected with wires. 2] To operate the reproducer at a distance from voice patterns. 3] A container filled with carbon granules that change their resistance according to air pressure. It is called a microphone. 4] An electromagnet that causes a metal disk to move in step with the changes of electric current in the electromagnet. It is called a headset or speaker. 5] A reproducer.

2-4:

1] Electrical energy that can travel through space. 2] It doesn't require a medium. It travels at the speed of light in a vacuum. 3] Removing air from a glass jar with a bell in it. You can still see the bell (light is still transmitted), but you can no longer hear the bell (sound is no longer being transmitted). 4] The range of usable electromagnetic radiation. 5] The range of frequencies that produce a visual stimulus to humans. 6] Its frequency.

2-5:

1] Electromagnetic radiation. 2] Electromagnetic radiation that contains some pattern of the intelligence being transmitted. 3] The light of the flashlight. 4] The intensity of the light. 5] To change or modify some measurable property of a carrier. 6] The amplitude (intensity), frequency, or phase. 7] Amplitude modulation. The amplitude of the carrier is changed by the modulating signal. 8] Frequency modulation. The frequency of the carrier is changed by the modulating signal. 9] Phase modulation. The phase of the carrier is changed by the modulating signal. 10] To restore the modulating signal by processing the modulated carrier signal.

2-6:

1] By assigning different frequencies to different stations. 2] Having transmitters all transmitting at the same time but at different frequencies. 3] Using different colors of light. 4] The tuner.

2-7:

1] One of the important things is that the measurement will not damage the system. 2] Record them. 3] Compare the measurement to some standard. 4] Troubleshooting, reporting the discrepancy, giving the system to someone else to repair.

Chapter Problems (odd-numbered)

1] True. 3] True. 5] True. 7] B. 9] D. 11] A. 13] B. 15] A. 17] Medium. 19] Frequencies. 21] Amplitude, frequency, and phase. 23] Using the advantages of electromagnetic energy to overcome the limitations of information sources by causing the information source to change some measurable quantity of the carrier. The carrier is necessary to overcome information source disadvantages. 25] Having several transmitters transmit at the same time at different frequencies. By the frequency. The tuner in the receiver selects the signal.

Review Questions Chapter 3

3-1:

1] Any device that converts electromagnetic radiation into electrical signals. 2] Converts electromagnetic radiation into electric currents. 3] An antenna will convert any electromagnetic radiation into electric currents.

3-2:

1] An electrical circuit that selects one frequency and rejects all others. 2] To select one radio station and reject all others. 3] The resonant frequency of the tuner. 4] By changing the amount of capacitance or inductance in the tuner.

3-3:

1] To convert electrical signals into something understandable. 2] An electromagnetic that effects the movement of a metal disk. 3] The vibrations of the metal disk in the headset. 4] It can't keep up with the rapid changes of the carrier.

3-4:

1] An electrical device or circuit that detects the information imposed on a carrier wave. 2] A device that allows current to flow in only one direction. It can be used as a detector. 3] It is rectified or converted into pulsating DC. 4] A capacitor.

3-5:

1] Four. 2] Antenna, tuner, detector, reproducer. 3] Converts the signal into electrical currents, selects one station and rejects all others, detects the information on the carrier wave, converts the electrical energy into sound.

3-6:

1] An electronic instrument that shows the amplitude and frequencies of the sine wave components in a waveform. 2] A variable tuner connected to an amplitude detector. 3] As vertical lines on the face of a CRT. The vertical axis is usually in volts, and the horizontal axis is in hertz. 4] The frequency and amplitude of the harmonic content are displayed. 5] Contains components of a pure sine wave. 6] It predicts the sine wave components of a periodic waveform. 7] A square wave.

3-7:

1] A functional way of representing a complex electronic system. 2] By the detail. 3] When circuit detail is not necessary. When circuit detail is necessary. 4], 5], and 6] Refer to Table 3-1.

Chapter Problems (odd-numbered)

1] False. 3] True. 5] True. 7] B. 9] C. 11] A. 13] B. 15] E. 17] Spectrum. 19] Sawtooth. 21] 2 MHz at 5 mV, 4.5 MHz at 6.5 mV. 23] $f_1 = 100$ Hz at 1.91 V, $f_2 = 200$ Hz at 0.95 V, $f_3 = 300$ Hz at 0.64 V. 25] See Figure A-37.

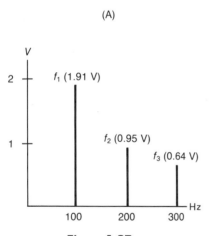

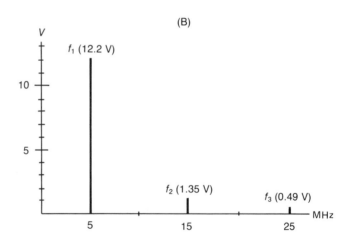

Figure A-37

Review Questions Chapter 4

4-1:

1] Two sine waves, with the higher frequency "riding" on the lower frequency. 2] The resulting signal is different from the original signal. In order to produce a resulting signal that is different from the original signals. 3] A carrier and modulating frequency are combined with the diode to produce pulsating DC. This waveform is amplitude modulated and restored to an AM waveform by the flywheel effect of a parallel resonant circuit. 4] To change or modify.

4-2:

1] An AM wave where the amplitude does not reach zero. An AM wave where the amplitude just reaches zero. An AM wave where the amplitude reaches zero and stays zero for a measurable time. 2] The ratio of the peak value of the modulating signal to the peak value of the unmodulated carrier. The percent modulation is the modulation factor multiplied by 100%. 3] The intelligence of the modulating signal becomes distorted. 4] Take the difference of the maximum peak value and the minimum peak value and divide this by the sum of these measured values.

4-3:

1] The original two frequencies along with the sum and the difference. 2] The sum and difference frequencies. The lower and upper sidebands are the difference and sum frequencies, respectively. 3] Many upper and lower sidebands resulting from the large harmonic content of a square wave. 4] The difference between the upper and lower sidebands. It is determined by the tuned circuits in the transmitter. 5] The highest-frequency component of the modulating signal. 6] 10 kHz.

4-4:

1] A circuit that increases the amplitude of a signal. A transistor amplifier. 2] A ratio of the output signal to the input signal. Gain can be less than 1. 3] They are all ratios of an electrical output quantity to an electrical input quantity. The units of each quantity. 4] Any undesired signal. Weather conditions, industry, electrical components, and devices. 5] The ratio of the amount of signal to the amount of noise. The ratio of the input signal-to-noise-ratio to the output

signal-to-noise-ratio. 6] One or unity. This means that the amplifier did not introduce any noise to the signal.

4-5:

1] A measure of the weakest signal that can be put into a useful form by the receiver. 2] By using RF amplifiers and cascading other amplifiers. 3] Connecting one amplifier to another in such a way that the signal is amplified first by one amplifier, then by the other, and so on. This greatly increases the strength of the received signal. 4] By multiplying the gains of each amplifier together. 5] You can add the gains of each amplifier and need not multiply. 6] The amount of power in dB referenced to 1 milliwatt. To simplify power measurements.

4-6:

1] The ability of a receiver to select one station and reject all others. 2] The bandwidth of the tuned circuits. 3] If the bandwidth of the receiver is less than that of the received signal, then not all of the sidebands will be received. If the receiver bandwidth is too large, then the receiver will lose selectivity. 4] The loss of information because the receiver bandwidth is too narrow. 5] It could receive more than one station at a time. 6] Selectivity is the ability of a receiver to select one station and reject all others. Sensitivity is the ability to receive weak signals.

4-7:

1] It comes from the word "heterodyne," which means to mix. The "super" part of the word was used for commercial purposes to help sell the concept. 2] A local oscillator and the received signal feed into a mixer. The resultant difference output is amplified by a fixed, tuned intermediate amplifier(s); the output signal(s) of the amplifier(s) is (are) detected, and the resultant detected signal is amplified and operates the transducer (usually a loudspeaker). 3] Automatic gain control is used to help maintain a constant volume while tuning across the band to receive weak or strong signals. 4] Excellent selectivity and sensitivity. 5] Local oscillator is tuned to the IF frequency above the incoming signal. The mixer combines this with the received signal to produce a fixed output signal. The fixed, tuned IF amplifiers always amplify the same frequency. 6] Amplifies the difference frequency between the local oscillator and received signal. 7] Twice the IF frequency plus the frequency of the received signal. The image frequency can be reduced or eliminated by RF amplifiers.

4-8:

1], 2] See Table 4-5. 3] Observing the output of each stage. Observing the output of the IF amplifier. The oscilloscope can be used. 4] An instrument that produces an RF carrier signal and a modulating audio waveform. Usually the RF carrier frequency can be changed by the technician. 5] Injecting a signal of the proper frequency and amplitude (usually with a signal generator) to the input of a stage and observing the results on the transducer. Injecting an audio frequency to the input of the audio amplifier and listening for an audio tone from the speaker.

Chapter Problems (odd-numbered)

1] The result is two sine waves, with the higher frequency "riding" on the lower frequency. 3] The resultant signal is different from the original signal. Useful to produce a resultant signal different from the original signals. 5] The process where one frequency modifies another frequency, as in the process of amplitude modulation where a lower-frequency signal modifies or modulates the amplitude of the higher frequency or carrier. 7] (A). 9] 0.333. 11] (A) 33.3% (B) 100%. 13] See Figure A-38.

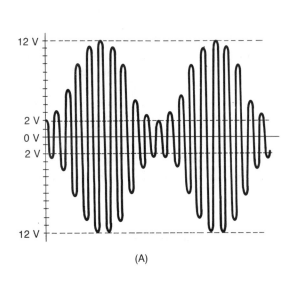

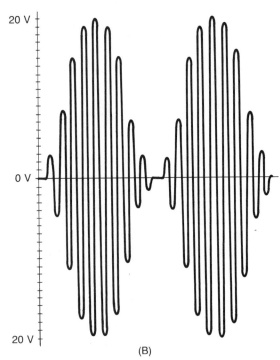

(A)

(B)

Figure A-38

15] (A) 43% (B) 75%. 17] 550 kHz, 538 kHz, 562 kHz, 12 kHz. 19] 24 kHz. 21] See Figure A-39.

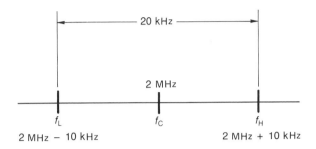

Figure A-39

23] (A) $A_V = 5$ (B) $A_P = 666.7$. 25] (A) 266 (B) 9 200. 27] 1.5. 29] 1 008. 31] 24.8 dB. 33] (A) Double power gain = 6 mW (B) Eight times the power gain = 24 mW (C) Half the power gain = 1.5 mW (D) 0.062 5 times the power gain = 0.188 mW. 35] (A) 14.8 dBm (B) 20.8 dBm (C) 8.8 dBm (D) 4.78 dBm. 37] (A) (B). 39] (A) 1.035 MHz (B) 1.255 MHz (C) 1.45 MHz. 41] (A) 1.570 MHz (B) 3.010 MHz (C) 2.46 MHz.

Review Questions Chapter 5

5-1:

1] Refer to Table 5-1. 2] (A) RF power amplifier (B) RF voltage amplifier/RF driver. 3] High-level modulation is more efficient, but it increases the system cost. 4] The oscillator. 5] The range of frequencies that they amplify. 6] To increase the oscillator frequency.

5-2:

1] In the sidebands. 2] No information is contained in the carrier. 3] When the modulation factor is 1. 4] One quarter. 5] They both contain the same amount of information. 6] The power content of the carrier is independent of the percent modulation. 7] Because the transmission bandwidth is less. 8] A range of unused frequencies between two transmitting frequencies. Guard bands use up frequency allocation space.

5-3:

1] There are no sidebands. 2] The reception of the station becoming weaker. It is caused by atmospheric

conditions. 3] (A) The sidebands, not the carrier, are transmitted. (B) Either the upper sideband or the lower sideband is transmitted in SSB, not both sidebands. (C) Part of the carrier and all of one sideband are suppressed in vestigial sideband, as opposed to all of the carrier and one sideband in SSB. 4] (A) Carrier and sidebands are transmitted. (B) Only sidebands are transmitted. (C) Only one sideband is transmitted. (D) Only one sideband and part of the carrier are transmitted. 5] Single-sideband transmission. The other resulting frequencies from the modulating process are removed. You would get unmodulated pulsating DC. 6] The receiver needs to reinsert the missing carrier. 7] The ability to have more transmitting stations within a given band of frequencies (such as TV transmission).

5-4:

1] The carrier is suppressed. 2] Suppresses the carrier and produces only the sidebands. 3] The envelope of a DSB transmission is twice the frequency of the modulating signal. 4] The difference between the upper and lower sideband is equal to twice the frequency of the modulating signal. 5] Reinsertion of the carrier in the receiver. 6] Bandwidths are the same.

5-5:

1] Either upper or lower. 2] In order to transmit low modulating frequencies. 3] Using a low-frequency RF signal to mix with the sidebands. This causes the sidebands to have a greater frequency spread from the carrier and makes the carrier easier to filter. 4] To provide a conversion frequency. 5] The sidebands are now further removed from the carrier, thus not as high a Q is needed in the filters. 6] Two filters. One before the converter and the other after.

5-6:

1] Visual and synchronizing. 2] A highpass filter. 3] Reduces transmission bandwidth, reduces power requirements, easy to detect the carrier. 4] The receiver must make up for the part of the carrier that is suppressed. 5] They are tuned to emphasize the visual frequencies above 750 kHz.

5-7:

1] SSB receivers must reinsert the carrier before detection can take place. 2] Stability. 3] Automatic frequency control. 4] A reference frequency used at the receiver for carrier reinsertion. 5] Beat frequency oscillator. A variable frequency oscillator that is referred to the pilot carrier. 6] Two mixers. The first produces a signal for the IF amplifier; the second reinserts the carrier (BFO).

5-8:

1] Wasted transmitter power. 2] Simulates the transmitting antenna but doesn't transmit any signals. This is used when making adjustments on the transmitter. 3] A coupling device that extracts a small amount of energy for measurement. 4] An unmodulated sine wave. 5] Amplitude changes will appear on the sine wave.

Chapter Problems (odd-numbered)

1] RF carrier oscillator, frequency multipliers, audio amplifier, RF power amplifier. 3] Input(s): RF oscillator and/or audio amplifier; output: antenna. 5] Having the carrier wave modulated at the power amplifier stage. 7] Not counting guard bands: 7. 9] Five. 11] (A) 7 (B) 15. 13] 303 W. 15] Carrier power is 4.44 W. Power in each sideband is 0.278 W. 17] The same power will be in the carrier. The total transmitted power is 7.5 W. 19] 66.3 W. 21] 82%. 23] 55%. 25] 72 W. 27] There is no carrier. All of the transmitted power is in the transmitted sidebands. 29] No carrier power, 6 kW in each sideband. Transmitting power is zero with no modulating signal. 31] Zero because there are no resultant sidebands. 33] Carrier power is 6.25 kW. Power in each sideband is 875 W. 35] A lowpass or highpass filter, depending on the sideband being transmitted. 37] Twice the

frequency. 39] In order to have a reference frequency for proper sideband detection. 41] Atmospheric conditions causing the phase relations of the sidebands to change. 43] (A) Standard AM (B) Suppressed carrier (DSB). 45] See Figure 5-14. 47] 500 kHz + f_M, 500 kHz − f_M. 49] 12 dB.

Review Questions Chapter 6

6-1:

1] Amplitude, frequency, and phase. 2] (A) Changes in frequency. (B) Changes in amplitude. 3] Can cause the amplitude to change. 4] By a circuit that removes the amplitude changes of the signal. 5] A circuit that removes amplitude changes. AM receivers cannot use them because the information is represented as amplitude changes.

6-2:

1] FM: carrier frequency is modulated. AM: carrier amplitude is modulated. 2] Changing the distance between the plates causes the frequency of an oscillator to change. By the changing air pressure caused by the voice wave. 3] Amplitude of modulating signal causes frequency to change; frequency of modulating signal affects rate of change. 4] How far the FM carrier is from its resting frequency. The frequency of the FM carrier without any modulation. 5] Carrier swing is twice the frequency deviation. By knowing the modulation index and the deviation of the FM wave. 6] How often the FM signal deviates from its resting frequency. 7] Deviation of the FM wave and frequency of the modulating signal.

6-3:

1] A range of frequencies where no transmission takes place. 25 kHz on either side of the transmitting signal. 2] Because of the number of sidebands produced by an FM signal. 3] FM bandwidth changes in a very complex way that depends on the modulating signal amplitude as well as its frequency. 4] The ratio of the maximum deviation of the FM carrier and the maximum modulating frequency. 5] Five. 1.67. 6] The ratio of the actual to the maximum allowable frequency deviation of the FM carrier expressed as a percent. 7] It contains a small enough modulation index so that the bandwidth is approximately equal to twice the value of the modulating frequency.

6-4:

1] Poor frequency stability. 2] To cause the carrier frequency to change. In an AM transmitter, the carrier amplitude was made to change. 3] A circuit that increases the input frequency. It allows the use of a lower-frequency oscillator. 4] Causing the output of one frequency multiplier to be fed into the input of another frequency multiplier. 5] It is the product of the amount of frequency multiplication and the frequency of the master oscillator. 6] It increases the modulation index for higher frequencies. This helps to reduce noise that causes frequency distortion.

6-5:

1] IF frequency, limiter, deemphasis network, AFC, wider bandwidth requirements. 2] Clips the peaks of the FM waveform, eliminating the amplitude changes caused by noise. The deemphasis network compensates for the effect of the preemphasis network used at the FM transmitter. 3] Automatic frequency control. Helps keep the local oscillator at a stable frequency. 4] How the modulating frequency affects the carrier. 5] PM can have greater efficiency, is more noise free, and requires a simpler transmitter.

6-6:

1] Stereo gives the listener a sense of direction. 2] Such a system is not economical and uses bandwidth. 3] Signal addition: combining two signals in phase. Signal subtraction: combining two signals that are 180° out of phase. 4] A stereo signal must be received on a monaural receiver as a monaural signal, and it must be received on a stereo receiver as a stereo signal. 5] The sum and the difference of the two channels are transmitted along with a pilot carrier. Only the sum of the two frequencies are reproduced by the monaural FM receiver. 6] Because a carrier is needed at the receiver to mix with the two channel frequencies. 19 kHz is used because it is easier to extract this subcarrier in the receiver. 7] By adding to get one channel and subtracting to get the other channel.

6-7:

1] An oscillator circuit whose output frequency can be rapidly changed (swept) by a changing voltage source (usually a sawtooth waveform). 2] Because of the instrument's ability to rapidly change its output frequency. 3] An oscillator whose output frequency

can be controlled by an external voltage. 4] Causes it to change frequency at a linear rate, return to its original frequency, and repeat the process. 5] Gives an output voltage where the amplitude is proportional to the frequency of the input. 6] Has an output voltage proportional to the difference between the two input voltages. 7] An automatic level control that helps maintain a constant output voltage for different frequencies. 8] To test the bandwidth of a circuit(s). An oscilloscope can replicate the frequency-response curve of the circuits under test.

Chapter Problems (odd-numbered)

1] FM, AM, PM. 3] Less noise, greater bandwidth. 5] FM: carrier frequency is modulated. AM: carrier amplitude is modulated. 7] To remove amplitude variations. 9] Two metal plates, one of which is flexible enough to respond to sound. 11] FM carrier frequency without modulation. How much the carrier frequency changes from its resting frequency. 13] (A) 30 kHz (B) 87.985 MHz − 88.015 MHz. 15] How rapidly the FM carrier changes its frequency. 17] The carrier deviation and modulating signal frequency. 19] 1.0. 21] (A) 40 kHz (B) 100 MHz (C) 20 kHz (D) 6.67. 23] 20%. 25] 40 kHz. 27] 1.0. 29] (A) 3.0 (B) 60 kHz. 31] Approximately 30 kHz. 33] (A) 2.5 (B) Approximately 100 kHz. 35] 4 kHz. 37] About 9 kHz. 39] f_C = 96 MHz, f_D = 12 kHz. 41] Used in FM transmitters to help reduce the effect of frequency distortion caused by noise. 43] To convert frequency changes to amplitude changes. 45] To help maintain a steady local oscillator frequency. 47] How the modulating signal affects the carrier. 49] For FM, a wider range of audio frequencies is presented to the listener. 51] Two signals in phase. Two signals 180° out of phase. 53] By adding signals to get one channel and subtracting the signals to get the other. 55] 19 kHz. 57] An instrument with a linear changing output frequency that produces a repetitive linear frequency change. 59] See Figure 6-29.

Review Questions Chapter 7

7-1:

1] No. It is sent in sequence and "traced" across the screen. 2] Sound, video, and synchronizing information. 3] The transmitting frequency assigned to a TV station. 6 MHz. In order to transmit all required information. 4] Channels 6 and 7. 5] See Figure 7-2. 6] Electrons pulled from electron gun by high voltage on picture tube; resultant electron beam strikes phosphor coating on face of picture tube causing it to glow. 7] By applying horizontal and vertical sawtooth waves to cause the electron beam to sweep across the CRT face. Two oscillators are required. 8] Synchronizes the frequency of the horizontal and vertical oscillators. Cuts off the electron beam during retrace. 9] Sync pulses riding on blanking pulses with video information.

7-2:

1] Must be fast enough to not cause flicker yet slow enough to fit within the bandwidth requirements. 2] Keep the bandwidth requirements low. 3] Transmitting two TV fields interwoven (interlaced) into one frame. Helps keep down bandwidth requirements. 4] A frame (525 lines every $\frac{1}{30}$ of a second) consists of two fields (262.5 lines every $\frac{1}{60}$ of a second). 5] The total lines per frame and the total frames per second. 6] Smallest amount of information that can be displayed. Number of lines on the screen. 7] Ratio of the screen width to its height $\left(\frac{4}{3}\right)$. The effect of system imperfections on the final received signal (75% unaffected). 8] 6-MHz bandwidth with picture carrier using vestigial sideband transmission; sound carrier 4.5 MHz above picture carrier with a 50-kHz bandwidth. 9] By using vestigial sideband transmission.

7-3:

1] UHF mixer, UHF oscillator, VHF RF amp, VHF oscillator, VHF mixer. 2] Sound (41.25 MHz) and picture (45.75 MHz) carriers. 3] To separate the vertical and horizontal sync pulses. 4] Two oscillators in the tuner: one covers VHF, the other UHF requirements. Horizontal oscillator at 15 750 kHz and vertical oscillator at 60 Hz. 5] Amplitude of video signal controls intensity of the picture.

7-4:

1] Compatibility with black-and-white receivers. Stay within 6-MHz bandwidth. Simulate a wide variety of colors. 2] Additive is mixing colors from light sources; subtractive is mixing colors from reflective light. 3] The three different transmitters would each transmit a different primary color. The system is impractical because of its cost and bandwidth requirements. 4]

A triad of three electron guns each exciting a different primary color phosphor on the screen. The color phosphor consists of the three primary additive colors: red, green, and blue. 5] Because each electron gun energizes only one color of phosphor (red, green, or blue). No. 6] To minimize interference with the rest of the composite signal. So the carrier frequency does not appear on the TV screen. 7] To synchronize the phase of the color reference oscillator. During the horizontal blanking pulse immediately following the horizontal sync pulse (the "back porch"). 8] Color signal amplifiers, burst separator and sync circuits, color reference oscillator, color demodulator circuits.

7-5:

1] Recommendations made to the FCC by the BTSC. 2] Multichannel TV sound allows the transmission of stereo sound. 3] Professional channels; another sound channel plus other options. 4] All capabilities of the system are utilized. 5] Transmit audio in a different language. 6] Communications between TV stations and their remote units.

7-6:

1] Video, sound, and control information. 2] VHS and Beta. 3] Because of the high-frequency recording requirements of a video signal; it isn't practical to make the recording tape move fast. 4] The vertical sync. 5] 30 Hz. Used to make sure the control signal is in step with the received video. Used to ensure a scanning speed of 30 Hz.

7-7:

1] A computer to process information, input devices to input information, output devices to record and display information. 2] Troubleshooting an amplifier. To convert the amplifier signals to a form that is useful to the computer. 3] Reduction of human error. Cost. 4] Convert analog information to digital information. 5] The amount of involvement of a human operator determines the level of CAT. The highest level of CAT is where there is no human intervention in any part of the repair process.

Chapter Problems (odd-numbered)

1] Sound, video, and sync. 3] Sound uses FM, video uses AM. 5] Commercial FM is between TV channels 6 and 7. 7] Received signal is split after detector; sound to sound section; video to picture section; sync to sync separator (separates vertical and horizontal sync.) 9] (A). Creates electron stream (B) Attracts electrons toward face of tube. 11] The phosphor coating on the picture tube face. 13] 525 (including those not seen during retraces). 15] Video information, horizontal sync pulse, horizontal blanking pulse, vertical sync, and blanking. 17] The greater the amplitude, the darker the picture. 19] Interlacing and persistence of the picture tube and human eye. 21] Two fields of video information are presented every $\frac{1}{60}$ of a second. The fields produce alternate scanning lines to form a complete picture. It is used to keep within the bandwidth requirements and prevent flicker. 23] $f_S = L_\tau F_\tau$. Scanning frequency equals the product of the total lines per frame and the total frames per second. 25] Horizontal retrace = 10.2 μs. Vertical retrace = 500–750 μs (depending on how many lines are actually transmitted). 27] The smallest amount of information that can be displayed on the screen. 252 (counting retrace). 29] Ratio of screen width to screen height is $\frac{4}{3}$. 31] Quality of the TV picture after imperfections. 75%. 33] 4.2 MHz. 35] The sound carrier is 4.5 MHz above the picture carrier. 37] 12 in the VHF band and 14–83 in the UHF band. 39] See Figure A-40.

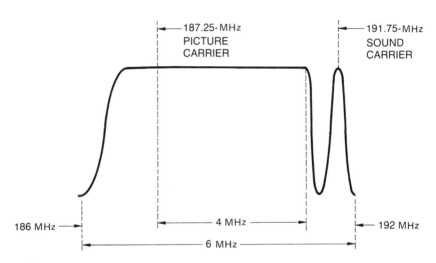

Figure A-40

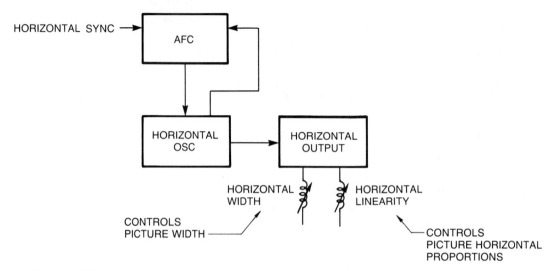

Figure A-41

41] Helps determine the sensitivity and selectivity. 43] IF amplifiers. 45] Automatic Gain Control. Controls the gain of the IF amplifiers to help maintain a constant signal strength inside the TV receiver. 47] It affects the difference between black and white on the picture tube. Controls the gain of the video amplifier. 49] Separates the vertical and horizontal sync pulses. Would cause problems in horizontal and vertical sync. 51] See Figure A-41.

53] Compatibility with black-and-white receivers. Within 6-MHz bandwidth. Simulate a wide variety of colors. 55] 30% red, 59% green, 11% blue. 57] Each gun is directed to its own color of phosphor; one at green, another at blue, and the third at red. Thus the names red, green, and blue gun. 59] The Y (white), R − Y, and B − Y signals. 61] Color subcarrier oscillator. The color sync burst. 63] The color signal. The vector sum of R − Y and B − Y signals. 65] By combining the R − Y and B − Y signals with the Y signal to get red and blue. This is then subtracted from the Y signal to reconstruct green. 67] Sound transmitted by the BTSC system. Stereo, sec-

ondary audio, and a professional channel form one of the many options. 69] By transmitting the sum and difference frequencies of the two stereo channels. This is the same kind of system used in stereo FM and makes both systems compatible with existing monaural sound receivers. 71] To record video, sound, and control information. 73] The TV vertical sync pulses. 75] The amount of human input needed. The higher the level, the less human input.

Review Questions Chapter 8

8-1:

1] Continuous monitoring is a constant observation; sampling is a periodic observation. 2] When both provide the same information. When the data change fast enough so that sampling misses information that would be obtained by continuous observation. 3] Decrease the sampling rate. Increase the frequency of the waveforms. Increase the response time of the recorders. 4] Separating different information by time. Separating different information by frequency. 5] Pulse amplitude, pulse width, pulse position, and pulse code modulation. 6] Information continuously varies some aspect of the pulse. Digital pulse information converts the information into a discrete code. 7] Pulse amplitude modulation (PAM).

8-2:

1] The process of taking a periodic sample and transmitting the samples. 2] It gives more time to transmit other kinds of information. 3] Sampling frequency of a pulse modulating system must be equal to or exceed twice the highest signal frequency to convey all information in original signal. 4] Distortion that occurs when sampling rate is less than that allowed by the sampling theorem. A moving picture of a wheel that appears to rotate the wrong way. 5] To ensure that a practical lowpass filter would pass the required frequency components. 6] Range of frequencies between the modulating signal and lowest-frequency component of sampling harmonic used. It is used to allow a practical lowpass filter to restore the modulating signal. 7] Similar to the action of an AM detector, where the capacitor's charging and discharging action restore the original modulating signal.

8-3:

1] The amplitude of the pulse is changed by the modulating frequency. 2] In dual polarity, the PAM signal goes above and below a zero reference. In single polarity, the signal never crosses the zero reference. A coupling capacitor connected to a resistive voltage divider. 3] The switch can be opened or closed by applying a small DC voltage. 4] Sampling different information in time. Used to transmit more than one information source over the same carrier. 5] Subject to noise (same as AM). 6] By time-sharing signals connected to individual switches. A counter and a decoder can be used. 7] Using an analog switch with a binary counter and decoder. A synchronizing clock signal is required between transmitter and receiver.

8-4:

1] PAM is modulating the amplitude of the pulse, and PDM is modulating the duration or length of the pulse. 2] The time it takes to complete the pulse is modified. 3] Symmetrical trailing-edge and leading-edge modulation. 4] Produces one of two voltage states, depending on the amplitude of an input signal

compared to a reference. 5] Using a triangular wave as the sampling waveform. 6] Using a sawtooth wave with the steepest slope leading for leading-edge PDM or trailing for trailing-edge PDM. 7] Produces a waveform consisting of positive and negative spikes. 8] Removes (clips) waveform spikes. Positive peak clipper removes positive spikes; negative peak clipper removes negative spikes. 9] A circuit that generates a linearly increasing voltage. 10] Differentiate PDM wave and use resulting positive and negative peaks to control a ramp generator. Pass resulting wave through lowpass filter.

8-5:

1] Position of the pulse relative to a reference is changed by the modulating frequency. 2] PDM is another form of pulse time modulation. 3] Pulses unaffected by the modulating signal. Used to determine the relative position of the modulated pulse. 4] Have PDM control the action of a one-shot F/F. 5] An F/F with one stable state. 6] PPM and reference pulses control R-S F/F. Resulting output is PDM. Use PDM demodulation.

8-6:

1] Amplitude of modulating signal is converted into a digital code. 2] Amplitude of modulating signal is converted into a digital code. 3] Sound is recorded as a digital code. 4] Slicing the waveform into small units. 5] Sampling the amplitude at equal intervals and assigning a code based on quantization levels of the signal. 6] It can be processed by software to remove noise or introduce changes. 7] A simple A/D converter that easily converts an input signal into a binary code. 8] An error or distortion introduced when modulating signal is not exact value of resulting binary code. 9] See Figure 8-30. 10] To accommodate very small and very large signal variations. The process increases the quantized steps for small signals and decreases for large signals.

8-7:

1] Transmitting a string of pulses where the pulse polarity indicates if the modulating wave amplitude is increasing or decreasing. 2] Simplified encoding and decoding. Reduction of transmitted bits, more rapid response to changing signal. 3] A counter that can increment or decrement its count. 4] Essentially a D/A converter. Staircase direction depends on an up or down count. 5] Staircase generator and modulating signal fed into comparator. Output controls a clocked D F/F. 6] Signal controls action of an up-down counter, which feeds a D/A converter and lowpass filter.

8-8:

1] Information is shared between frequencies. Information is shared between times. 2] The medium reserved for a particular signal. 3] The sampling time and the system bandwidth. 4] A reference signal used to restore the missing carrier for each channel. 5] Easily remove the carrier. 6] By an oscillator controlled by the pilot carrier.

8-9:

1] A perfect" square wave with zero rise time. No. 2] Preshoot, finite rise time, nonlinearity, overshoot and rounding, settling time, ringing, droop, pulse width. 3] The amount of time it takes the pulse to go from 10% of its maximum value to 90% of its maximum value. 4] The amount of time between the 50% amplitude points of the pulse. 5] More harmonics of the same phase. An impossible infinite bandwidth of equal amplitude. 6] Differences between harmonics decrease.

Chapter Problems (odd-numbered)

1] Some aspect of a pulse is modified by the modulating signal, or a pulse code represents the modulating signal. 3] More efficient use of a communication system. Possible loss of information. 5] Information continuously varies some aspect of the pulse. Digital pulse information converts the information into a discrete code. 7] PAM: Amplitude of wave changes. PDM: Duration of wave changes. PPM: Relative position of wave changes. 9] TDM: Separating information in time. FDM: Separating information with frequency. 11] Sampling frequency of a pulse modulating system must be equal to or exceed twice the highest signal frequency to convey all in-

formation in original signal. 13] 20 kHz. 15] Symmetrical and nonsymmetrical. 17] Some aspect of the pulse is varied in time. 19] To create a guard band for using a practical lowpass filter. 21] An RC lowpass filter. 23] The 4016 contains four electronic switches that are electrically controlled. 25] A voltage of either $+5$ V (switch is closed) or 0 V (switch is open). 27] Schematic should contain two bilateral switches and one F/F. 29] Pulse duration modulation. The duration of the pulse changes, not the amplitude. 31] The same as FM over AM. 33] See Figure 8-18. 35] See Figure A-42.

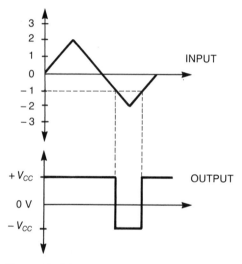

Figure A-42

37] Output signal is at maximum. 39] Use a sawtooth wave with the steepest slope leading for leading-edge PDM or trailing for trailing-edge PDM. 41] A circuit that generates a linearly increasing voltage. 43] PDM: The duration of the pulse is modulated. PPM: The relative position of the pulse is modulated. 45] It controls the timing of a one-shot F/F. 47] The quantized wave is sliced into separate sections to be sampled. 49] See Figure A-43.

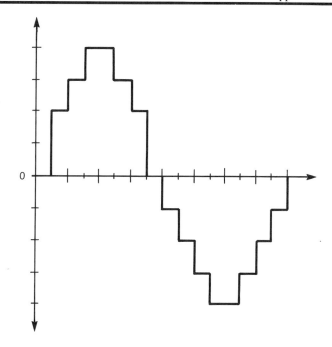

Figure A-43

51] 7, 8, 6, 4, 1, 0, 2, 4. 53] Converts a digital code to an analog signal. 55] An error or distortion introduced when modulating signal is not exact value of resulting binary code. 57] Transmitting a string of pulses, where the pulse polarity indicates if the modulating wave amplitude is increasing or decreasing.

59] A waveform that increases or decreases in discrete steps (looks like a staircase). 61] See Figure 8-35. 63] TDM: Time-division multiplexing. FDM: Frequency-division multiplexing. 65] See Figure A-44.

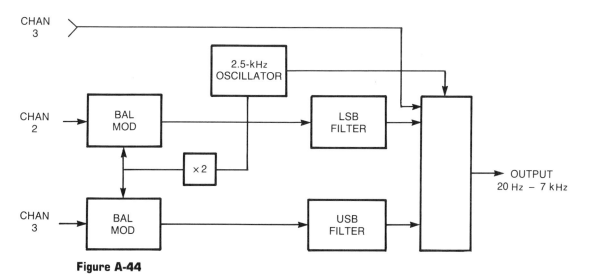

Figure A-44

67] See Figure A-45.

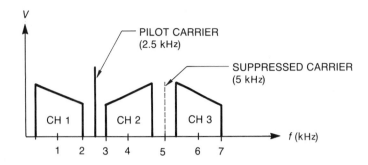

Figure A-45

69] More harmonics of the same phase.

Review Questions Chapter 9

9-1:

1] Converts a digital code into analog information. 2] Digital decoder, bilateral switch, and resistive voltage divider. 3] Value of the 4-bit binary input determines which of the 16 outputs will be active. 4] To produce a DC voltage source of equal voltage increments. 5] No. Because the 4–16-line decoder allows only one switch at a time to be active.

9-2:

1] Process of creating a specified waveform. 2] By the "shape" of their waveform. 3] Decoder, control matrix (ROM), voltage source, and bilateral switch. 4] Allows a predetermined output. To prevent shorts between outputs of the decoder. 5] Read only memory. Determine the shape of the output waveform.

9-3:

1] Process of converting an analog quantity to a digital quantity. Digital sound. 2] Amplitude of analog signal is converted to a digital code during a single clock pulse. 3] To store the results of A/D converters. 4] Amplitude of analog signal is compared to stairstep by a comparator. 5] Amount of time taken to make stairstep comparison. Amplitude of analog signal, clock speed.

9-4:

1] Has upper and lower threshold points that are not equal, and they cause the device to operate between an on and an off condition. Used to clean up a digital pulse. 2] Line driver used to prepare signal for transmission; line receiver prepares signal for reception. 3] Coax, ribbon cable, and shielded twisted pair. Ribbon cable. 4] Single ended uses coax cable; differential transmission uses twisted pair. 5] (A) One-way transmission (B) Two-way transmission (C) Two-way simultaneous transmission. 6] Maximum number of signals (carrier states) transmitted in 1 second.

9-5:

1] Serial is one bit at a time; parallel is all bits at the same time. Serial for long transmissions. 2] Whether or not there is a timing relationship with the clock pulse. 3] Converts parallel data for serial transmission and serially received data for parallel processing. Universal asynchronous receiver transmitter. 4] A standard interface specification. States an interfacing connector and allocates a use for each connection. 5] 4. Transmitted and received primary data; transmitted and received secondary data.

9-6:

1] Mark represents a 1; Space a 0. 2] A Mark and Space are distinguished by two different frequencies. 3] Mark and Space are distinguished by phase. Difficult to detect without a reference. 4] A synchronizing pulse is transmitted with the data. Mark and Space distinguished by phase. Easier to detect. 5] Amplitude of signal is proportional to digital value. Increasing bits per second without increasing the baud rate. 6] Relationship is BPS = N(baud rate).

9-7:

1] Signal processing circuit controlling an A/D converter with a digital readout. 2] To convert the electrical quantity into a form useful to the A/D converter. 3] Circuits inside the DMM automatically adjust range requirements of the instrument to accommodate value of input measurement. 4] Quantity to be measured is converted into heat; resulting heat is converted to a voltage with a thermocouple. 5] By using a constant-current source with known current ranges. 6] DMM has greater accuracy but is susceptible to noise.

Chapter Problems (odd-numbered)

1] True. 3] False. 5] True. 7] B. 9] C. 11] A. 13] C. 15] A. 17] Several. 19] Synchronous. 21] (A) 2, 3, 14, 16 (B) 4, 5, 6, 8, 12, 13, 19, 20, 21, 22, 23 (C) 15, 17, 24 (D) 1, 7. 23] Produces two frequencies: 1 270 Hz represents a Space, and 1 070 Hz represents a Mark. 25] By allowing different modulation levels to represent two or more bits. 27] Analog meter uses a meter movement with a moving indicator across a fixed scale. Digital meter uses a readout device that presents a numerical display as would be seen on a pocket calculator. 29] Comparators are used to control input voltage. It can be done with a binary up-down counter operating through a decoder and a bilateral switch.

Review Questions Chapter 10

10-1:

1] 3×10^8 m/s. Yes, in other mediums. No, they cannot travel faster. 2] The wires connecting the source to the load. The voltage and current are everywhere the same. 3] It takes time for the pulse to travel the length of the transmission line. No. 4] Inductance and resistance in series; capacitance and conductance in parallel. 5] The relationship between a unit-length inductance to a unit-length capacitance. An infinite length. 6] Balanced means both wires carry RF energy 180° out of phase. Unbalanced means one wire is at ground potential while the other carries RF energy. TV twin-lead in; coax cable.

10-2:

1] When terminated in its characteristic impedance. 2] The reflection of some of the power from the load back to the source. They indicate that the line is not terminated in its characteristic impedance. 3] At the open end, the voltage is a maximum and the current is a minimum. At the shorted end, the current is a maximum and the voltage is a minimum. 4] Incident means from the source, and reflected means returning to the source. 5] When the line is terminated in its characteristic impedance, there are no longer any standing waves.

10-3:

1] When the load is not matched to the impedance of the transmission line. 2] The greater the amount of mismatch, the less will be the power absorbed by the load. 3] Power not absorbed by the load. Total power transmitted by the source. 4] Flat line is terminated in characteristic impedance; resonant line is not terminated in characteristic impedance. 5] When terminated in its characteristic impedance. 6] (A) At a quarter-wavelength (B) At half-wavelength. 7] (A) At half-wavelength (B) At quarter-wavelength. 8] Series resonance: When the source sees maximum current and minimum voltage. Parallel resonance: When the source seems maximum voltage and minimum current.

10-4:

1] That portion of a communication system used for radiating waves into space or receiving them from space. 2] Antenna ends look like an open, so impedance at source looks very low (73 Ω). Same endpoint characteristics as a quarter-wave transmission line. 3] At endpoints, current is a minimum and voltage is a maximum. At source, current is a maximum and voltage is a minimum. 4] Ratio of the total power dissipated to the square of the effective antenna current. 5] Strongest perpendicular to the antenna and vertically zero at its ends. 6] Contains mutually perpendicular magnetic and electrostatic fields. Horizontal: Horizontal electric field. Vertically: Vertical electric field. 7] See Figure 10–22. 8] Ratio of the antenna transmitting frequency to its bandwidth.

10-5:

1] Produces a pattern of four lobes. 2] Construction is the same; wavelength depends on transmission or reception frequency. Radiation pattern of $\frac{1}{2}$ wave is four major lobes and two minor lobes. 3] By using two different frequencies: a lower frequency where the antenna appears as a full-wave dipole, and a higher frequency where the antenna appears as a $\frac{1}{2}$ wave. 4] See Figure 10-26. (Almost a cylinder pattern with antenna parallel to axis of cylinder). 5] The angle from the radiation pattern between two points on either side of maximum radiation where the field strength drops 3 dB. 6] To transmit a commercial radio signal toward population densities. 7] A comparison of the output strength of the radiation of an antenna in a particular direction and a reference antenna that radiates an omnidirectional wave (such as a Hertz). 8] A driven collinear array, where the elements are placed in a line (see Figure 10-32).

10-6:

1] Ground wave, sky wave, and line of sight. 2] A wave that travels along the earth's surface. Wave energy would be absorbed. 3] Radio horizon is $\frac{4}{3}$ greater than actual horizon. 4] Characteristics of the ionosphere. Affected by hourly, daily, and yearly changes. 5] Skip distance is the distance wave travels over the earth. Multiple hops caused by sky wave being reflected back to earth and reflected back to the ionosphere, and so on. 6] From multiple-path transmission where one signal interferes the other. Can be caused by the two signals being out of phase with each other.

10-7:

1] All impedances should match. 2] To help match impedances when there would otherwise be an impedance mismatch. 3] The same value of the source impedance. 4] The same value as the load impedance.

Chapter Problems (odd-numbered)

1] 166 ns. 3] 1 m. 5] 11.1 pulses. 7] 32.26 cycles. 9] 77.5 Ω. 11] 71 pF. 13] (A) VSWR = 2.0 (B) ISWR = 2.0 (C) SWR = 2.0. 15] SWR = 2. 17] (A) SWR = 3 (B) K_{ref} = 0.5 (C) P_{ref} = 25% (D) P_{abs} = 75%. 19] Open stub = 43.1 cm. Shorted stub = 20.4 cm. 21] Length = 16.3 cm; location from V_{max} = 37.3 cm. 23] 125 kW. 25] 20 Ω. 27] 11.8 A. 29] Q = 40. 31] 15 m. 33] 1.425 m. 35] 87.7 MHz. 37] 2.5 dB. 39] 1.58 kW. 41] 1 385 ft. 43] Ground wave. 45] 30 MHz. 47] F layer. 49] From multiple-path transmission where the phase of one signal can interfere with the other.

Review Questions Chapter 11

11-1:

1] For the efficient transmission of frequencies above 1 GHz over short distances. 2] At high frequencies, majority of conduction takes place along the surface of the conductor. 3] The waveguide width determines the lowest frequency that can be transmitted in the waveguide. 4] TE and TM modes. 5] The lowest possible frequency that can be propagated in the waveguide of given dimensions. 6] Submodes are in the categories of TM_{mn} and TE_{mn}, where m is the number of half-wavelengths along the a dimension and n is the number of half-wavelengths along the b dimension.

11-2:

1] The lowest frequency that can be accommodated in a waveguide of given dimensions. 2] Factors are the waveguide width and the number of half-wavelength fields. 3] Cutoff wavelength is larger than the free-space wavelength. 4] The guide wavelength is a function of the free-space wavelength and the cutoff wavelength. 5] For connecting a rotating element, such as an antenna to the communications system. 6] For circular, the m indicates the number of full periods of the field, and the n is the number of half-period variations of the field. 7] It can operate at lower frequencies than dictated by its outside dimensions, thus saving space and weight. 8] Constructed from spiral-wound ribbons of brass or copper covered outside with a soft material such as rubber. Used in laboratory situations or wherever continuous flexing is required.

11-3:

1] By inserting a short piece of wire to act as an antenna. The number and location determine the mode of waveguide operation. 2] By using waveguide bends and twists. 3] Through the location of a piece of metal in a waveguide slot. Q is determined by the diameter of the metal; the smaller, the higher the Q. 4] Using a carbonized material or resistive vane or permanent/removable end plates or by adjustable flap/vane attenuators.

11-4:

1] Means "small wave," which refers to its short wavelength. Start at 1 GHz. 2] Hybrid-T junction. Some of the arms are not electrically connected to each other. 3] A piece of wire bent into a C shape acts as a resonant circuit. 4] Resonant cavity can be thought of as being constructed from an infinite number of hairpins. 5] Contains a calibrated movable plunger used to tune the cavity.

11-5:

1] Exhibit the property of light. 2] Allows the use of directional antennas. 3] Essentially an extension of the open end of a waveguide. 4] Ability to focus electromagnetic waves. 5] Line-of-sight transmission with repeater stations. 6] Minimal atmospheric disturbance, some privacy, and no transmission line maintenance.

11-6:

1] Comes from the word "radio" and can be read backwards or forwards, thus describing the basic principle of radar. 2] A pulse of electromagnetic energy is transmitted from the radar system, reflected from an object, and received by the radar system. The time it takes for this to happen is a measurement of the distance of the object from the radar system. 3] Consists of a transmitter, directional antenna, and receiver. 4] A pulse of RF energy followed by a period of receiver time during which no transmission takes place. 5] An electronic switch that allows the same antenna to be used as a transmitting and receiving antenna.

11-7:

1] Line-of-sight transmission. 2] Earth communications and extraterrestrial geographies. 3] Source of continuous power, directional antennas, correcting thrusters, environmental sensors, communications equipment. 4] Geosynchronous or geostationary. 5] Satellite systems contain many transponders allocated to different frequencies. 6] Uplink is communications to the satellite; downlink is communications from the satellite.

11-8:

1] The approximate resonant frequency. 2] A device that separates the forward wave from the reflected wave. 3] Sweep generator, VSWR bridge, signal generator, marker adder, RF detector, and scope. 4] RF generator, VSWR bridge, and RF voltmeter.

Chapter Problems (odd-numbered)

1] True. 3] False. 5] True. 7] A. 9] A. 11] C. 13] A. 15] Horn. 17] Parabolic. 19] Duplexer. 21] 12 cm. 23] $m = 1$. 25] 2.55 cm.

Review Questions Chapter 12

12-1:

1] The study of light transmission in a controlled path accomplished with light conductors. 2] Light amplification by simulated emission of radiation. 3] The photophone, which consisted of a speaking device that caused a mirror to vibrate and a receiver that converted the light vibrations back to voice. Some problems were atmospheric disturbances and availability of a reliable light source. 4] Wide bandwidth, immunity from electromagnetic interference, immunity from interception by external means, inexpensive materials, corrosion resistance, and immunity to atmospheric changes.

12-2:

1] Light can be thought of as an electromagnetic wave or as particles called photons. 2] Confined to frequencies from 3×10^{11} Hz to 3×10^{16} Hz. 3] Optical spectrum includes frequencies from 3×10^{11} Hz to 3×10^{16} Hz, and the visible spectrum is that part of the optical spectrum that can be seen by the eye. 4] The range of frequencies within the optical spectrum. 5] Light beam is the actual cone of light transmitted from the source. Light ray is an imaginary line that shows the direction of travel for the light ray. 6] Reflected ray strikes a surface and returns back into space. Refracted ray is a light ray that is bent as it passes from one medium to another. 7] The ratio of the sines of the angles of the incident and refracted waves is equal to the ratio of the velocities of light in the mediums causing the refraction. 8] The angle of incidence that causes the angle of refraction to be exactly 90°.

12-3:

1] The critical angle determines if a light ray will stay within the transmitting medium or be refracted from it. 2] Refract and leave. Reflect within the glass rod. Conduct along the surface. 3] See Table 12-2. 4] Step-index has a uniform index of refraction. Graded-index has a changing index of refraction (largest in the center). 5] The refraction of the light changes within the material in such a manner as to encourage the beam to stay within the transmitting medium.

12-4:

1] By causing light to refract out of the transmission medium. 2] Multiple strands contain many small strands in a bundle. Single strand contains just one optical fiber. 3] Caused by rays of light arriving at the end of a fiber at different times, resulting in a potential loss of information. 4] Use of optical cladding and a small-diameter fiber. 5] Model and material dispersion. 6] From about 1.3 to 1.5 microns.

12-5:

1] A solid-state device that emits light when forward biased. 2] A measure of the optical fiber's light-gathering capabilities. 3] The sine of the angle is related to the square root of the difference of their squares. 4] Improper alignment, connecting ends not parallel, separation resulting in light scattering. 5] Scattering: Imperfections within the optical fiber. Bending: Caused by a kink in the optical fiber.

12-6:

1] It is in phase of one wavelength and travels in a narrow beam. 2] Radiometric measurement: Full spectrum of electromagnetic energy within the optical spectrum. Photometric measurement: Measurement of visible light. 3] Energy is the ability to do work. Power is the amount of work per unit time. 4] Power per unit area. 5] A solid angle of 1 steradian is equal to a surface area of the square of the radius. 6] Measurement of the strength of the light source.

12-7:

1] A central nucleus consisting of positive particles called protons orbited by negatively charged electrons. 2] A shell represents the distance from the nucleus where an electron is most likely to be found. The further the electron is removed from the nucleus, the higher its energy level. 3] Energy is related to the ratio of Planck's constant and wavelength. 4] Spontaneous: Cannot predict when electron will return to its ground state. Stimulated: Energy is released in precise and predictable amounts. 5] External energy (such as light) is pumped into the lasing medium rais-

ing energy levels of electrons. Mirrored surfaces amplify stimulated emission. One of the surfaces has a "partial" mirror allowing some of the laser light to escape.

12-8:

1] Medium, exciter, feedback, and output. 2] Medium: laser rod. Exciter: light source. Feedback: mirrors. Output: partial mirror. 3] Radiation emitted from the PN junction. 4] Reflectivity from the diode–air interface. 5] Consists of an amplifier operating a solid-state laser connected to optical fibers as the transmitting medium. Receiving end consists of phototransistor operating an amplifier that reproduces the original transmitted signal.

12-9:

1] The eye and skin. 2] Ultraviolet radiation. No. 3] Erythema. Several hours after exposure (similar to a "sunburn"). 4] Federal 1040 and ANSI Z136. 5] Classes I, II, III, and IV. 6] I: No known harm. II: Can cause eye damage if viewed directly. III: Will produce eye damage if viewed directly. IV. Danger to skin and eye directly or by diffused reflections.

Chapter Problems (odd-numbered)

1] Light amplification by stimulated emission of radiation. 3] Light waves in phase. 5] Einstein predicted; T. Mainman demonstrated. 7] Communications, surveying, medical applications, accurate energy source, and optical scanning. 9] See Figure 12-3; 3 × 10^{11} Hz to 3 × 10^{16} Hz. 11] Light ray is a line representing the direction of the light beam. Light beam is the actual light pattern. 13] They are equal. 15] Reflection: Light waves returning back to space from a polished surface. Bending of light from one medium to another. 17] Ratio of the sine functions of the angles of the incident and refracted waves is equal to the ratio of their velocities. 19] It will not change since the ratio of the velocities doesn't change. 21] 54°. 23] (A) 32.8° (B) 41.8°. 25] See Table 12-2. 27] See Figure 12-18. 29] Mode: Various light paths in optical fiber. Multimode step-indexed fiber: An optical core covered by a jacket that has a smaller refractive index than the core. 31] By use of a jacket with a smaller refractive index than the core. 33] See Figure 12-24. 35] Model dispersion and material dispersion. 37] See Figure 12-28. 39] 0.72. 41] It is in phase of one wavelength and travels in a narrow beam. 43] Radiometric measurement: Full spectrum of electromagnetic energy within the optical spectrum. Photometric measurement: Measurement of visible light. 45] (A) 5 W (B) 240 μW. 47] (A) 625 mW/cm^2 (B) 30 μW/cm^2. 49] 487 μW/(st-cm^2). 51] See Figure 12-40. 53] See Figure 12-43.

Review Questions Chapter 13

13-1:

1] One of the major reasons is due to its low cost as an IC. 2] Three applications are AM, FM demodulators, and frequency synthesizers. There are many other applications. 3] Phase detector, lowpass filter, amplifier, and VCO. 4] See Table 13-1. 5] Free-running: When frequencies of input and VCO signals are different. Capture: VCO begins to change frequency to reduce frequency difference with input signal. Phase-lock: When input frequency and VCO frequency are the same.

13-2:

1] To produce an output voltage proportional to the phase difference of two signals. 2] Analog and digital. 3] Two signals of the same waveform are out of phase when they have the same frequency but their maximum and minimum points do not occur at the same time. This is true for sine waves and square waves. 4] The output of an XOR is +5 volts only when the inputs are different. Hence, when the inputs are in phase, the output is 0 volts. As this phase relationship changes, the average DC voltage varies accordingly. 5] For a digital phase detector, the output voltage is a maximum when the input signals are 180° out of phase and becomes zero as the phase difference approaches 0°.

13-3:

1] A commonly used IC found in many timing applications. 2] A Set-Reset flip-flop, two comparators and a switching transistor. 3] The outputs of the two comparators control the inputs of the *S-R* flip-flop. The *Q* output of the flip-flop controls the base voltage of the transistor, causing it to be in cutoff or saturation. 4] The value of the external resistors and capacitor(s). 5] By externally controlling the reference voltages to the comparators.

13-4:

1] Amplify DC voltages. 2] The inverting input produces an output signal 180° out of phase with the input. The noninverting input produces an output signal in phase with the input. 3] Inverting amplifier: Feedback resistor from output to inverting input; input signal to inverting input. Noninverting amplifier: Feedback resistor from output to inverting input; input signal to noninverting input. 4] So the output signal may swing above or below zero volts. 5] The values of external resistors connected to the amplifier.

13-5:

1] The ratio of the amount of output voltage per degrees of phase shift. 2] Increases the sensitivity of the PPL. 3] Voltage-to-frequency conversion in hertz per volt. 4] Consists of a phase detector, op amp, and VCO. 5] By the values of external components.

13-6:

1] Changes in the carrier frequency when used as the input signal to a PLL will cause a corresponding change in the output voltage of the phase detector. This is amplified by the op amp, thus converting frequency information into amplitude information. 2] A communications receiver that uses a PLL. See Figure 13-22. 3] A PLL. 4] A PLL where the input frequency (FSK) will cause the output of the PLL to be at one voltage or the other. 5] A device that can either create or detect FSK (usually contains a PLL).

13-7:

1] To produce a predictable output at a specific frequency. 2] Zero volts. 3] The values of external components (*R* and *C*). 4] See Table 13-2. 5] The circuit can use a two-input NOR gate whose inputs are each connected to a separate tone decoder, each one having a different select frequency.

13-8:

1] Digital selection of a given frequency. Multiple higher frequencies from a stable single source frequency. 2] A PLL with the VCO output connected to frequency dividers the output of which is connected as the phase detector reference. 3] The frequency dividers. 4] By using a frequency divider at the input to the frequency synthesizer.

Chapter Problems (odd-numbered)

1] TV sync, Touch-Tone decoders, and FSK decoders. 3] See Table 13-1. 5] 0°. 7] The band of frequencies centered on the free-running frequency that the PLL can lock on to. 9] The VCO control voltage is directly proportional to the phase detector output voltage. 11] (A) 14.3° (B) 360° (C) 180°. 13] 45° out of phase. 15] 30°. 17] See Figure A-46.

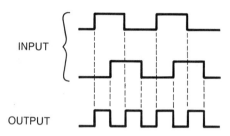

Figure A-46

19] (A) 2 V. (B) 2 V 21] Period = 10.4 ms. Duty cycle = 66.6%. Peak voltage = 5 V. 23] See Figure A-47.

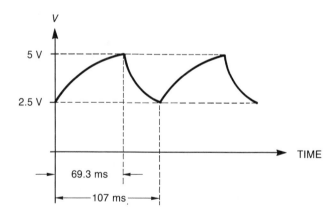

Figure A-47

25] See Figure A-48.

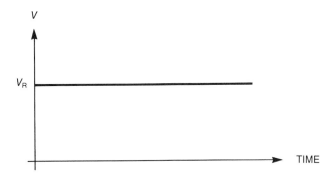

Figure A-48

27] (A) −6 (B) 1.2. 29] (A) 25 V in phase with input (B) 11.36 V out of phase with input. 31] 1.89 V/ rad. 33] 1.35 V. 35] 125 kHz. 37] 8.5 Hz. 39] See Figure 13-21. 41] See Figure 13-24. 43] From Figure 13-27, the output of the NOR gate will be +5 V only when both tones are present (1 209 Hz and 697 Hz). 45] Output frequency is $\frac{1}{8}$ that of the input frequency.

Review Questions Chapter 14

14-1:

1] According to the range of frequencies they will amplify. 2] An amplifier will give a power gain. A transformer can never give a power gain. 3] The range of frequencies from the lowest to the highest that the amplifier is capable of amplifying at a power gain that is more than half of its maximum power gain. 4] Vertical scale is linear; horizontal scale is logarithmic. 5] See Table 14-1.

14-2:

1] Transistors, FETs, and ICs. 2] Because when damaged, the whole unit must be replaced. 3] NPN and PNP. 4] The gate voltage controls the channel current. 5] Plate, control grid, cathode, and heater. 6] It can contain most of the parts required for a complete communications receiver.

14-3:

1] Classes A, B, and C. 2] The amount of time the device is conducting compared to the input signal. 3]

See Table 14-2. 4] Tuned circuit. To restore the rest of the input signal by the flywheel effect. 5] Only one transistor conducts at a time.

14-4:

1] Input and output signals are 180° out of phase, low input impedance, very high output impedance, current gain of unity or less. 2] Input and output signals are 180° out of phase, produces voltage and current gains. 3] Source follower. 4] 180°, 0°, 180°. 5] Common collector family.

14-5:

1] Small-signal amplifier has a power rating of $\frac{1}{2}$ watt or less. Large-signal amplifiers more than $\frac{1}{2}$ watt. 2] Low-power examples: RF amplifier for receiver, audio amplifier for transmitter. High-power examples: Audio amplifier for receiver, RF amplifier for transmitter. 3] A signal small enough to ensure that the amplifier operates within its linear portion. 4] Determine the operating characteristics of the amplifier. 5] Use of tuned circuits and decoupling networks.

14-6:

1] Yes. The LM 1866 is an example. 2] Automatically controlling the gain of the radio so that tuning to different stations keeps the volume constant. 3] Replace the whole chip. 4] Liquid crystal display. 5] Display is viewed in a mirror from light passing through an LCD screen.

14-7:

1] To forward bias the transistor and determine its operating point. 2] The relative values of these resistors along with V_{CC} determine the value of V_B. 3] Within limits, R_E has no effect on V_B. 4] R_E helps determine the value of the transistor current (I_E and thus I_C); this in turn directly affects the value of V_C. 5] The larger R_C, the smaller V_C. The smaller R_C, the larger V_C.

Chapter Problems (odd-numbered)

1] True. 3] True. 5] False. 7] C. 9] D. 11] Q_3, Q_4. 13] Q_2. 15] Q_1. 17] Filament or heater. 19] Input. 21] V_E decreases if V_{CC} decreases, R_{B_1} decrease, or R_{B_1} increase. Just the opposite occurs for V_E increases. Value of V_E is effected by resistors R_{B_1} and R_{B_2}. 23] $V_C = I_C R_C$. R_C has no effect on V_E. 25] $I_{max} = V_{CC}/(R_C + R_E)$. $V_{max} = V_{CC}$.

Review Questions Chapter 15

15-1:

1] A circuit that produces its own waveform. 2] A source of energy, frequency-determining circuit, amplifier, positive feedback circuit. 3] Product of amplifier gain and feedback circuit must be unity. 4] Initial turn on: Rich in harmonics. System build-up: System gain greater than 1. Sustained oscillations: System gain becomes unity. 5] Compensating component in the circuit or amplifier operation between cutoff and saturation.

15-2:

1] Armstrong: Transformer feedback. Colpitts: Capacitive voltage divider. Hartley: Inductive voltage divider. 2] Phase shift utilizes a 60° phase shift each of three RC legs. Twin-T has network that produces 0° phase shift at resonant frequency. 3] Frequency stability. 4] Physical dimensions and method of cutting. 5] Parallel mode and series mode. Frequency of both modes determined by equivalent inductance and series capacitance. Parallel resonant mode frequency is also determined by parallel capacitance.

15-3:

1] Amplitude, frequency, and phase modulation. 2] Frequency and phase modulation. Angle modulation. 3] Distortion and resonant circuit. 4] Produces the sidebands of two mixing frequencies but not the carrier. 5] FM: Carrier frequency deviates. PM: Carrier phase deviates.

15-4:

1] Collector, series, and IC modulators. 2] Carrier applied to transistor base; modulating signal to transistor collector. Uses class C operation. 3] Two transistors are used: carrier applied to base of one; modulating signal to base of the other. 4] CA3080A. See Figure 15–23.

15-5:

1] A mixer that produces the carrier but suppresses the sidebands. 2] Diode-ring, active, and IC balanced modulator. 3] Carrier output is ideally balanced (assuming perfectly matched diodes). Addition of modulating signal causes imbalance in output producing sidebands. 4] Carrier is applied to balance the output. Modulating signal is applied so that output is not balanced and sidebands are produced with a suppressed carrier. 5] Usable up to 100 MHz; carrier suppression up to 65 dB.

15-6:

1] Angle modulation. 2] Solid-state phase modulator. Varactor diode modulator. VCO FM generator. 3] A single FET for the solid-state phase modulator. A varactor diode for the varactor diode modulator. A complete VCO in an IC. 4] Capacitor C_2 and the varying gain of the FET. 5] When reversed biased, it acts as a capacitor. The amount of reverse-bias DC voltage controls the amount of capacitance.

15-7:

1] When circuits not designed to be oscillators act as oscillators. 2] See Table 15-5. 3] Low-frequency oscillations caused by poor power supply regulation. 4] Proper shielding. 5] An area where all ground connections are made. Prevents oscillations by preventing ground loops.

Chapter Problems (odd-numbered)

1] See Figure 15-1. 3] Wires, optical links, and radiation. 5] $\frac{1}{3}$. 7] Initial turn on, buildup, and sustained oscillations. 9] Bring driven into cutoff and saturation. 11] (A) C_3, C_4, L_2 (B) L_2, L_3, C_3 (C) L_1, C_7 (D) L_1, C_3. 13] (A) 10 μH (C) 45 μH (D) 12.5 μH. 15]

(A) Twin-T (B) Phase shift (C) Wein bridge. 17] (A) 362 Hz (B) 13 Hz (C) 1.33 Hz. 19] Double the value of each capacitor. 21] (A) f_{par} = 85 256 Hz, f_{ser} = 82 228 Hz (B) f_{par} = 2 056 135 Hz, f_{ser} = 2 055 724 Hz. 23] (C) Q_{par} = 4 625, Q_{ser} = 4 609; BW_{ser} = 37.5 Hz, BW_{par} = 37.5 Hz (D) Q_{par} = 6 358, Q_{ser} = 6 324; BW_{par} = 66.3 Hz, BW_{ser} = 66.3 Hz. 25] (A) Clapp (B) Colpitts (C) Pierce. 27] It introduces distortion. 29] There is a 90° phase difference. 31] Modulating signal at B, carrier at A. 33] Series modulator. 35] IC modulator. Modulation level adjustment. 37] Balanced modulator. Used to produce a suppressed carrier. 39] Suppressed carrier with upper and lower sidebands. 41] (A) Phase modulator: Carrier at *a*, modulating signal at *b* (B) Carrier at *b*, modulating signal at *a*. 43] The modulating frequency changes the capacitance of the reverse-biased diode D_1. This changes the resonant frequency of the oscillator, producing an FM carrier. 45] When circuits not designed as oscillators act as oscillators. Poor power regulation, improper shielding, and signal grounding. 47] Improper ground connections that result in unwanted oscillations. See Figure 15-35. 49] See Figure 15-37.

Review Questions Chapter 16

16-1:

1] Automatic gain control. 2] AM, FM, and PM detectors. 3] Automatically controls the gain, so when tuning to different stations it isn't necessary to adjust the volume control for weak or strong stations. 4] To give an indication of the signal strength by use of a meter. 5] Delayed AGC: No AGC action until signal is a specific strength. Keyed AGC: AGC action in TV that takes place during horizontal retrace.

16-2:

1] Simple AM, suppressed carrier, and single sideband. 2] AC converted to pulsating DC that is filtered by a capacitor. This action reconstructs the original modulating signal. 3] A synchronous detector. 4] See Figure 16-7. 5] Suppressed carrier.

16-3:

1] To convert the frequency or phase changes into a reconstruction of the original modulating signal. 2] A parallel resonant circuit tuned just off the carrier

frequency of the FM signal. As the frequency changes, the voltage across the tuned circuit will change, thus reconstructing the original modulating signal. 3] For the ratio detector: Self-limiting and produces its own AGC. 4] Availability of economical ICs. 5] The LM 1800 uses a PLL.

16-4:

1] Lowpass, highpass, bandpass, and band-reject. 2] Active filter supplies power to the signal; passive does not. 3] Inductors are not needed; present a high impedance to the originating circuit. 4] Decade: A change in frequency by a factor of 10. Octave: A change in frequency by a factor of 2. 5] How a filter responds to a change in frequency.

16-5:

1] Lowpass has parallel input capacitor; highpass has series input capacitor to op amp. 2] Ratio of resonant frequency to bandwidth. 3] Using resistance or capacitive reactance as the numerator for the output voltage formula. 4] Keep the voltage gain less than $2Q^2$.

16-6:

1] High *Q*. 2] Series resonance: Minimum impedance. Parallel resonance: Maximum impedance. 3] Inductive: Just above resonance. Capacitive: Just below resonance. Resistive: At resonance. 4] Two matched pairs of crystals (see Figure 16-28).

16-7:

1] A signal generator and a voltmeter or scope. 2] The AGC must be deactivated. 3] The amplitude of the signal generator output does not change. 4] No change in the output amplitude of the circuit under test. Reduce the amplitude of the input signal to the amplifier under test. 5] A plot of amplitude (in dB) linearly plotted on the vertical axis versus the corresponding frequency (in hertz) plotted logarithmically along the horizontal axis.

Chapter Problems (odd-numbered)

1] False. 3] False. 5] True. 7] C. 9] A. 11] C. 13] B. 15] C. 17] Flattest. 19] Decade. 21] (A) Refer to Figure 16-14 (highpass filter): f_B = 637 Hz (B) Refer to Figure 16-14 (lowpass filter): f_B = 4.825 kHz. 23] See Figure 16-22: R_1 = 31.8 kΩ, R_2 = 6.4 Ω, R_3 =

63.7 kΩ. 25] See Figure 16-24: $R_1 = R_2 = 85$ mΩ, $R_3 = 10.6$ kΩ, $R_4 = 21.2$ kΩ. 27] Refer to Figure 16-27: $f_{ser} = 411$ Hz, $f_{par} = 29$ kHz. 29] See Figure 16-30.

Review Questions Chapter 17

17-1:

1] Steady output voltage, low output impedance, overload protection. 2] A constant output voltage for any value load. Implies that it can deliver infinite current when load is zero ohms, an impossible condition. 3] No-load voltage is larger than full-load voltage. 4] The difference between no-load and full-load voltages.

17-2:

1] Adjusting the value of one resistor in a circuit consisting of two series resistors (the other resistor acts as the load) and a voltage source, keeps the voltage across the load resistor constant as its value changes within a known range. 2] The base current can be thought of as controlling the amount of resistance between the collector and the emitter. The net result of this is transistor action, where the amount of base current controls the amount of current between the emitter and the collector. 3] It maintains a constant voltage drop when operated at its zener point. It is used as a voltage reference. 4] The transistor controls the amount of current delivered to the load. The op amp using the zener voltage as a reference monitors the regulator output voltage and in turn controls the amount of current in the transistor. The net effect is to maintain a constant output voltage for a wide range of load conditions.

17-3:

1] The operation is similar to a series regulator, except the variable resistor is in parallel with the load. 2] The shunt regulator resistor must decrease in value to help maintain a constant voltage across the load. 3] Similar to the action of a series regulator, except the transistor is in parallel with the load. 4] The control circuitry which controls the amount of current in the shunt transistor is similar. This action in turn helps maintain a constant output voltage for a wide range of load conditions.

17-4:

1] Three fixed: Positive, negative, and dual-tracking. Those with variable output voltage(s). 2] The better the heat-dissipating capabilities of the case, the greater the maximum power dissipation. 3] With the proper connection of a variable resistor. 4] If one output voltage changes due to load conditions, so will the other. 5] Thermal shutdown, current limiting, current foldback, remote shutdown.

17-5:

1] The product of the transistor voltage drop and current. Since these values may be significant, so is the resulting power dissipation. 2] The period is equal to the sum of the on time and the off time. The ratio of the area under the pulse and the period determines the average value. 3] The transistor operates between saturation (ideally 0 volts drop) and cutoff (ideally no current). Since power dissipation is the product of these two values, the power dissipation is ideally zero. 4] To filter the pulsating DC and restore pure DC. 5] 20 kHz to 50 kHz. Smaller-value filter capacitors are required compared to 60-Hz operation, resulting in greater reliability.

17-6:

1] Adjust the pulse width to accommodate changing load requirements. 2] By the voltage level of the comparator. The lower the output voltage, the longer the comparator will be on and the wider the pulse width will be. 3] To increase the operating efficiency. 4] DC must be converted to a changing current in order to take advantage of transformer action. The new voltage is filtered back to DC.

17-7:

1] Unregulated, series regulated, shunt regulated, switching regulator, and battery supplied. 2] See Table 17-3. 3] Break open the circuit at key points and reapply power. When symptom no longer appears, the problem is in the disconnected section of the power supply. 4] The problem may be in the load, causing the output of the regulated power source to shut down or undergo current foldback. 5] Dry cells, nicads, and lead–acid.

Chapter Problems (odd-numbered)

1] A steady source of voltage, small output impedance, and an overload safety mechanism. 3] See Figure 18-1. 5] (C) V_{NL} = 1 kV, V_{FL} = 0.99 kV (D) V_{NL} = 85 V, V_{FL} = 85 V. 7] Figure 17-36(B). The internal resistance is closest to the value of the load. 9] (C) About 1% (D) Better than 1%. 11] Series regulator. 13] (A) Decrease R_S (B) Increase R_S. 15] Maintains a constant voltage. Used as a reference voltage. 17] Refer to Figure 17-19. 19] 20 V. 21] Prevent damage to the load and voltage regulator. 23] Controls the amount of voltage across the load (shunt regulator). 25] See Figure 17-13. The transistor controls the amount of current in R_{int}, thus controlling the voltage drop across R_{int} and R_L. The control circuit consists of D_Z and the differential amplifier. 27] Three fixed types: positive voltage, negative voltage, and both positive and negative voltage. Variable output voltage. 29] Depends on the type of regulator selected. The regulator that could dissipate the most power would be housed in a TO-3 case. The one that dissipates the least power would be housed in a TO-92 case. 31] See Figure 17-16. 33] If one output voltage changes due to load conditions, so will the other. 35] Thermal shutdown, current foldback, and remote shutdown. 37] (C) 1.25 V (D) 10.28 V. 39] (A) 0.5 (B) 0.71. 41] See Figures 17-26 and 17-27. 43] (B) 4 V. 45] See selected technical manual. 47] See Figures 17-31 and 17-32. In order to get transformer action. 49] Dry cells: Make voltage measurements under loaded conditions for accuracy. Nickel–cadmium: Potentially explosive gases given off during recharge. Lead–acid: Dangerous high current possible.

Review Questions Appendix A

A-1:

1] They can represent circuits that have two conditions: on or off. 2] Decimal number system has 10 different symbols; binary number system has two different symbols. 3] The subscript 2 as in 10_2. 4] 8, 4,

2, 1. Largest number is 15. 5] 10_2, 0110_2, 1100_2. 6] 3_{10}, 5_{10}, 11_{10}. 7] 255_{10}.

A-2:

1] To control signal processing and time events. 2] On: Closed, 1, true, high, +5 V. Off: Open, 0, false, low, 0 V. 3] Two switches in series. Two switches in parallel. 4] The output of an AND gate is true only when all the inputs are true. The output of an OR gate is false only when all inputs are false. 5] Changes the logic level. 6] The output of a NAND gate is false only when all inputs are true. Output of a NOR gate is true only when all inputs are false. 7] XOR is true only when the inputs are different. XNOR is false only when inputs are different.

A-3:

1] A circuit used to store a single binary number. 2] An on–off switch being turned on and off for the same amount of time. 3] An F/F whose output is controlled by two inputs called Set and Clear. 4] An on–off timing signal. One cycle of the clock pulse. 5] The F/F can change state only in step with the clock pulse. 6] An F/F that has one input (the D or Data input). It has two outputs that are always different (complements). 7] Leading is the rising edge; trailing is the falling edge. 8] The F/F will change state only on the edge of the clock pulse. It allows a more precise timing. 9] The J input can be thought of as the Set input; the K input as the Clear input. When both inputs are active, the F/F will toggle. 10] They override the actions of the J-K and clock inputs. Clear sets Q false; Set causes Q to be true. 11] It has no effect on the output state of the F/F.

A-4:

1] A group of flip-flops. 2] The contents of any group of flip-flops treated as a unit. 3] Bit: Single binary digit. Nibble: Four binary digits. Byte: Eight binary digits. 4] Loading an entire register in 1 CP. Parallel transfer or loading. 5] Entering 1 bit per CP into a register. 6] Takes less time but requires more connections. Serial loading requires fewer connections but requires more time. 7] Data is loaded within 1 CP and transferred out 1 bit per CP. 8] Data is loaded in 1 CP at a time and transferred out within 1 CP.

9] Transferring data between computers. 10] A register connected in such a way that it simulates a binary number that increments or decrements during each clock pulse.

A-5:

1] FCC requirements. 2] A binary counter and a precision timer (such as a one-shot). 3] The accuracy of the precision timer. 4] It converts a binary number to a decimal number to make the number easier to read. 5] Seven controlling inputs activate a specific segment. The segments are arranged to suggest a number from 0 to 9. 6] A light-emitting diode. Used as a low-power indicator device.

Appendix Problems (Odd-numbered)

1] True. 3] False. 5] False. 7] B. 9] C. 11] B. 13] E. 15] C. 17] False. 19] Clock. 21] 0101. 23] 4. 25] A circuit simular to Figure 8-30 with only one cycle of the unknown frequency sampled. The results of the binary counter would represent the amount of time it took for one cycle to complete.

INDEX